Advanced Television Systems

Advanced Television Systems: BRAVE NEW TV

Joan Van Tassel

Focal Press
Boston   Oxford   Johannesburg   Melbourne   New Dehli   Singapore

Focal Press is an imprint of Butterworth-Heinemann

Copyright © 1996 by Butterworth-Heinemann and Joan Van Tassel

 A member of the Reed Elsevier group

Library of Congress Cataloging-in-Publication Data

Van Tassell, Joan M.
    Advanced television systems : BRAVE NEW TV / Joan M. Van Tassel.
      p. cm.
    Includes bibliographical references and index.
    ISBN 0-240-80243-8 (pbk.)
    1. Television. I. Title.
  TK6630.V33 1996
  384.55—dc20   95-51361
                                         CIP

British Library Cataloguing-in-Publication Data
A catalogue record for this book is available from the British Library.

The publisher offers discounts on bulk orders of this book.
For information, please write:
  Manager of Special Sales
  Butterworth-Heinemann
  313 Washington Street
  Newton, MA 02158-1626

10 9 8 7 6 5 4 3 2 1

Printed in the United States of America

This book is dedicated to my mother,
Lucille Newton-Van Tassel, whose generous spirit
has filled my life with gifts
and
James Bromley, lifelong friend and
engineer par excellence.

Table of Contents

List of Figures

List of Tables

Acknowledgments

So many people have given generously of their time and energy to make this work possible. The list must begin with Julia Marsh of Technology Futures, Inc., a brilliant writer in her own right and the Ideal Editor in her position on New Telecom Quarterly. I told her about the book I wanted to write at lunch during the InterMedia show in San Jose and the next thing I knew, she had taken it in to Focal Press on my behalf.

At Focal Press, I was then lucky enough to draw Marie Lee as my editor there and I remember a fabulous lunch with her at Uglesich's in New Orleans, where we settled the final arrangements over soft-shelled crab po' boys. My thanks go to Karen Speerstra for believing in the project. Product Manager Joan Dargan is a pleasure to know and work with and I thank her for all she did for the book. Tammy Harvey and Valerie Cimino brought helpful comments and encouragement in their editing of the early versions of the first chapters.

All of the ideas and information presented in this book come from others. The errors are mine.

My deepest appreciation goes to Professor Ron Rice (Rutgers University) for his careful and thoughtful review of the manuscript. His work helped me sharpen many facets of the book and directed me to the concepts not thought through and ideas not stated.

Two individuals, Jim Bromley and Steve Rose, spent hours that accumulated into many days explaining the technical side of advanced television to me. Jim Bromley and I have talked about technology for the past fifteen years and I am still amazed at the depth of his knowledge. Steve Rose is such an enjoyable person the time just flew. From the beginning, we've multiplexed on the same wavelengths and the tapes of our discussions are filled with laughter and excitement about the emerging information infrastructure.

The book began with an assignment from Paula Parisi at *The Hollywood Reporter* to compile a complete listing of interactive TV test sites. She taught me so much about reporting on emerging technologies—she's a savvy journalist, a terrific writer, and an outstanding editor. I'm also grateful to the folks at THR, Editor Alex Ben Block, Miles Beller, Scott Hettrick, David Grunwald, and fellow co-conspirator, Jim Kearney, for their support.

A book like this couldn't have been completed without the help of PR, media relations, corporate communications, and publicity staff people. Mike Schwartz (CableLabs); Mary Barnsdale (AT&T); Brigitte Engle (General Instrument); John Taylor (Zenith); Tammy Lindsay (Time Warner); David Harrah (then at IBM); Ginger Fisk and Larry Plumb (Bell Atlantic); Nancy Buskin (Viacom); Steve Lange and Dave Banks (U.S.West); James Carlson (Jones Intercable); Linda Brill (DIRECTV); Linda Alexander (MTV); Ujesh Desai (Silicon Graphics); Jim Dougherty (Pacific First Networks); Sue Dean (TCI); Karl Buhl (Microsoft); Ellen East (Cox Cable); Linda Healy (PacTel); Kevin Doyle (BellSouth); Dick Jones (GTE); Bob Ferguson (Southwestern Bell); Mike Brand and Lynn Brown (Ameritech); Sylvan Leclerc (Groupe Videotron); Ellen Van Buskirk (Sega); Mike Gwartney (Family Channel); Anita Corona (IN); Paul Sturiale and Marty Lafferty (EON); Tom Hagopian (ESPN); Jim Boyle (YCTV); and Virginia Gray (Southern New England Telephone Multimedia).

This book enabled me to have some fascinating conversations about the emerging communications system. I hope I have been able to capture just a small portion of the excitement, energy, and intelligence that came from the people I interviewed. A full list of the interviews conducted for the book follows the list of references. However, some people stand out in my mind so clearly, I must mention them. Joseph Widoff (Advanced Television Test Center); Dr. Joseph Flaherty (CBS); Robert Rast and Dr. Woo Paik (General Instrument); Jim Chiddix (Time Warner); Marcia DeSonne (National Association of Broadcasters); Jonathan Seybold (guru); Ken Locker and Tamiko Thiel (Worlds, Inc.); Bob Doyle (digital video expert extraordinaire); Michael Liebhold and Alan Brightman (Apple); Gary Arlen and Peter Krasilovsky; Ed Horowitz (Viacom); Lee Rosenberg (William Morris Agency); Steven Spielberg; William Samuels (ACTV); and Gary Teegarden (U.S. West).

The publications that track new television technology are invaluable and I cited them over and over again: *Broadcasting & Cable; Multichannel News; WiReD; Communications Technology; CED; The Hollywood Reporter; Inter@ctive Week; Convergence; On Demand;* IEEE publications;

and *TV Technology*. Matt Stump, Peter Lambert, Mark Berniker, Peter Krasilovsky, Leslie Ellis, Roger Brown, Dana Cervenka, Bob Doyle, and Christine Perey all stand out as knowledgeable observers who write with great clarity about these topics.

Behind every author stands a long line of people who believed in her and encouraged her to stay the course. I will always be grateful to the Annenberg School for Communications at the University of Southern California for the opportunity to study there and to Ambassador Walter Annenberg whose extraordinary generosity made it possible. My mentors Peter Clarke and Susan Evans and professors Everett Rogers and Bill Dutton were four points of light. Professors Peter Monge and Janet Fulk have my sincerest thanks for the tremendous help they have always given me. Carolyn Spicer made my research in the Comm Center a pleasure.

My colleagues were always ready to help and, best of all, to listen: August E. Grant (University of Texas); Noshir Contractor (University of Illinois); Diane McFarlane (Yorktown State University); Bill Richards (Simon Frazier University); George and Debbie Manross (CSU, Fullerton); and Carl Ferraro (SUNY/New Paltz).

During much of the time I spent writing the book, I taught at Pepperdine University. I appreciate their support and the comfort of that beautiful campus. I'm particularly indebted to Lynne Gross, Milton Shatzer, Bert Ardoin, Bob Woodruff, Fred Casmir, Don Shores, Juanie Walker, Louella Benson, David Lowry, Nan Bartlett, Esther O'Connor, and Provost Stephen Lemley for their warmth, understanding, and help.

My assistant throughout this project has been Debbie Reed. She has worked harder for her college degree than just about anyone I've ever seen, in spite of the legislature of the state of California. I've been able to track the effects of the unconscionable destruction of the California system of higher education through its deleterious effects on her life and dreams. She has triumphed over the odds. Debbie is outstanding at research, editing, and hand-holding—and I am grateful to her for each of those marvelous skills. I consider myself to have been extremely fortunate to have obtained her help.

My friends and family have listened to my complaints and comments with kindness, patience, and some humor. I'm thankful for pals like Susan Trieste, Nancy Jones, Mary McGuire, Bill Cooley, and Leigh. My uncle and aunt, Robert and Irene Newton, gave staunch support. Finally, my brother Gordon and sisters Nancy and Elaine are simply beautiful people. We've all come a long way and it's a privilege to share a life with them. Our father, Gordon E. Van Tassel, would have been proud.

Introduction

Encountering something new is almost always difficult, especially when the same territory previously seemed simple and familiar. The three-network world of a decade ago was such a well-known and natural media environment that, in retrospect, it took little effort to understand and encompass it in its entirety.

In the late 1970s and early 1980s, I worked as a documentary and magazine show segment producer for Los Angeles television stations. I started shooting with 16 mm film equipment, then switched to ¾" video, and finally to ½" betacam equipment. My first pieces were cut (literally) on a Moviola; within three years, everything was edited electronically on ¾" video offline and 1" video online. Audio post-production changed from a track mix to a sweetening session.

Video post-production changed most of all. In the days of 16 mm film, the editor marketed the "ins" and "outs" of special transition effects like dissolves and wipes and sent them out for the finished "opticals." Video post let the director see these effects immediately and change editing decisions so directors and producers assumed much more direct control over their work. Editing on video cut down post-production time by as much as 50%. It made sophisticated effects as easy as pushing thirty or so buttons. It was done by computers.

I never expected to know anything about computers. I documented the plight of the homeless, the ill, and the socially marginalized. During sweeps periods, I shot my share of pieces on aerobics, soap operas, and food trends. I learned about swingers, diseases of the week, homicide cops, and bathing suit styles. But computers?

At the Annenberg School of Communications at USC in 1983, studies of computers had been underway for some years. Everyone used computers for word processing, number crunching, presentation formatting, desktop publishing, email, and conferencing. Television was already dead tech there.

As a graduate student, I began working on the Annenberg Videodisc Project developing a disc for cancer patients about some side effects of treatment. I was not grateful. Disdaining the clumsy software and limited reach of the medium, I was dragged metaphorically kicking and screaming into the world of advanced television, which I considered vastly inferior to the commercial television I knew.

So I understand the visceral response people have to the description of advanced television: high definition, digital, compressed, interactive, connected, switched, global television. For many, it is a mouthful of unfamiliar, unwelcome syllables that sound vaguely threatening. Nevertheless, it is the future.

The purpose of this book is to give each of those adjectives meaning to the interested reader. People who work in broadcasting, cable, satellite delivery services, telecommunications, the computer hardware and software industries, the power utilities industry, television and commercial production, program development, and service and content marketing will all discover aspects of the emerging television landscape they didn't know about.

Part One treats the adjectives, one by one—high definition, digital, compressed, interactive, connected, switched, and global—describing what it is, why it is important, its current state, and how it is changing. If there is a technical aspect, I try to give the reader a broad picture of the technology without getting bogged down in the engineering detail.

Part Two discusses these attributes as they are assembled into the wired and wireless infrastructures that deliver television to viewers. These two chapters, especially the one on wired systems, are among the most technical in the book because this is the arena where the decisions are currently being made.

Part Three covers the business side of advanced television: content creation, technology and content testing, and business models. Beyond the facts, I try to provide the tenor of the enormous undertaking in which the participating organizations and individuals have involved themselves. It is a wonderful and terrifying task to bring a new system of communication into being!

Hopefully, at the end of reading this book, readers will be able to attend the many expositions and conventions that feature the emerging television technologies and understand most of what is presented. They will see these topics repeatedly in the business section of their newspapers and be able to separate the hot from the hype. They will talk to the technical folks and ask the right questions—even when the answers are all techspeak.

Most of all, this book is intended to let readers look over the horizon line at the next twenty-five years of the revolution of communications that began with the 20th century. The Annenberg School for Communication gave me that vantage point and it has been my pleasure to pass it on.

Joan Van Tassel
November 9, 1995
Malibu, California
jvantass@pepperdine.edu

High Definition Television (HDTV): The First Step

Introduction

High definition television (HDTV) is the name given to an array of improvements to the television picture that make images clearer, brighter, and more detailed and raise the sound to the level of compact disc quality. HDTV occupies a premier place in the emerging television system of the future—even if it is never actually adopted. Its real legacy may not even be the superior television picture its creators envisioned. Rather, the lasting benefit of HDTV is likely to be the accelerated development of digital television, hastening its appearance by as many as ten years. As we shall see in this and succeeding chapters, digital television is already being deployed across the media landscape with important implications for all the communications industries.

Even in the longer term, digitized TV will probably have greater impact than the high definition picture that inspired it.[1] Digitized video can be transported across cable, computer, telephone, and wireless networks. This universality promotes the convergence and interconnection of these different telecommunication architectures. This new technology has encouraged international standards-setting bodies to attempt the difficult goal of establishing a single global standard for a digital television system.

The race for HDTV has been an extraordinary effort of truly massive proportions. It required enormous financial investments since the late 1960s, commitment from more than twenty national governments, and nearly thirty years of intense research, conducted by a host of commercial consortia, hundreds of business organizations, research laboratories, universities and thousands of scientists, researchers, and skilled technicians.

HDTV became a kind of industrial Holy Grail that would endow its discoverers with economic supremacy. By the late 1980s, the endeavor that began as a clever commercial goal to improve television technology became inextricably intertwined with national and regional pride, economic autonomy, and the spirit of competitive pursuit. In countless news reports of that time, HDTV was evoked as proof of cultural domination or subordination, as confirmation of wise government policy or, contrarily, as foolish anarchy.

During an era of extravagant corporate news stories, HDTV was one of *the* hot topics. The high cost, high-tech, high stakes global race to establish an HDTV system was long and hard, the arena strewn with the detritus of fallen ideas and exhausted prototypes. More than once, the pace of technological change compelled companies and their research and development departments to execute abrupt changes in competitive strategy or to abandon the quest altogether. Even today, after almost three decades of struggle, the research still continues, generating a continuous stream of new techniques and technologies—and new heroes and losers.

The turbulent development of high definition television and other aspects of the new television system prompted Joseph Flaherty, the much-respected vice president of technology at CBS, to comment: "There are three ways to lose money. The most pleasurable is romance. The fastest is gambling. And the surest is technology."[2]

Even before implementation, HDTV has cost billions of dollars just for research and development. Japan has spent an estimated $1.3 billion. The Europeans have spent nearly that much, with estimates ranging from $994.5 million to $1.066 billion. The U.S. has spent far less, approximately $200 million, expended by private industry with no federal funding. The private effort in the U.S. contrasts with the combination of governmental and private funding provided to both the Japanese and European HDTV research teams.

No matter whose checks were supporting the effort, HDTV was always a game for high rollers. It cost $17 million just to build the Advanced Television Test Center (ATTC) near Washington D.C., a facility designed to test the prototype HDTV systems. Although the test results from the ATTC will be used by the U.S. Federal Communications Commission (FCC) to set standards, the ATTC was paid for by private U.S. broadcasting, cable companies, and the high-tech organizations that actually developed the HDTV systems for testing.[4] Much more

Figure 1.1 April 5, 1995: Testing the latest version of the "Grand Alliance" Digital HDTV System at the ATTC. Source: ATTC. Photographer: David Poleski.

will be spent retooling the present television system, with estimates of the cost of conversion for industry and consumers running into billions of dollars. So it's reasonable to ask: What will come out of all this effort and all this money, to say nothing of all this hype?

HDTV: Creating a Better Television Picture

The current U.S. television format is called NTSC, which stands for National Television Standards Committee, the group that defined the technical standards for black-and-white TV in 1941 and for color TV in 1954. HDTV is the first major change in TV technology since then. (See Textbox 1.1, below.)

HDTV offers many benefits to viewers and brings several movie-like properties to the small screen. The most obvious visible characteristic is the changed aspect ratio, or the relationship of screen width to height. HDTV replaces today's almost-square screen (4:3, or 4 units of width to 3 units of height) with a 16:9 ratio (16 units of width to 9

A camera works by rapidly scanning an image and translating the color and brightness of a scene into electrical impulses, called a video signal. Retranslated back into the color and brightness of the original scene, these signals are then displayed on a screen. The quality of the picture depends on how many data points the camera picks up, how many times the scene is scanned in a given period of time, and how accurately the display device reproduces the scanned scene.

When the picture is displayed on the TV set, the scanning starts at the top left corner and moves horizontally across the screen to the right, creating a single "scanning line." At the end of the line, the scanner shuts off and snaps back to the left hand side of the screen. Then it turns back on to emit light along a second line, this one slightly lower than the first. The time during which the scanner is off is called the "horizontal blanking interval."

Along each line, NTSC television emits a continuous stream, an ever-changing mosaic of light and dark, equivalent to about 300 points of information. Digital HDTV increases the number to about 1,408 points of information, or "pixels" (a contraction of "picture elements"), providing much greater horizontal resolution.

From top to bottom, there is room for 525 lines on an NTSC TV screen, with about 480 of them used for the picture. Engineers call the 45 lines that aren't used to carry video signals the "vertical blanking interval." Just as HDTV provides greater horizontal resolution, it also increases the vertical resolution to 1080 lines, with 90 lines not used for video signals.[5] The result of the greater resolution is that the HDTV picture contains about 5 times as much information as an NTSC frame.

NTSC employs an interlaced system, meaning that the 240 even lines are scanned from top to bottom, 30 times each second. Then the 240 odd lines are scanned from top to bottom. Each scan is called a field; a double-scan, including both the even and odd lines, is called a frame. Thus, the NTSC system transmits about 480 active lines at a rate of 60 fields, or 30 frames, per second.

Even though half the picture is shown at a time, the viewer doesn't see a sparse picture because of "persistence of vision," which means that the human eye continues to see an image for a fraction of a second after it has disappeared. Another reason people see a complete picture is that the phosphor (the incandescent material in the screen) continues to glow after the electrical impulse that excited it decays, long enough for viewers to retain an impression of a whole image.

Textbox 1.1 NTSC and HDTV television.

units of height), approximating the rectangular shape of the screens in movie theaters, as shown in Figure 1.2.

Research shows that this wider screen takes advantage of the fact that the eyes see more along the horizontal plane than the vertical plane, resulting in a picture that seems more life-like and realistic.[6] The 16:9 ratio also makes the act of viewing a wide screen more like ordinary vision. When a viewer watches today's NTSC 4:3 television, the eyes move to the right and left only about 10 degrees. By contrast, the wider aspect ratio of HDTV causes the eyes to move from side to side at about 30 degrees, a scan that approximates everyday visual behavior.[7]

Undoubtedly the most important improvement HDTV offers is superior image quality, especially on large-screen television sets. The HDTV picture offers greater resolution than NTSC, which means it shows more detail and displays richer, deeper, more complex colors. Digitizing the image also results in higher quality. It eliminates some of the problems of NTSC; there is no shimmer, jitter, dot-crawl, and false color display. Since large TV sets are the fastest-growing segment of the television set market, there is considerable demand for the higher quality picture that HDTV can deliver.[8]

One way to appreciate the difference higher resolution will bring is to experiment with the distance you sit from your TV set. In order to see the picture without lines, the sitting distance must be approximately seven times the screen height. If the screen is eighteen inches high, you need to be ten and a half feet away. HDTV allows the viewer

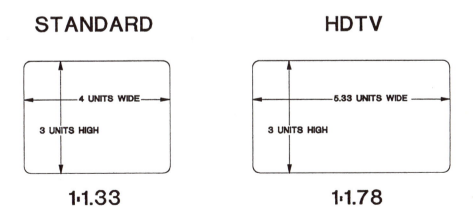

Figure 1.2 Aspect ratios of NTSC and HDTV screens.

STANDARD NTSC   HDTV

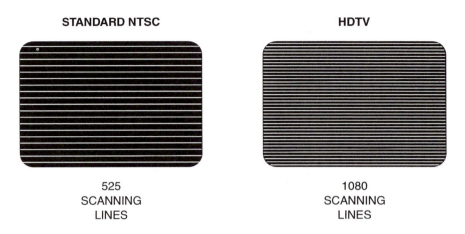

525   1080
SCANNING   SCANNING
LINES   LINES

Figure 1.3 Comparing the resolution of the NTSC and HDTV pictures.

to sit much closer to the screen, at a distance of only three times the screen height. In the case of an eighteen-inch screen, one could sit four and a half feet away and see a line-free picture.

Television sets display pictures with continuous lines, where all the odd lines light up, then all the even lines—called an interlaced scan picture. Textbox 1.1 provides a detailed explanation of the display of NTSC and HDTV television signals.

It is clear that the advent of digital HDTV would fundamentally change TV sets on the outside to accommodate the new rectangular picture and inside to process the digital signal. On the screen, HDTV sets will display detailed images, easy-to-read text, and sharp graphics, far beyond the capabilities of today's sets.

HDTV brings with it a whole new sound

If seeing is believing, hearing is all-involving. Many people don't realize that most film budgets allot nearly as much money towards the sound track as they do for the picture. Seamless audio is often as important as a believable picture for the audience to become psychologically involved, or to suspend disbelief, as the movie industry refers to this viewing phenomenon.

Every HDTV standard will include improved audio. The U.S. has adopted 5.1 channel Dolby AC-3 audio technology to provide compact disc-quality and stereophonic surround-sound which expand the enjoyment of television entertainment dramatically. This innovative new technology offers better sound separation, the potential for wider range dynamics, and efficient encoding for up to six high-quality audio channels.[9]

A World-Wide Effort to Develop HDTV

The path to HDTV was different for each of the three great economic powers: Japan, Europe, and the U.S. The Japanese started first, followed by the Europeans, while Americans waited until very late in the game to begin research and development. Surprisingly, that delay proved to be quite advantageous, as shown in the next sections covering the dramatic events that led to U.S. supremacy.

HDTV development, 1970–1986: Japanese diligence, Western complacence

The Japanese started work on HDTV in 1970 under the direction of Dr. Takashi Fujio of Nippon Hoso Kyokai (NHK), Japan's state broadcasting system. From the very beginning, the research aimed to create a better picture with greater detail and more vivid color. A major goal was to heighten the psychological effects of the medium.[10]

High definition television came first as the focus of research because it would directly affect the viewing experience. From a technical perspective, just getting there was a daunting challenge. From an economic perspective, the successful realization of the goal would sell more toasters, toothpaste, and tacos—and more televisions.

The Japanese effort was based on industry-wide cooperation, coordination, and government encouragement. In the early 1970s two government offices, the Ministry of Posts and Telecommunications and the Japan Development Bank, began subsidizing and parceling out research to 11 Japanese companies. Sony and Ikegami developed cameras and the recorder; NEC, Mitsubishi, and others worked on projectors, tubes, and processing devices.[11] Through the Development Bank, the Japanese government began subsidizing the research. Since the psychological, social, and economic benefits they hoped would flow from HDTV rested on the delivery of a higher resolution image, the concept guiding the effort was simply to increase the number of continuous scanning lines from 525 to 1,125. The improvement over NTSC would be significant, but only an extension of the existing system, not a radical departure from it. Specifically, the Japanese did not envision the possibility of producing a digital television signal.

Nevertheless, the Japanese were quick to realize that any improvement at all would offer the opportunity to develop and market an entire array of technologies in order to produce, process, transmit, receive, and display the superior video and audio.[12] To demonstrate its quality, Sony provided Francis Ford Coppola with HDTV equipment for his film, "One From the Heart," which he utilized for storyboarding scenes and as a database.[13]

In Europe and the United States, these developments went virtually unheralded. Among the three networks, only CBS' savvy and in-

---

Bandwidth refers to how much of the electromagnetic spectrum (EMS) a signal requires. The EMS is a range of frequencies of waves of light energy (or electricity) that move through the air. The waves are measured in three ways: 1) by how many times they arc, or "cycle," in one second of time, called "frequency"; 2) how high and low they arc, termed "amplitude"; and 3) how long they are, named "wavelength."

The frequency of the cycles and the length of the wave are related: The faster they complete cycles, the shorter the wavelength; the slower the cycles, the longer the wavelength. If the EMS is displayed horizontally, as a person moves their finger from left to right, the number of cycles (frequency) increases and the wavelength decreases.

One cycle per second is called a "hertz" (Hz), named after Heinrich Hertz, who first measured the EMS' waves of light and energy. At the low end of the EMS are radio waves, also called sine waves. A wave that cycles ten times per second would have a frequency of 10Hz and would be very long indeed.

In reality, radio waves "oscillate," or cycle, at much higher frequencies. The AM radio band starts at 540 kilohertz (kHz), or five hundred and forty thousand hertz; both AM and FM radio are in the kilohertz portion of the EMS. Television signals are measured in megahertz (MHz), or millions of cycles per second (Hertz). Gigahertz (GHz) are radio waves used for microwave transmission that cycle a billion times per second. Located in the upper reaches of the radio portion of the electromagnetic spectrum from 1 to 100 GHz, microwaves carry out such diverse tasks as carrying banks' financial data to cooking food in millions of households!

At the center of a radio or television signal is the "carrier wave," which is transmitted on a single frequency, a tiny fraction of the EMS. Traveling with the carrier wave are "modulation signals" that occupy additional frequencies on either side of the carrier, termed its "sidebands," which act on the carrier wave

---

Textbox 1.2 Understanding bandwidth.

fluential Senior Vice President of Technology, Joseph Flaherty, seemed aware of the implications of the Japanese advances. In 1981, he led CBS to apply to the FCC for permission to utilize three satellite channels and a medium-sized cable system (less than 90,000 subscribers) for experimentation with new television technologies, including HDTV.[14]

In the application for the satellite channels, CBS stipulated that they needed 27 MHz of bandwidth, substantially more than the usual 6 MHz of bandwidth allotted for the NTSC signal. Figures 1.4 and 1.5 and Textbox 1.2 provide a basic explanation of the term, "bandwidth."

---

to vary its strength. The variation that modulation signals generate allows the carrier wave to be a precise representation of the information that is sent: the sound of music, a voice, or a television picture.

"Amplitude modulation" means that information in the sidebands is reflected by variations in the amplitude of the carrier wave. "Frequency modulation" means that information is carried by variations in the number of cycles of the carrier wave. Pictures use more bandwidth than sound because there is more information in them than there is in audio. Thus, television requires a bandwidth of 4.2 MHz just for the NTSC picture, AM radio uses 10 kilohertz, and stereo FM radio takes up 150 kilohertz.

For example, 55.25 MHz is the placement of the carrier signal for any television station designated as Channel 2. Channel 2's AM video signal takes up 4.2 MHz of bandwidth on one side of its carrier signal (the upper sideband) and .75 MHz on the other side (the vestigial sideband), giving a total of 4.95 MHz. This 4.95 MHz occupies the frequency range from 54.50 MHz to 59.45 MHz.

Channel 2 also has a frequency-modulated (FM) audio signal 4.5 MHz above the visual carrier frequency, at 59.75 MHz. The audio modulation adds sidebands to this signal that extend 25 kHz on either side of the sound carrier signal. This gives the sound signal a bandwidth of .050 MHz extending from 59.70 MHz to 59.80 MHz. The total bandwidth is a fraction less than 6 MHz because a little is left on each side to avoid signal interference with other channels, called "guardbands."

Bandwidth applies to transmission over wires, as well as over the air. Wires are referred to as "pipes," "conduits," "lines," and "cables," pretty much interchangeably. A given wire has only so much signal-carrying capacity—this is also called its bandwidth. For example, the copper wire that links telephones has a bandwidth of 3 kHz, while the typical cable TV "pipe" delivers 550 MHz, or about 90 channels of 6 MHz NTSC TV.

Textbox 1.2 *Continued*

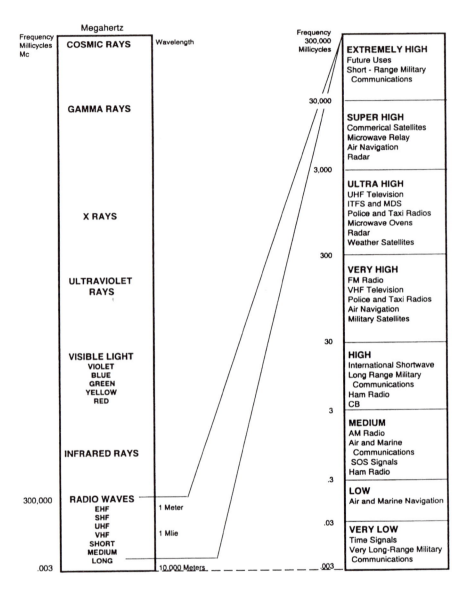

Figure 1.4 The electromagnetic spectrum.

In the early 1980s, experts believed that transmitting HDTV would require a minimum of 27 MHz, more than 4 times as much bandwidth as the NTSC signal. Depending on the specific design, HDTV might take up as much as an enormous 72 MHz! However, at that time, it wasn't technically feasible to transmit such a large signal

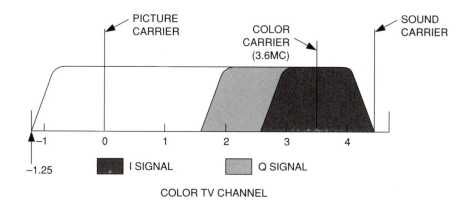

Figure 1.5 The bandwidth of a single television channel.

using a terrestrial transmitter. (That is, a transmitter located on earth, as opposed to being sited on a satellite or other airborne platform.) For that reason, the Japanese developed a satellite-based transmission system for HDTV, which did not conflict with the centralized approach of Japanese broadcasting.

However, in the United States, unlike Japan and most of the other countries of the world, local broadcast is an historically important element of television.[15] In addition to network-supplied or purchased-syndicated programming, local stations present local news, weather, and sports. They cover Main Street parades, school board elections, and council meetings, and the audience comes to revere some of the personalities who bring them this programming. To accommodate local interests, stations use terrestrial transmission, sending their signals over the air from transmitters (usually built on the highest hill around) to receiver antennas within line of sight.

In spite of executive interest at CBS, satellite transmission posed a considerable problem because it could not be adopted within the framework of a local television system. Throughout the early 1980s, CBS was the country's number one network whose business was deeply embedded in the heritage of local broadcasting. It was out of the question for CBS to single-handedly transform the entire television system, in spite of the lure of an improved picture.

It is less clear why the Europeans, with their centralized television systems, failed to respond sooner to the Japanese advances in HDTV research. However, while Europe and the U.S. continued to bask in blissful indifference, the Japanese accelerated their efforts at a furious pace. The Japanese called their system "HiVision" and, throughout the 1980s, they continued to develop equipment that would displace the entire range of the world's television technology—cameras, video recorders, switchers, routers, transmitters, and receivers rolled out of the laboratories of Japan's consumer electronics giants.

HiVision delivered a stunningly better picture that attracted crowds wherever it was demonstrated. By 1986, the Japanese were poised to extend their commercial domination of the television equipment industry, already significant, using entirely proprietary designs and technology.

As a first step towards adoption of the HiVision production standard, NHK sent design plans and specifications to the Comite Consultatif International de Radio (CCIR), the technical evaluation apparatus for the International Telecommunications Union (ITU). The ITU, part of the United Nations, is the most important global standards-setting organization, comprised of more than 150 member nations. Its mission is to propose, develop, revise, and administer worldwide technical standards and radio frequency spectrum allocations.[16]

At first, it looked like the Japanese would have clear sailing. The United States and Canada both supported the NHK standard, called 1,125/60. (This nomenclature meant that the HiVision signal had 1,125 lines, scanned 60 times per second.) The North American position was articulated by the Society of Motion Picture and Television Engineers' standard, SMPTE 240M, a U.S. version of the 1,125/60 format.

HDTV development, 1986–1987—The era of confusion

While the Europeans had heretofore appeared complacent and unconcerned, the Japanese filing with the CCIR galvanized them into action. Led by vociferous French objections, Europe balked at the idea of allowing Japan to take over the next generation of television.[17]

Rather than accept Japanese technology and the threat they felt it posed to their own consumer electronics industry, the European Community (EC) decided to subsidize the development of a domestic HDTV television system. In 1986, the EC funded Eureka 95 with an initial $180 million.[18] A consortium of European universities, research institutions, and electronics firms, including Philips, Thomson, and Bosch, shouldered the mission of developing HD-MAC, standing for High Definition-Multiplexed Analog Components. The proposed HD-MAC standard would call for 1,250 lines, transmitted at the same 50 frames-per-second rate as Europe's PAL television system, and delivered via satellite. (PAL, which stands for Phase Alternating Line, is a technique for creating a color television signal that was adopted by many European countries.)

The U.S. giant awakes Like ripples in a pond, the European response opposing the acceptance of a Japanese HDTV standard provoked a re-examination of policy within the United States. Fifty-eight broadcast-related companies urged the Federal Communications Commission to consider the issues involved in establishing high definition television in the U.S. In July, 1987, the FCC opened an inquiry to explore the possible introduction of an advanced television service.[19]

The FCC had observed the storm of protest over HDTV that engulfed the regulatory authorities in Europe. Recognizing the potential for controversy and the need to proceed carefully, the FCC created an organization to study the difficult technical, economic, and social issues involved in HDTV adoption. Composed of 25 volunteer members, the group was named the Advisory Committee on Advanced Television Service (ACATS).

Intense public discussion about the parameters of the new television system began almost immediately. The key issues during this early stage included compatibility, scalability, and interoperability.

The debate over compatibility concerned whether or not sets would be capable of receiving both the new HDTV signal and the NTSC signal. This idea hearkened back to 1954. When color TV was introduced, the engineers of the NTSC designed a signal that allowed people with color sets to receive a color picture, while those who had only black-and-white sets could still get a good picture. In the late 1980s, many observers hoped consumers would be able to keep their existing sets after the introduction of HDTV; however, this prospect seems increasingly unlikely.

Scalability was one idea for obtaining backwards compatibility between advanced HDTV and current NTSC sets. In a scalable system, a television set would accept and display any number of lines; the more lines the set could receive, the better the picture it would display. An NTSC set would accept and display 525 lines, while an HDTV set would accept 1,250 or 1,125, or whatever the maximum number of lines set for the future HDTV standard.

The issue of interoperability addressed HDTV sets as one component in an overall system. For example, in a stereo sound system, all CD players connect to and function with all amplifiers and speakers so that consumers can assemble equipment from different product lines. Similarly, interoperability dictates that all TV sets will work with all audio speakers, settop boxes, and VCRs, using standard connectors to link the components.

While interest in the U.S. in HDTV was high, in 1987 (and indeed for several years more) it seemed inevitable that a Japanese-designed satellite-based transmission system must surely triumph. Japan held a commanding two-decade lead that, to all appearances, was simply insurmountable. The only possible way to overtake the Japanese would be to develop a digital high definition alternative to the analog HiVision. At the 1987 National Association of Broadcasters' convention, knowledgeable people made public predictions to the effect that, unfortunately, a digital television design was at least 10 years away.[20]

1988–1992: The era of competition

In spite of the difficulties, by the end of 1987 the seemingly unequal battle was joined and an enormous global corporate race for HDTV was on. In the United States, the David Sarnoff Research Center, Zenith Electronics, and BellLabs started work. In France, Thomson scientists began an intense effort, as did Dutch engineers at Philips.

Little did anyone realize, not even those who were to accomplish great deeds in the future, that Home Box Office, the premium cable channel programming giant, had already taken the steps that would lead directly to a revolution in television technology. For the moment, such developments were nowhere in sight. Confusion and uncertainty prevailed.

In early 1988, the FCC laid out a plan for deciding how to adopt an advanced television system in the United States, but no one was sure the plan would work. In contrast to the cooperative Japanese effort, the FCC's procedure suited American business culture: The decision would be reached through a competitive process, pitting corporate giants against one another in order to reach the best possible system. The role of government was neither to coordinate nor subsidize; rather, it was to ensure a fair and level playing field for contenders.

By the end of 1988, the FCC had received 23 HDTV proposals, submitted by their developers, called proponents. It was a bewildering alphabet soup of competing systems, including IDTV (Improved Definition Television), EDTV (Enhanced Definition Television), and ACTV (Advanced Compatible Television). The different designs ranged widely in how much they would improve the NTSC picture and how they would transmit the signal.[21]

Still, it seemed nothing could stop the Japanese juggernaut. While Europe and the United States belatedly tried to organize themselves to develop their own high definition television, the Japanese did not hesitate to press their advantage. In February of 1988, they produced a live HDTV broadcast of the opening and closing ceremonies of the Olympics in Seoul, Korea, beaming pictures back to Japan via satellite. The public watched the broadcasts on 200 monitors displayed at 50 sites, including railroad stations and malls.

At the Shinshu University medical school, professors used HDTV pictures of brain surgery to teach operating procedures, an application made possible by the detail of the recorded HDTV image. At a museum, a "HiVision Gallery" allowed visitors to call up paintings and background information from individual viewing booths. The printing industry in Japan began designing equipment to allow them to make four-color separations and to print from HDTV images on tape. A restaurant started using HDTV pictures of its dishes in place of the plastic renditions usually on view in Japanese restaurants. Even a bookmaker used HDTV to show the horses racing that day.[22]

By October, 1988, NHK started to broadcast one hour a day in HDTV. The head of the Japanese Ministry of Posts and Telecommunications confidently predicted that there would be millions of HDTV sets in use by 1994 and 10 million by 2000.[23] As Japan looked at the slow progress in the West, mired in international politics, economic considerations, and technical issues, their optimism seemed more than justified.

Early in 1989, the Europeans entreated the U.S. to join them in blocking acceptance of the 1,125/60 standard still awaiting approval from the CCIR.[24] Within the U.S., opinion was divided. The broadcast networks disagreed among themselves as to which HDTV standards ought to prevail. Debate also raged within the National Technology Information Agency (NTIA), headed by a man who would soon go on to play an important role in the U.S. effort to develop HDTV, Alfred Sikes.[25] That spring, the House of Representatives and the Senate held hearings on the standards-setting issues, a sign of the heightened interest in and concern over HDTV.

In May, the Advanced Television Standards Committee (ATSC) reversed its support for the Japanese 1,125/60 HDTV standard. Established in 1983, the ATSC is a group composed of executives from the U.S. television industry. They come from broadcast networks, stations, cable companies, producers, electronics equipment manufacturers, and satellite companies. This influential group makes recommendations to the U.S. State Department on pending issues before international standards-setting organizations, such as the CCIR and the ITU.

The ATSC's view prevailed and the U.S. withdrew its support for NHK's standard, putting the issue of HDTV production standards back to square one.[26] Moreover, the European position hardened and it became clear they were unlikely to ever support Japanese-designed and controlled HDTV.

By mid-1989, the U.S. began to make substantial progress towards defining the country's parameters for HDTV. In July, the Advanced Television Standard Committee (ATSC) technical group approved production standards for HDTV:[27]

Aspect ratio:	16:9
Pixels per active line:	1,920
Pixel arrangement:	Orthogonal
Pixel aspect ratio:	1:1
Active lines per picture:	1,080
Interlace:	2:1 (present), 1:1 (future)
Unresolved:	frame rate
	size of vertical blanking interval

Then, in October of 1989, all hell broke loose when the American Electronics Association (AEA) published their forecasts of HDTV's potential effects on the U.S. economy and technological prowess.[28] At once, the report ignited a firestorm of controversy. The simmering conflicts over Japanese economic success and the fears it engendered broke out into open flames.

The entire tone of the public discourse concerning television's next generation changed from an arcane, technical, insider style to one of immediacy, urgency, and action. Within a month, there were ten congressional hearings scheduled and 9 bills concerning HDTV development on the docket.

The AEA is a trade association comprised solely of U.S.-based electronics firms. Their report considered the role that high definition television would play in the overall future market for U.S. high technology products and, by extension, in the U.S. economy. The AEA document reported one development in an ominous tone: "In mid-1988, several managers from a top U.S. semiconductor firm were able to preview the 40-chip set for HDTV receivers that has been developed by a Japanese firm. Validation that these chips already existed—a concern the U.S. executives had feared—was greatly overshadowed by the realization that their engineers probably did not even know how to engineer the chips. The level in sophistication in consumer electronics had taken another quantum leap forward."[28]

The report predicted that the semiconductor-gobbling HDTV system would give the Japanese a competitive advantage that would allow them to produce integrated circuits cheaper. This development was dangerous, concluded the report, because lower prices would enable Japan's companies to undercut American chips used in computers. Their formula was that if the U.S. held a 10% market share in HDTVs, then the country would keep 21% of semiconductor production and 35% of the personal computer market. By contrast, if the U.S. held a 50% market share in HDTVs, the nation's share of the semiconductor market would be 41% and its portion of the personal computer market would be 70%.

The AEA report concluded with a call for government to support the American electronics industry through subsidies, tax incentives, changes in patent procedures, and relaxation of anti-trust laws. Following the AEA bombshell, a rash of articles appeared in both the popular and trade press, detailing Japanese superiority and prescribing what the U.S. should do to cure its trade ills.

That same month, October, 1989, at the annual convention of the Society of Motion Picture and Television Engineers (SMPTE), HDTV was the star of the show, which featured a live transmission from Tokyo to the convention's location in New York. Rank Cintel, LTD, a British manufacturer of equipment used to transfer film to videotape, called a telecine, announced they were developing film-to-HDTV video transfer equipment.

Not all references to the coming HDTV system were positive. At that same SMPTE convention, broadcasters complained bitterly and loudly about the anticipated high costs of HDTV implementation for their stations, as outlined in the Ross Report. Bob Ross, Director of Broadcast Operations and Engineering for Group W station, WJZ-TV, Baltimore, had prepared and presented his findings earlier at the National Association of Broadcasters' convention in the spring of 1989. The Ross Report showed that an HDTV station, built from scratch, would cost $38.5 million, $24.5 million more than an equivalent NTSC station.

While politicians agonized privately over whether there should be government support for the U.S. HDTV research effort and journalists stirred the public pot, European and American electronics companies continued active development on HDTV. The David Sarnoff Research Center had made sufficient progress on their Advanced Compatible TV (ACTV) to broadcast New York's St. Patrick's Day Parade over WNBC-TV, although only the researchers themselves had the HDTV set required to display the high definition video. Zenith and AT&T were also both well along in their development. By November, 1989, the FCC had 26 remaining proposals: 21 designs for some form of improved definition television, 4 concepts for terrestrial transmission of HDTV, and 1 plan for satellite transmission.

In March of 1990, the Advisory Committee (ACATS) met to promulgate basic determinations to guide the development of HDTV. They reached three crucial decisions:

1. They would give preference for high-definition designs, rather than enhanced or improved definition systems;
2. They would re-visit technical questions again in 1992, in case radical new technologies had been discovered in the meantime;
3. They chose a simulcast, rather than an augmented transmission plan, meaning that HDTV would have its own independent signals, rather than merely adding information to the existing NTSC signals. However, the committee directed that the simulcast signal not exceed 6 MHz bandwidth.[29]

The little group that could—and did   These ACATS decisions were especially important to a small group of little-known engineers who worked for the out-of-the-mainstream VideoCipher division of General Instrument Corporation in San Diego, California.[30] For the GI VideoCipher group, NHK HDTV was a danger to the profitable business they had built around the existing satellite television transmission system because it demanded the replacement of GI's satellite hardware and software by Japanese-designed equipment.

As the global race to develop an alternative HDTV design that could compete with NHK's HDTV was reaching mach speed, two engineers who had been trained at the Massachusetts Institute of Technology (M.I.T.), Jerry Heller and Woo Paik, headed up General Instrument's VideoCipher division. They were laboring at a tangent to the HDTV efforts, on an innovation they called digital multi-channel NTSC that they thought would appeal to broadcasters. However, NHK's satellite-delivered HDTV menaced this new product as well as their lucrative VideoCipher business.

Jerry Heller and Woo Paik shared a background that enabled them to follow the HDTV race, even though they weren't actual participants. The two had worked together since 1978, part of an "invisible college" that originated at the Massachusetts Institute of Technology. Invisible colleges, which permeate the upper ranks of American society, are composed of people who are related through an association with a university or research tradition.[31]

The M.I.T. invisible college that placed Jerry Heller and Woo Paik together at General Instrument had its roots in 1968, when Irwin Jacobs, an M.I.T. professor, supervised Jerry Heller's Ph.D thesis there. When Jacobs formed a company called Linkabit to do defense-related digital communication research, he hired Heller. (Linkabit proved to be a fecund developmental incubator, ultimately resulting in more than 19 derivative companies, including Qualcomm, Jacobs' most recent achievement.)

In 1978, Jerry Heller went back to M.I.T. and hired Woo Paik to work for Linkabit. Heller and Paik worked there together for the next 2 years, then transferred to M/A-COM when that company bought out Linkabit in 1980. Robert Rast, now Vice President of HDTV Business Development at General Instrument, describes Heller as the "visionary" and Paik as "the one who makes it happen."

In 1985, Heller received a Request for Proposals from the Home Box Office (HBO) premium cable network, asking for ideas about how HBO could scramble their signal. The cable giant was losing as much as $10 million a month in potential revenue to satellite dish owners who were bootlegging the unscrambled HBO signal as it was downlinked to cable head ends for transport to subscribers' homes. HBO executives decided to encode their signal to prevent unauthorized (and unpaid) viewing of their movies.

Heller and Paik responded to HBO's request because of their considerable backgrounds at both M.I.T. and Linkabit with digital signals and their knowledge of the advanced computers needed to process them. Although they didn't have direct experience with television, they brought to M/A-COM specialized knowledge about the rapid processing of massive amounts of data, similar to the quantity of information in a television picture. Based on the Heller-Paik proposal, HBO invited M/A-COM to meet with them before awarding the contract.

Heller sent Paik to the January, 1985 meeting with HBO. "I had to pretend I was some kind of expert in the scrambling area," recalls Paik. "Of course, I had some ideas about what could be done, but I'd never actually worked with television signals before, so my knowledge was theoretical. After we landed the contract, I went out and bought

Do It Yourself TV Repairs by Robert Middleton, for $6.95 at a local drug-store because I needed the actual numbers for color television process-ing."

M/A-COM won the HBO signal-scrambling contract from a crowded field of at least 20 other high-tech companies. The scheme Heller and Paik devised for HBO, called VideoCipher, became the ba-sis of an extremely profitable business for M/A-COM. One of their competitors, General Instrument Corporation, watched these develop-ments unfold and made a successful bid to buy that division, and only that division, from M/A-COM.

Now working at GI, the two engineers trained their sights on de-veloping a product they called "multi-channel NTSC." They reasoned that since sending a television signal over satellite transponders was very expensive, if they could figure out a way to make the NTSC sig-nal more efficient, television networks could save money by piggy-backing more channels on each transponder. In order to reduce the size of the 6 MHz NTSC signal, they digitized it, invented ways to shrink the amount of information in it, and sent it over the satellite. Then, back on earth, they de-digitized the signal and reconstituted it into a complete 6 MHz NTSC signal.

By early 1990, Heller and Paik had simulated the solution and built a prototype. They invited potential customers, including several network executives, for a demonstration of their digital multi-channel NTSC product. Since broadcasters followed the HDTV race closely, these knowledgeable network people informed the GI team that they believed the design elements used in multi-channel NTSC would ap-ply directly to the high-stakes HDTV sweepstakes.

May 31, 1990 was the deadline for proponents to submit HDTV plans for evaluation by ACATS and testing at the Advanced Televi-sion Test Center. Inside General Instrument's VideoCipher Division, Woo Paik was pushing hard for an HDTV project, at least partly be-cause he was certain he could digitize and compress a true high-defini-tion television signal. Heller agreed that Paik's ideas were the only way an HDTV signal could be transmitted over the air in 6 MHz, thus meeting the preferences expressed by the ACATS in March. Barely a month before the deadline, Heller gave Paik the go-ahead—but not, he warned Paik, at the expense of multi-channel NTSC.

Paik promptly canceled his appointments and, leaving his col-leagues at work on the multi-channel NTSC project, went home for a week of isolated, uninterrupted thinking and writing. Despite the nov-elty of having their hard-working father home during the day, Paik's three children tiptoed around, staying far from the private study at the end of the house.

One week later, just days before the deadline, Paik had successfully designed the most advanced television system ever invented, a design that would put the United States into the forefront of HDTV, years ahead of the Japanese. He carried a yellow-lined pad into the General Instrument facility in north San Diego; written on it were the descriptions, formulas, block diagrams, and charts that laid out the basic design for digital, compressed HDTV that could be sent over-the-air by terrestrial transmitters.

Looking at Paik's work, Jerry Heller was so convinced that the design would work that he sat down and designed some of the boards himself. The week before the deadline, Heller and Larry Dunham, another GI executive, engaged in daily discussions about whether they should submit their system for HDTV testing precertification, as the

Figure 1.6 Dr. Woo Paik standing next to the digital HDTV system he invented.

FCC termed the first step. By the end of the week they hadn't resolved the question, so the two men continued talking over the weekend. On Sunday, they called the president of General Instrument, Frank Hickey, in Chicago and requested the $130,000 they needed for submission. Hickey agreed and a check was cut on Monday.

The day of the deadline, Heller and Dunham flew to Washington, D.C. and hand-delivered the check and application letter to the Advanced Television Test Center (ATTC) in Alexandria, Virginia, minutes before the close of business. Joe Widoff, Deputy Executive Director of Finance of the Advanced Television Test Center, remembers the shockwaves that General Instrument's all-digital proposal sent through the advanced television development community: "In one day, the whole HDTV landscape changed."[32]

The changed HDTV landscape   The reaction from the media was immediate, and Heller recalls being deluged by telephone calls from reporters. The response from the industry was less sanguine. In late June, Woo Paik traveled to Alexandria and presented more details about the design to ACATS. Skeptical of GI's claims to having developed a workable digital television system, the representatives of the broadcast, cable, and manufacturing concerns questioned Paik closely. Only at the end of several hours were their doubts allayed.

Digitizing the TV signal answered all the issues raised at the beginning of HDTV development: compatibility with NTSC equipment, scalability (as a means of accomplishing compatibility), and interoperability with other system components. The General Instrument digital coup added an immensely important new issue to the discussion of HDTV—"extensibility." Extensibility means that a digital TV format could extend the medium's images to computer and telephone networks which process digital data.

When digitization answered these overarching questions, the whole ground of public discussion about HDTV adoption shifted from technology to implementation. Specific standards such as spectrum use, transmission and transport standards, frame rate, pixel size, scanning type, and cost came to the fore.[33] Although still daunting, new figures showing the cost of implementation would be $12 million per station (rather than the $24.5 million predicted in the earlier Ross Report), helped allay broadcasters' fears.[34]

By August of 1990, the test bed was in place at the Advanced Television Test Center in Alexandria, Virginia and testing of the proposed HDTV systems was set for April, 1991. Unable to match GI's all-digital HDTV system, many proponents dropped out, leaving only six prototypes in the testing process. With strong encouragement from ACATS, most of the proponents who were left embarked on programs to develop their own all-digital systems.

For the Japanese, General Instrument's breakthrough was such a bitter pill to swallow that they did not respond quickly to the threat it posed. In any case, they may have been skeptical of the feasibility of the GI system in the real world, a belief still held by some. Having invested more than 20 years and a billion dollars in their products, Japan continued to back Narrow-MUSE, a version of HiVision for the U.S. Narrow-MUSE was a hybrid analog/digital system, which added digital processing to an analog signal.

Likewise, the Europeans still supported their own hybrid system, HD-MAC, although the EC countries experienced considerable difficulty getting the member nations to agree to it. Eventually, they reached accord on a two-step plan. First, they would launch D2-MAC, an improved-definition solution offering a better picture than the current system, in a 16:9 format, accompanied by 4-channel digital sound. Then they would introduce HD-MAC, true high-definition television.[35]

In the U.S., the proponents readied their operating prototypes for testing in April, 1992, at the Advanced Television Test Center. The complex task, as well as the high costs involved, drove corporations to make alliances. In February, 1992, General Instrument teamed with M.I.T. and in March, Zenith Corporation joined hands with AT&T.[36]

Action at the Advanced Television Test Center The start of formal testing was by far the most important aspect of the HDTV story in 1992, despite several delays caused by the need to redesign the test site to accommodate the changeover to digital prototypes. The GI breakthrough had forced other proponents to convert to a digital design, and 4 of the 6 systems were now digital.

The ground rules for testing decreed that each proponent bring in their own equipment to the Advanced Television Test Center where technical experts would evaluate the picture across approximately 30 dimensions over a ten week period. Proponents couldn't make any changes to their designs and breakdowns counted against their time. General Instrument, Zenith and AT&T, and the Thomson-Philips-NBC consortium each experienced at least one problem during the rigorous workout for both humans and machines.[37]

In March, General Instrument successfully transmitted the first-ever over-the-air digital television signal at noon on March 23, 1992.

This transmission finally laid to rest any doubts about the viability of digital HDTV. To most observers, the FCC's encouragement of a digital solution was vindicated. A few weeks later, General Instrument duplicated its transmission to the 20,000 broadcasters attending the annual National Association of Broadcasters in Las Vegas.[38]

The triumph of digital television as a production standard inevitably led to consideration of the next step, transmission standards. Like the earlier standard, this would be a hard-fought, step-by-step battle.

In early 1992, the FCC formally laid out its implementation plan for HDTV transmission, allowing a fifteen year transition from NTSC to HDTV. From the time the commission set the transmission standard, current television broadcasters would have two years to apply for an HDTV license. In addition to the 6 MHz of VHF spectrum they currently use, they would be entitled to an additional 6 MHz in the UHF portion of the spectrum. They would then have three years to build the HDTV transmitter. For the remaining years, broadcasters would simulcast NTSC programming on their existing VHF channel, along with available HDTV programming on their additional 6 MHz UHF channel. At the end of the fifteen year period, the broadcasters would have to turn over their VHF 6 MHz allotment back to the FCC.[39]

One long-range consideration that tilted the FCC towards the simulcast approach was that the agency would be able to reallocate the released VHF 6 MHz for mobile communications. This change would make U.S. spectrum use compatible with the international assignment of those frequencies to mobile use.

Media coverage of these developments towards setting the HDTV standards was so intense that some called 1992 "the year of HDTV." The implementation plan introduced by the FCC marked the beginning of the end of the standard-setting process for the new television system.[40]

Three elements emerged that would recur in the future discussions and negotiations over HDTV. The first realization was that cooperation could take place between the rival proponents who had previously competed for FCC approval. The second development was the replacement of theory by hardware, allowing more accurate assessments of implementation requirements and costs. The third circumstance arose from this increased knowledge: an increased resistance of broadcasters to HDTV, whether or not it was digital.

Behind the scenes, Richard Wiley, the head of the FCC's Advisory Committee on Advanced Television Service (ACATS), began encouraging the former competitors to form a "Grand Alliance." He argued that just as all the systems had become digital, their designs would come to resemble one another more and more as testing continued. This increased similarity would make it more and more difficult for ACATS to recommend any one system to the FCC. In addition, an alliance would allow the proponents to pool their resources, to share both the risks and the rewards. Wiley, a telecommunications attorney, devoted his considerable diplomatic talents towards arranging an alliance that would benefit each member.

1993–present: The era of cooperation

As the standard-setting process for HDTV neared conclusion, Richard Wiley's corporate diplomacy among communications giants came to fruition in the Grand Alliance he had envisioned. However, even as the creative side of the equation came together, the possibility for actual implementation of HDTV grew increasingly uncertain.

Economic bedfellows   The testing of the HDTV systems was extremely expensive and evaluation of the four systems would continue for at least one more year. Without an agreement, the result would be one winner and two losers (GI had two entries). Worse, the losers would be likely to initiate costly, time-consuming lawsuits to challenge the process itself. By contrast, an accord would make everyone a winner and offer each a share in proceeds from licensing.

Given these incentives, preliminary negotiations to form a Grand Alliance began in 1992, but they formally ensued in February of 1993. In May, just as another round of testing would have started, the bargaining in Washington, D.C. gained in intensity. Wayne Luplow, vice president of Zenith, Robert Rast from General Instrument, and Peter Fannon, Executive Director of the Advanced Television Test Center, all canceled their participation on a panel at the Information Display Show in Seattle to rush back to Washington for the meetings.[41]

In order to forge the agreement, the participants postponed the hardest decisions about unresolved technical issues. According to one report, Richard Wiley intervened personally to resolve some of the disagreements that threatened to prevent the alliance. Despite the difficulties, Wiley was successful and the announcement of the formation of the Grand Alliance took place in the last days of May, 1993.[42]

The Grand Alliance television set   Once settled, the member organizations divided the tasks among themselves; ironically, they parceled out the tasks in a style reminiscent of the Japanese HDTV effort twenty years before. The alliance reached agreement over the scan mode and frame rate by recommending hardware that would accept both progressive and interlaced scanning and 24, 30, and 60 frame rates. All parties agreed that the best solution would be a receiver that could display 1,125 progressively-scanned lines at 60 frames per second. However, no commercially-available monitor existed with this capability, and estimates of when such receivers might be available ranged from 1996 to 1999.[43]

There was a further complication to developing an HDTV display—the CRT problem. The technical term for today's television receiver is the cathode ray tube, configured with the glass-faced picture tube we all know. In order to get the maximum benefits from HDTV, viewers need a large screen, measuring at least about 30 inches along the diagonal. Since large-screen televisions are the fastest-growing segment of the receiver business, the overall trend fits nicely with deriving the maximum benefit from the new HDTV technology. There is only one problem: A large CRT television set won't work.

A 30-inch CRT is power-hungry, hot, and heavy—it could weigh as much as a monstrous 600 pounds! An even worse difficulty is that as the number of lines increases from 525 to 1,125, the screen becomes unacceptably dark because of low luminance.

Faced with these difficulties, the Grand Alliance and the FCC put a priority on developing affordable, practical HDTV sets. The U.S. government had early recognized the importance of consumer displays. In 1988, the Defense Advanced Research Projects Agency (DARPA) had established a $30 million program of grants and loans for research and development of high resolution displays and the underlying electronics to support them.

The most likely candidate to replace the current television screen is called "active matrix liquid crystal display," or amlcd, originally invented by U.S. researchers. Once again, American companies are playing catch-up to a two-decade Japanese lead. Many observers believe that in order to make a large-size screen, a design called "tiled amlcd" will provide the solution. Tiling refers to new techniques that allow manufacturers to piece small sheets together into larger panels, in place of difficult-to-produce sheets of silicon. One company, Optical Imaging Systems, in Troy, Michigan, uses this method to combine amlcd panels into a seamless image display at a reasonable cost.[44]

At RAF Electronics, based in San Ramon, CA, Robert Fiske has developed a high-resolution projection system the company hopes will retail for less than $1,000. Robert Fiske was one of the original inventors of amlcd technology, but he couldn't find U.S. investors. According to RAF, the new projection system displays images three times as bright as existing systems with four times the resolution. The manufacturing of the product has also been simplified to permit a relatively low introductory price.

Texas Instruments also demonstrated a new prototype for an HDTV display at the Advanced Research Projects Agency (ARPA) High Definition Systems Conference in Washington, D.C. in February, 1994. The company claims their system will deliver digital, high-definition pictures on 12-foot screens at a reasonable price within the next two years.

The Grand Alliance television signal   In February, 1994, the Grand Alliance announced the test results of two alternative ways of creating an HDTV signal for transmission. The two proponents were Zenith Corporation, using a modulation method called VSB, standing for "vestigial sideband," and General Instrument's QAM, the acronym for "quadrature amplification modulation." The ATTC technical staff evaluated the signals for robustness, range, and the potential for and vulnerability to interference.[45]

Zenith's VSB was the winner. In mid-September of 1994, after three months of field tests in Charlotte, North Carolina, the Advisory Committee reported that VSB "provided satisfactory reception where the NTSC service is presently available, and in many instances where NTSC reception was unacceptable . . . [it] performed well under real-world conditions of multipath and other propagation phenomenon."[46] Further tests conducted in January, 1995, confirmed these results.

Broadcasters' resistance grows   Broadcasters continued to evince concern about the costs of implementing HDTV studio and transmission equipment. By 1993, estimates of conversion had lowered further to $1 million per station. However, in an era of declining audience share and scarce advertising dollars, this expenditure still represented a substantial investment, especially for stations in small markets.

Many broadcasters were also unhappy with the selection of the VSB modulation scheme recommended by the Grand Alliance and accepted by ACATS. They wanted the FCC to test another transmission method called COFDM, standing for "Coded Orthogonal Frequency Division Multiplexing." This transmission method is the basis for all the transmission designs under consideration in Europe; the Japanese are also testing it.[47]

Broadcasters liked COFDM for several reasons. It is the least expensive of all the transmission schemes. Used in France since the late 1980s for audio transmission, its economics are well-understood. COFDM also lends itself to transmission from small geographically-dispersed cells, instead of one giant, central transmitter. Dispersed transmission means the signal can be propagated into valleys, canyons, and otherwise inaccessible locations.[48] Finally, COFDM held out the potential for a world-wide transmission standard.

At that time some believed that the testing of COFDM as the broadcasters wanted would add at least another year to setting the standard for HDTV. Some cynics suggested that this demand was a mere ploy designed to delay acceptance of HDTV, since broadcasters don't yet see how to recoup their conversion investment.[49]

The position of broadcasters was strengthened by the endorsement of COFDM by the Digital Audio Visual Infrastructure Committee. (DAVIC is a private international standards-setting group.) In mid-1995, broadcasters got their wish. A window of opportunity opened for COFDM testing. The Grand Alliance partners had taken longer than expected to build their prototype HDTV system, so ACATS scheduled a "bakeoff" between the already-accepted VSB and COFDM for July, 1995. COFDM was convincingly demonstrated to be inferior to Zenith's VSB transmission scheme. However, some observers believe that QAM will ultimately become the transmission system that is actually adopted because it is better understood and less expensive than the Zenith proprietary VSB scheme.

Spectrum flexibility is another area where HDTV proponents and broadcasters are on a collision course. As broadcasters reviewed their assets, they realized that the additional 6 MHz of bandwidth they would receive for implementing HDTV might help fund the conversion.[50] Recall that the FCC proposed to allocate existing broadcasters a second 6 MHz channel assignment in the UHF portion of the spectrum for simulcast HDTV transmission along with their VHF 6 MHz NTSC assignment. This outright gift took an ironic turn as broadcasters started planning how to use this second 6 MHz allocation of spectrum.

Local stations realized that they could deliver digital services, reproduce their regular NTSC channel four to six times to broadcast their programs at different times, or develop altogether new programming channels. In one example, Fox Broadcasting announced it was considering an all-news channel.[51] In order to effect these various money-making schemes and develop additional revenue streams, broadcasters organized their powerful lobbying apparatus.

In the fall of 1994, Congress considered an omnibus bill that would have updated the Communications Act of 1934, which provided the legal structure for FCC regulation. Although the bill wasn't enacted, broadcasters were successful in adding language giving them "spectrum flexibility" with the UHF 6 MHz intended for HDTV simulcast. They proposed that instead of using the UHF bandwidth solely for HDTV, which would require significant capital investment, they would have the freedom to use the extra 6 MHz for other purposes. These offerings, argued broadcasters, would help pay for HDTV—if they ultimately decided to implement it.[52]

Horrified at the mere suggestion that stations might not present HDTV programming, the companies developing HDTV appealed to ACATS, to the FCC, and ultimately to the Congress itself.[53] Everyone realizes that the adoption of HDTV depends on the availability of programming. Without something to watch, why would consumers buy sets? ACATS acquiesced to a limited version of broadcasters' demands for spectrum flexibility, agreeing they should be allowed to use the additional UHF 6 MHz for other services when HDTV wasn't shown.[54] However, they also suggested to the FCC that if broadcasters didn't intend to use the additional bandwidth for HDTV at all, then the FCC should open up the additional 6 MHz allocation to all comers, not just existing broadcasters.

The issue of spectrum flexibility continues to be a source of friction between the Grand Alliance and broadcasters. In mid-1995, a compromise was reached. The Grand Alliance backed down on its demand that the UHF channel for advanced television (ATV) could never be used for other digital services. Some members of Congress had suggested that if broadcasters were not going to use the UHF channel for HDTV, it should be auctioned to them, rather than given to them for free. Faced with the threat of paying for the extra 6 MHz, the broadcasters hastened to restore the 1992 agreement and announced they would use the spectrum primarily for HDTV. In April, 1995, NBC disclosed that the network planned to provide widescreen HDTV as early as fall 1997.[55]

In late 1995, the planned spectrum giveaway became something of a scandalous consumer issue, with the FCC placing a value on the frequencies between $10 and $70 billion. Liberal groups such as the Center for Media Education teamed with conservative groups like the Americans for Tax Reform, saying broadcasters should pay for the spectrum, and calling the free spectrum allocation "cyberpork." The cellular industry joined the fray, since the spectrum for advanced TV services deprives them of potential spectrum.

Japan and Europe abandon analog HDTV   Among communication experts, dissatisfaction in Europe with D2-MAC and HD-MAC had been brewing for two years and, finally, in January of 1994, the European Community formally replaced the unsuccessful "MAC directive" (92/38/EC promulgated in May, 1992) with COM(93)-556. The new ruling adopted the 16:9 format with flexible transmission standards. Digital standards remain to be set.[56]

Europeans agree that migration to a digital system will be the next step, including a high definition version. In 1994, a team of technical experts began meeting to define the digital video standards. They reported their findings to the EC Digital Video Broadcasting Group (DVBG), a panel composed of broadcasters, manufacturers, satellite operators, and regulators. The DVBG made their recommendation to the European Community Telecommunications Council, the body that will actually define European digital standards.[57]

Initially DVBG recommended widescreen 625-line PAL-SECAM, foregoing HDTV. However, in a technical symposium in Montreux, Switzerland, in June, 1995, the Americans threw down the gauntlet. Joseph Flaherty, Vice President of CBS, told the Europeans if they stuck to a 625-line standard and failed to adopt HDTV: "It won't be a bad picture, but no picture. You'll have a revolution in the streets." U.S. determination caused the influential European Broadcasting Union to schedule reconsideration of the matter in September, 1995.[58] However, in January, 1996, European digital standards remained unresolved.

The Japanese waited even longer than the Europeans to signal their surrender to digital technology. The abdication began in February, 1994, with a senior government official observing that HiVision was obsolete and that Japan would move towards a digital standard. The remark caused such an uproar in Tokyo—especially angering NHK, the public television system that developed HiVision—that the official was forced to recant.

However, a few months later the NHK director of engineering, Shuichi Morikawa, conceded that the organization's research had shifted to more advanced digital systems. To make the point, NHK's laboratory opened itself to a reporters' tour to demonstrate a variety of digital research projects.[59]

Although Japan accepted the U.S. advanced system for export, internally the country will continue to broadcast HiVision analog HDTV. In September, the Ministry of Posts and Telecommunications designated Hamamatsu City as the "HiVision City," the first of thirty-eight such municipalities. The Japanese will build a city-wide video

HDTV network for use at concerts, international conferences, and academic meetings.[60] The MPT also announced they would license NHK and 9 private stations to conduct trials of rectangular 16:9 enhanced definition television (EDTV). The ministry launched these terrestrial broadcasting tests in the summer of 1995.[61]

Implementation Issues for HDTV

In the U.S., the implementation of HDTV will make a complete break with the existing NTSC television system. In Europe and Japan, the status of advanced television is somewhat more uncertain because the United States leaped ahead to a technologically-superior digital design system, leaving their analog efforts far behind. At the moment, Europe and Japan plan to make a gradual migration from their current systems to advanced TV.

As research and political accommodation laid out the parameters of compatibility, scalability, interoperability, and extensibility, specific implementation issues came to the fore: spectrum use, frame rate, display design, and conversion costs. By mid-1995, technical agreements had been reached to resolve these issues. Speaking at the Montreux Television Symposium, Richard Wiley, Chairman of ACATS reported that two standards would be promulgated:

> For desktop computer applications: 720 lines, 1,280 pixels, progressively scanned at 24, 30, and 60 frames per second;
>
> For entertainment television applications: 1,080 lines, 1,920 pixels, progressively scanned at 24 and 30 frames per second, and interlaced scanned at 60 fields per second.[62]

By late 1995, the FCC had moved to settling the final details of implementation. In mid-November, ACATS had finished all field testing and gave the FCC a formal recommendation to accept the Grand Alliance proposal. At that time, the agency still needed to set a channel allotment policy, establish a table of channel allotments, and officially assign channels to broadcasters.[63]

While technical answers exist to some aspects of these problems, their overall resolution lies in the political sphere and, finally, in the marketplace. Ultimately, it is the consumer who will determine whether HDTV will diffuse—or stagnate as an interesting technological road not taken. The best assessment of demand for HDTV is a forecast by the American Electronics Association that predicts that ac-

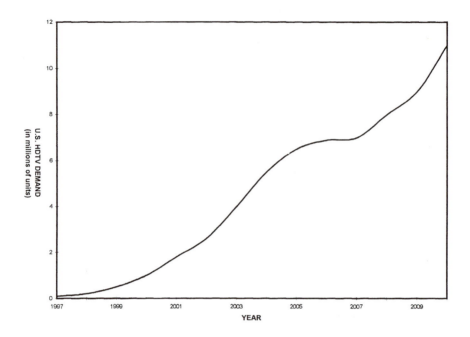

Figure 1.7 AEA HDTV demand forecast.

ceptance will be slow and steady. They anticipate that 11 million HDTV sets will be in use by the year 2010.[64]

The issue of a single international standard has now come full circle. In 1985, when the Japanese applied to the CCIR for approval of the 1,250/60 standard, there was great hope that the world would agree on one television system. As events unfolded, those dreams were dashed on the rocks of regional and national economic self-interest. Now, given the extraordinary advances in digital signal processing, there is renewed emphasis on simple conversion among systems even if they do not utilize identical standards. A formula promulgated in directive CCIR-601 lays out the basic premises for such conversion and will continue as the CCIR moves through the process of setting HDTV standards.

Conclusion

The U.S. entered the HDTV race to develop a better television picture in 1987. It was an era when Americans felt an uncomfortable, unaccustomed sense of technological inadequacy and economic defeat. Dire predictions about the country's future abounded. In 1991, the Council on Competitiveness—a coalition of chief executives from business, labor, and higher education—warned that America's technological edge had eroded in one industry after another. On September 19, 1994, that same council reported that its membership believes that the U.S. has shown a dramatic improvement in several areas critical to developing an advanced communication network. They added that "new markets and applications are helping the United States to recoup its position in electronic components."[65]

By 1997, ten years after initiating an effort to develop HDTV, the U.S. will have taken the first steps towards actually building that advanced communications system, transporting moving images that are compatible with telephone and cable networks and computer processing.

Changes in modes of communication exert a wide influence on many activities within a society. The telegraph was the first virtually instantaneous communication technology, and it changed the way generals fought wars, the way robber barons manipulated markets, and the way people learned about the world around them. Later, the telephone, radio, and television allowed people to hear and then to see the world, bringing immense social change with these new perceptions.

One implication of an interconnected communication system that carries pictures, sounds, and data in two directions with equal facility is that everyone is potentially a broadcaster. This revolution will also spark changes in the ways we work, play, learn, and live.

The key advance that makes this possible is digitization, or the coding of moving images into bits. The dictum among information engineers, "bits are bits," means that data, whether it is voice, graphics, moving images, or text files are all the same to a computer, as well as a network. In the next chapter, we will consider the meaning of the familiar word, digital, and the processes and consequences of digitization.

Notes

1. Andrew Lippman, "HDTV sparks a digital revolution," *BYTE* (December 1990):297–305.

2. Dr. Woo Paik, General Instrument, interview by author, on August 23, 1994 in North San Diego County, CA.

3. The estimate of Japanese spending is from J. Farrell and C. Shapiro, *Brookings Papers: Microelectronics 1992* (Brookings Press, 1992), cited by Junhao

Hong, "High Definition Television." *Communication Technology Update: 1993–1994* 3rd ed., eds. August E. Grant and Kenton T. Wilkinson (Austin: Technology Futures, Inc., 1993):23. Figures for European costs are taken from a Reuters News Service report, June 17, 1994, found on NEXIS, an electronic information service of Mead Data Central. A good source for information about Japanese development of HDTV is R. Akhavan-Majid, "Public service broadcasting and the challenge of new technology: A case study of Japan's NHK" paper presented at International Communication Association, Miami, FL, May 21–29, 1992.

4. Joseph Widoff, Deputy Executive Director, Advanced Television Test Center, in telephone conversation with author, September 15, 1994.

5. Federal Communications Commission Advisory Committee on Advanced Television Service, *ATV System Recommendation, draft-SP Version,* (Washington: FCC):10–11.

6. Dr. Lynne S. Gross, *The New TV Technologies,* 3rd ed. (Dubuque: Wm. C. Brown, 1990):188.

7. William J. Cook, "There's a battle ahead sparked by a revolutionary technology—even a death star," *U.S. News & World Report* (September 10, 1990):75.

8. Gross, *New TV Technologies,* 192–193.

9. "DMX Music Service first DBS user of AC-3 audio," *PR Newswire* (January 17, 1994).

10. Takashi Fujio, "High-Definition Television Systems," *Proceedings of the IEEE* 73:4 (April 1985):646–655.

11. Andrew Kupfer, "The U.S. wins one in high-tech TV," *Fortune* 60:4 (April 8, 1991):123.

12. Ibid.

13. B. Winston, "HDTV in Hollywood," *Gannett Center Journal* 3:3 (1989):123–137.

14. Information about the CBS application for satellite channel allocations is in Jay C. Lowndes, "14 seek direct broadcast rights," *Aviation Week and Space Technology* (August 10, 1981):60. A story covering CBS' request for a waiver to operate a cable system is found in Ernest L. Holsendorph, "CBS Cable Bid Cleared by FCC," *New York Times* (August 5, 1981):D-1.

15. Susan Tyler Eastman, *Broadcast/Cable Programming: Strategies and Practices,* 4th ed. (Belmont: Wadsworth):29–30.

16. Richard I. Nickelson, "The Evolution of HDTV in the Work of the CCIR," *IEEE Transactions on Broadcasting* 35, no. 3 (September 1989):250–258.

17. Lippman, op. cit., 302.

18. Farrell and Shapiro, op. cit.

19. Richard E. Wiley, "High Tech and the Law," *American Lawyer* (July 26, 1994):6.

20. "HDTV: Broadcasters look before they leapfrog," *Broadcasting* 117, no. 11 (September 11, 1989).

21. Lippman, op. cit., 298–299.

22. "HDTV Developments in Japan," *Financial Times* (May 9, 1989), found on NEXIS, an electronic information service of Mead Data Central.

23. "HDTV Live Broadcasts," *Japan Economic Newswire* (February 23, 1988), found on NEXIS.

24. "HDTV cooperation asked," *Television Digest* 29 (May 22, 1989):9.

25. "HDTV production standard debated at NTIA," *Broadcasting* 116 (March 13, 1989):67. The role of HDTV in the U.S. economy is well-discussed in M. Dupagne, "High-definition television: A policy framework to revive U.S. leadership in consumer electronics," *Information Society* 7:1 (1990):53–76.

26. John Burgess, "U.S. withdraws support for studio HDTV standard; Japanese suffer setback in global effort," *The Washington Post* (May 6, 1989):D-12.

27. *Engineering Report,* (Washington, D.C.: National Association of Broadcasters, September 4, 1989):1.

28. *Development of a U.S.-based ATV industry* (Washington, D.C.: American Electronics Association, May 9, 1989).

29. J. Stilson and P. Pagano, "May the best HDTV system win [an interview with R. Wiley]," *Channels* 10 (August 13, 1990):54–55.

30. The story of the development of a digital HDTV design and its submission to the ATTC comes from Robert Rast, Dr. Woo Paik, and Jerry Heller in individual personal interviews with the author, August 1994.

31. Derek Price et al., "Collaboration in an Invisible College," *American Psychologist* 21 (1966):1011–1018.

32. Joseph Widoff, interview by author, June 1, 1992 in Alexandria, VA.

33. "SMPTE: Seeking a universal, digital language," *Broadcasting* 121 (November 4, 1991):62.

34. "Refined HDTV cost estimates less daunting," *Broadcasting* 118 (April 9, 1990):40–41.

35. Daniel Tyrer, "The high definition television programme in Europe," *European Trends,* no. 4 (4th quarter, 1991):77–81.

36. "HDTV transmission tests set to begin next April," *Broadcasting* 119 (November 19, 1990):52–53.

37. Mark Lewyn, Lois Thierren, and Peter Coy, "Sweating out the HDTV contest," *Business Week* 33, no. 6 (February 22, 1993):92–93.

38. Peter Lambert, "First ever HDTV transmission," *Broadcasting* 122, no. 10 (March 2, 1992):8. For a good contemporaneous account of the Las Vegas

transmission, see Carl Patton, "Digital HDTV: on-the-air!," *ATM* 1 (Advanced Television Markets), no. 5 (April 1992):1–2. The first simulcast occurred over WRC-TV in Washington, D.C., detailed by Fritz Jacobi, "High definition television: at the starting gate or still an expensive dream?" *Television Quarterly* 16, no. 3 (Winter 1993):5–16.

39. Hans Fantel, "HDTV faces its future," *New York Times* (February 2, 1992):H17.

40. Ibid.

41. Brian Santo and Junko Yoshida, "Grand alliance near?" *Electronic Engineering Times* (May 24, 1993):1, 8.

42. Brian Robinson, "HDTV grand alliance faces tough road," *Electronic Engineering Times* (May 31, 1993):1, 8.

43. Robert Rast, Vice President, HDTV Business Development, General Instrument, interview by author, August 23, 1994 in North San Diego County, CA.

44. Fred Dawson, "The state of the display: flat-panel screens coming soon to a PDA or computer near you," *Digital Media* 3, no. 9/10, (February 1994):11. For a good discussion of plasma flat panel display technology, see Steven W. Depp and Webster E. Howard, "Flat-Panel Displays," *Scientific American* (March 1993):90–97.

45. Peter Lambert, "ACATS orders issue of broadcast multichannels," *Multichannel News* 15, no. 9 (February 28, 1994):3.

46. System Subcommittee Working Party 2, Document SS/WP2-1354, (Washington: Advisory Committee on Advanced Television Service, September 1994).

47. Marcia DeSonne, interview with author, February 1995, Los Angeles, CA.

48. Mike Sablatash, "Transmission of all-digital advanced television: State of the art and future directions," *IEEE Transactions on Broadcasting* 40 (June 1994):2.

49. Marty Levine, "Critical time nears for setting HDTV standard," *Multichannel News* (June 12, 1995):12A.

50. Robert Rast, Statement of Robert M. Rast, Vice President, HDTV Business Development, General Instrument Corporation Communications Division, (Washington, D.C.: FDCH Congressional Testimony, March 17, 1994), archived on NEXIS, an electronic information service of Mead Data Central. For TV station point of view, see Doug Halonen, "FCC: Who pays for advanced TV?" *Electronic Media* (March 1995):1, 75.

51. Edmund Rosenthal, "FBC studies multiplexing strategies," *Electronic Media* (February 28, 1994):26.

52. Jenny Hontz, "Infohighway bill passes Senate panel," *Electronic Media* (August 15, 1994):1.

53. Dennis Wharton, "HDTV org threatened by flexibility," *Daily Variety* (August 8, 1994):32.

54. Peter Lambert, "Abel says multichannel options could pay for HDTV," *Broadcasting* 122:44 (October 26, 1992):44–45. See also, Richard Wiley, "Entertainment for tonight (and tomorrow too)," *The Recorder* (July 26, 1994):6–10.

55. Harry A. Jessell, "Broadcasters come together over HDTV," *Broadcasting & Cable* (April 17, 1995):6. See also Jube Shiver Jr., "Congress urged to make broadcasters pay license fees for new technology," *Los Angeles Times* (November 8, 1995):D1, D4.

56. "Commission background note on digital TV," *Reuters News Service* (December 6, 1993). For country-by-country transmission scheme selection, see: Commission of the European Communities, "Wide-screen television lifts off," *RAPID* (press release IP:94-21, January 14, 1994), both archived on NEXIS, an electronic information service of Mead Data Central.

57. "Member states ready to impose EU norm on digital TV," *European Insight* (June 10, 1994):615.

58. Don West, "HDTV gauntlet thrown down in Montreux," *Broadcasting & Cable* (June 19, 1995):37–38.

59. "Will shift to digital HDTV, Japan firm says," *Los Angeles Times* (June 9, 1994):D3.

60. "Hamamatsu City designated HiVision City by Ministry of Posts," *COMLine Daily News Telecommunication* (September 6, 1994), archived on NEXIS.

61. "NHK and 9 private stations for trials of EDTV," *COMLine Daily News Telecommunication* (September 1, 1994), archived on NEXIS.

62. Don West, "Wiley talks technical," *Broadcasting & Cable* (June 19, 1995):39.

63. West, "HDTV gauntlet," op. cit., 38.

64. Barry L. Bayus, "High-definition television: assessing demand forecasts for a next generation consumer durable," *Management Science* 39:11 (November 1993):1319–1333. For the original AEA forecast, see: American Electronics Association, *"High Definition Television (HDTV): Economic Analysis on Impact,"* prepared by the ATV Task Force Economic Impact Team of the AEA, Santa Clara, CA, (November 1988).

65. "U.S. Staging Comeback in Technology," *Los Angeles Times* (September 19, 1994):D3.

2

The Digital Destiny

When Woo Paik and Jerry Heller of General Instrument first began considering how to transform their work on HBO's signal encryption solution into HDTV, they quickly realized the only answer was to change from an analog to a digital television signal. They reached this conclusion because the amount of information in the HDTV picture was simply more data than could be transmitted over the air in the FCC-allotted 6 MHz bandwidth. In this way, high definition television became the catalyst to an even more important change in our current television system: digital television (DTV).

"Digital" is a word that has become part of the popular vernacular. Yet many people use the word without knowing or much caring what the term means. For people who work in the communication industries, however, understanding the meaning of this now-common expression is of paramount importance because so much of the discussion of both the present and the future is conducted with the terms and concepts of the digital revolution and because it offers so many opportunities.

Digitization brings information handling to a new level of ease and efficiency. It allows the transformation of diverse original materials, such as music, voice, still images, moving video, and numerically-represented information into a universal, compact, transportable, processible, storable, and retrievable format.[1]

Digital processing brings awesome speed to the manipulation of masses of information, including computationally-intensive procedures that simplify and minimize the data. Digital storage and retrieval permit unlimited, virtually error-free recreation of the original material, as distinct from the successive, error-full copying procedures that non-digital media require for duplication.

Digitization has already revolutionized text processing and numerical computation. Now it is poised to transform the visual world of television. In the 1992–1993 fiscal year, U.S. companies spent $150 million on research and development of digital video for three applications: 1) professional equipment for the generation, manipulation, and editing of full-motion images; 2) images stranded on personal computers; and 3) delivery of images across telephone and cable networks, for such uses as video-on-demand, videoconferencing, and multi-player video games.[2] Analysts predict each of these markets will bring in millions of dollars, perhaps billions, to the companies that can create popular applications for them.

There are many advantages to digital television. Compression reduces the information in the signal, so more signals fit in the same bandwidth. This brings down the high cost of bandwidth for every kind of over-the-air delivery of television—broadcasting, satellite direct-to-home, and cellular. Digital technology also brings CD-quality sound and clearer images. It enables the wide-screen 16:9 format, providing an interim step between the current system and HDTV. Finally, digital television makes convergence at the hardware level possible: the coming together of computer, telephone, and broadcast equipment and networks.

Digitization underlies all the coming changes to the communication infrastructure, including television. Recent advances in processing power and speed will appear in video products and services in the next five years. Microchips are now available that can perform 8 billion operations per second for real-time video processing. The broadband PCI bus accelerates a computer's internal speed so it can transport the information load of images. New, inexpensive software-based video compression programs will allow even consumers to send images over their home telephones.

This chapter will first describe the nature of digital and analog signals and how they are transmitted. We then turn to the extent to which the television system has already become digital and the likely timetable of adoption. Finally, the chapter covers digital compression techniques for moving video which will allow distributors to deliver many more program services to consumers. The span of this important revolution demonstrates how important it is for industry practitioners to have a basic understanding of the process of digitization and the data compression it enables.

The Digital Revolution: Freeing the Message from the Messenger

In the 1970s, digital visionary Jonathan Seybold published a newsletter to printing concerns that covered progress in computerized typesetting. Seybold describes the sense of wonder he felt when he realized that digital computers had set a historical force in motion, not only because of computer technology itself, but because of the data structure it brought into being. Seybold argues that for the first time in history, translating messages into digital data frees them from the media that has heretofore served as their messengers.[3]

Messages and media have always been communicational Siamese twins. Now through digitization, paintings become independent of canvas and paint, music no longer necessarily resides on a vinyl record or a compact disc, and moving images come to life, unfettered by celluloid and magnetic tape. All these disparate messages are expressed by gazillions of 0s and 1s, tossed into the great bitbucket of computer memory.

As digital data, messages are infinitely plastic, transformable into other, new forms that allow new interpretations and expressions. Music becomes visual, translated into colors, notes, graphs, bars, or dots. Manipulating images is commonplace. Using a system costing less than $5,000, a competent computer user can process and retouch digitized images in ways only experts with hundreds of thousands of dollars of equipment could have done just a decade ago.

The unparalleled plasticity of digital data cannot be overemphasized. It permits changes and amendments to messages, usually accomplished more easily in the digital domain than in the primary format. Once stored, digital data is extremely compact, almost always smaller than the initial data. For example, a novel or the complete score for a concerto can be stored on a $3\frac{1}{2}''$ diskette. From storage, random access to any part of the message is possible: With little difficulty, any page of text, any frame of a series of moving images, or any note sequence of music can be called up. Figure 2.1 demonstrates the flexibility of digital data.

The transformability of digital data allows easy, instant, virtually error-free replication of images, regardless of the complexity of the original material. This near-perfect replication is due to the fact that the image is not copied or reproduced; rather, it is re-created anew each time from the stored 0s and 1s. Error-free recreation means that images can be processed and re-processed in post-production with no

Figure 2.1 Digitally creating the Virtual Studio. Source: Ultimatte.

loss of quality, a characteristic termed multi-generational perform-ance.[4]

The loss of picture quality is a severe problem with analog im-ages. Every time the sequence is recopied, the signal is a "generation" further from the originally-produced signal. For example, television directors take great care to avoid copying videotape because with each successive generation, the image becomes less and less clear—and more marred by unwanted "artifacts." Since digitized video recreates the original rather than copying it, the result is more "robust" (retains its quality) through multiple processes.

Tying many computers together into networks adds to the flexi-bility that is the defining characteristic of digital data. In a networked world, many users may have access to stored material, allowing any one of them to dissect, alter, or disseminate all or part of the original document. Messages can be transmitted from one to many others (po-tentially millions of people) with little more than a keystroke.

While these applications are apparently admirable, they also en-tail some negative consequences. The distinctions between different types of public works is confusing and may weaken intellectual prop-erty rights.[5] The ease of changing messages may result in unauthor-ized alterations to important works. Even more serious, documentary evidence may be lost, destroyed, or fabricated. With these caveats in mind, this chapter now turns to a consideration of digital television.

The Multiple Universes of Digital Video

There are several levels of digital television (DTV) already in use and it will continue to replace analog equipment. One way to classify DTV is by platform technologies: 1) Variable-quality video stranded on indi-vidual personal computers; 2) Low-to-moderate-quality video trans-ported over bandwidth-limited telephone networks; 3) Distributed high quality video transported over broadband cable and telephone networks and transmitted over the air.

Another technological perspective on DTV is to rank systems in order of functional complexity. This analysis leads to the following ranking:

Level 1: *Dumb*—Analog-digital or digital-analog converter that does not allow the user to exert any control over the equip-ment. Its functionality is entirely predetermined;

Level 2: *Digital control*—PC-based video edit controller that allows user to manipulate videotape and performs digital effects on a few frames at a time;

Level 3: *Digital video data and control*—Computer-based nonlinear editing system that produces digital data and allows manipulation of a continuous stream of fully digitized images;

Level 4: *Distributed digital video*—Server–client technology that allows real-time, multiple-access to video archives, video-on-demand, and other asymmetrical digital video services. Asymmetrical means that there is more downstream (server-to-client) capacity than upstream (client-to-server) capacity;

Level 5: *Networked two-way digital video*—Videotelephone, PC-or workstation-based desktop videoconferencing, shared video, multi-user games, and other symmetrical digital video services, where participants can both send and receive digital video in real time.

Any one of these functions can occur at variable levels of quality. When the CCIR established standards for digitizing the television picture, they decided to issue a family of standards specifying agreed-upon levels of quality, rather than identifying one single standard.

When these standards are integrated into digital video equipment, the products come in several different levels of quality and complexity.[6] The highest level is "studio quality," referring to first-generation video sent directly from studio cameras. "Broadcast quality" is at the lower level of this category. It describes images that have been processed, edited, and often layered with text or matted elements. Video gathered with field equipment, such as news cameras, are at the lowest level of studio-quality images.

The next level is "industrial quality," often used for corporate communications, professionally-videotaped weddings and ceremonies, and some news gathering applications. One wag observed that its quality is sufficient to tell whether the smile on the face of a videoconferencing participant is sincere—approximately the level of S-VHS quality.

"Consumer quality" refers to DTV of standard VHS quality, what viewers expect from a rented videocassette. Digital video products at all levels of quality and complexity have made extraordinary advances in the past decade.

Studio and broadcast quality digital video products

The driving force for costly professional equipment is the need for exceptionally sharp, clean images in feature films, broadcast television, and television commercials. These applications demand the highest quality, while subjecting the original material to the most image-degrading post-production processing, frequently involving multiple layers of effects.

In the late 1980s, pioneering high-end post-production companies and in-house operations began constructing studio-quality digital facilities—now they are becoming commonplace. The Home Box Office cable network was in the vanguard, opening their facility in 1989, as was director George Lucas' Industrial Light and Magic digital special effects company ("Jurassic Park," "Terminator 2").[7] In mid-1994, Sony built Image Works, an in-house, state-of-the-art, digital post-production studio. Disney created its own facility as well. At about the same time, computer giant IBM teamed with the digital special effects wizard Stan Winston ("Star Wars," "Terminator") to open Digital Domain, specializing in state-of-the-art digital post-production.[8]

During the past decade, the increased contributions of digital technology became clear. From pre-production storyboarding to post-production editing and compositing, digital processing is deeply embedded in today's movie and television production. Morphing and other high-impact effects added unique entertainment value to "Terminator 2." Cut-and-paste techniques made the characters in "The Coneheads" look convincing.

"The Beauty and the Beast" illustrated the cost-saving capability of digital assistance in animation. Sophisticated digital matting made it possible to replicate extras in "In The Line of Fire" and to cut that portion of the budget by as much as 80%. In pre-production for "Addams Family Values," producers David and Heather Nicksay used computerized storyboarding and scheduling to save considerable time in a very tight production schedule.

In the entertainment industry, the pros invoke an old, cynical adage, "Let's fix it in post!" whenever there are problems in production. The infusion of digital technology is changing that aphorism to, "Let's create it in post!"[9] George Lucas digitally created matte backgrounds for the television show, "Young Indiana Jones," and edited the entire program on digital equipment. The dinosaurs in Steven Spielberg's "Jurassic Park" demonstrated how digital image generation could create "characters" and effects not otherwise possible at all. And the double waterfall that heightened the drama of Meryl Streep's role in "The River Wild" was created by digitally duplicating the one waterfall that was actually there.[10]

Broadcasters are converting to digital, too. ABC invested $153 million in digital equipment in 1994 and 1995, starting with the replacement of analog VTRs by digital D-2 VTRs. Footage will be stored on the Media Pool tapeless server (from manufacturer BTS) and routed in and out of the network's digital editing rooms by a BTS routing switcher. ABC is also building digital control rooms, including digital monitors, switchers, special effects units, and audio equipment.[11]

While digital control and processing has triumphed, implementation of real-time, two-way networked video studio applications is spreading slowly. Two-way, networked video allows creative people to conduct real-time editing and image processing when they are in different geographical locations. Currently, creative people are at work on The Silicon Valley Test Track, HollyNet, and the Silicon Graphics Studio with their ability to allow high-speed, high-capacity, bi-directional collaboration.[12]

Digital video products at the industrial level of quality

Industrial-quality video equipment is a fast-growing market. It satisfies the need for small-screen applications and fills many corporate, promotional, and informational needs. The quality is sufficiently high for the images to withstand editing, matting, and layering with text, although with less resolution, color saturation and contrast than studio-quality pictures.

Just in the past few years, the increase in processing power of computer chips has allowed the introduction of industrial-quality digital video products at all levels of complexity. Workstation and even PC-based hardware, with companion software, is now sufficiently powerful to convert high-resolution analog TV signals to digital data and back into an analog format.

There are also many products that provide digital control. The Matrox Studio and the Fast Video Machine are computer-based edit controllers that manage two or three input VCRs, provide switcher functions, and output the result to a VCR for recording on tape.

In addition to digital control, products that perform digital processing now compete with analog equipment. The Avid Media Pro and Media ProSuite, the only all-digital, full-screen nonlinear editing systems, have successfully penetrated the industrial market. Many advertising agencies use the Avid to edit commercials for client approval that show exactly what the finished spot will look like, including effects, graphics, and text. In the past year, Avid has developed a full-screen, 30 frames per second editor, an advance over the one-quarter to one-half screen model previously available.

Full-featured enabling software programs, such as Aldus Free Hand, PostScript, and Adobe Photoshop, have become widespread, bringing an enormous range of special effects to the desktop, previously only executionable at multi-million-dollar post-production facilities.

There is some growth in industrial-level networked video equipment as well. Organizations have begun to distribute video over their client-server systems in the workplace. The last generation of file servers on local area networks carried text and low-resolution graphics; few transported audio, high-resolution graphics, or full-motion video. However, as organizations upgrade their systems for videoconferencing, they increasingly install higher-capacity servers and storage devices to handle industrial-quality digital video.

Videoconferencing is a more and more common use for industrial-level networked video, particularly systems designed for small groups ("huddleware") and individual two-way audio-video communication. The technology varies. It can be picture phones or telephones linked to computer screens, with images carried over telephone lines. Self-contained "rollabout" systems, marketed for small group use, include digital encoder and decoders (codecs), side-by-side camera, and monitor.

Digital video products at the consumer level of quality

The least developed area of DTV products is in the lowest commercially-viable level of quality, equipment intended for consumers that reproduces images comparable to those of ½" VHS videotape. There are a few circuit boards for consumer computers that will translate analog video signals into digital data.

There are no purely digitally-controlled video systems at the consumer level. One reason is that digital control of videotape machines requires the use of extremely expensive VCRs, priced well above the budget of most consumers. Consumer VCRs don't have the electronics that permit control, nor do they have the mechanical precision to execute frame-accurate editing.

Industrial-level products, such as Apple's QuickTime, the Matrox Studio, and the Fast Video Machine have some crossover into the consumer market for serious hobbyists with higher-than-average disposable income. In 1995, the price tag for a complete system may run between $10,000 and $50,000, depending on the equipment and features.

Figure 2.2 shows a Level 2 product which allows digital control over analog video. Figure 2.3 shows an example of a Level 3 product that allows the manipulation as well as control of digital video data.

For consumer equipment, digital control goes hand-in-hand with digital processing. One example in this product category is the Amiga-based Video Toaster. The Amiga computer provides digital control and is equipped with additional boards that grab video frames and perform substantial digital processing. Toaster features include previously-expensive transitions and special effects: wipes (one picture pushes another picture off the screen), dissolves (one picture dissolves into another picture), fly-ins and fly-outs (a picture comes flying in and expands to fill the frame, or squeezes down and flies out), character generation of text, and image quantization (turns the image into squares) and polarization (turns the image into contour-only).

Client-server distributed video systems have no consumer counterpart. The closest is the Internet's World Wide Web which allows

Figure 2.2 The Fast Video Machine System 3000 that provides control over videotape editing. Source: Fast Video.

Figure 2.3 Apple QuickTime lets users make digital movies. Source: Apple Computers, Inc.

computer users to access photographs and audio. It is possible to view video but only after installing special telephone lines. At present, there are efforts to digitize material from university libraries, including still images. The federal government has pledged $24 million toward digitizing materials in university libraries. Grants will go to Carnegie-Mellon, Stanford, the University of California at Berkeley and Santa Barbara, the University of Illinois, and the University of Michigan. People will access these "virtual libraries" from the Internet.[13]

So far, this chapter has covered existing digital video, its uses, and the products that deliver it to specific markets. The next section will examine the process by which analog video signals become digital video signals.

Count Your Digits: Analog Versus Digital

According to Webster's dictionary, the word digital comes from "digits," meaning fingers and toes, the oldest calculators of all. So in a sense, "digital" means counting, but it has also come to include other ideas such as "discrete," "binary," "sampling," and "high-quality, clear sound."

Another word we often hear applied to our current television system is "analog" (or analogue, as the British would write)—the opposite of digital. Analog is also at the root of the word "analogous," which means "similar to" or "comparable with."

Consider the analog and digital watchfaces in Figure 2.4. The analog watchface is analogous to the revolution of the sun. Allowing for the 2-to-1 compression (12 hours of day, 12 hours of night), we can roughly calculate the position of the sun from the big hand on the dial. If we were given this code, we wouldn't need to be able to read the Arabic numerals at all; we could estimate the time simply by the position of the big hand alone. By contrast, the digital watchface tells one time and one time only. If we don't read Arabic numerals, we're just out of luck since we don't have the clues that the analogic system gives us.

Another point to note is that the analog watch gives us the time as one point along a continuum of points, whereas the digital watch tells us the time at one time only. This difference demonstrates an important distinction between analog and digital information: Analog data is continuous; digital data is discrete.

Discrete does not describe a person who doesn't gossip (which is spelled discreet); rather, it means separate categories. Digitizing a continuous signal into discrete elements can be seen if we put a wave representing a wavering musical note on a graph. The analog and digital measurements are shown in Figure 2.5.

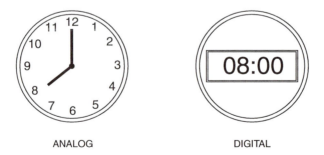

ANALOG   DIGITAL

Figure 2.4 Analog and digital: Two ways to measure and present the same underlying reality.

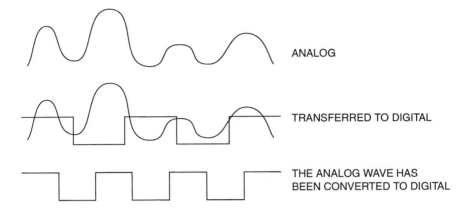

Figure 2.5 Sampling an analog signal.

Sampling: The first step to convert analog to digital data

Starting with the analog representation, the first step in digitization is to take measurements at discrete intervals, approximately twice the rate of the occurrence of the event. (This formula was discovered by Harry Nyquist, an early researcher of telegraph communication systems.) For example, if a musical note changed every half-second, then measuring the stream every quarter-second would capture all the information, allowing all the intervening material to be discarded.

This process of measuring only at pre-designated points is called "sampling." The measurement could be taken every nanosecond, every millisecond, every quarter- or half-second—the decision depends on the frequency of changes in the signal, the meaning of the data, and the purpose to which it is being put.

Figure 2.4 shows two important points that help to understand DTV. Analog and digital signals point to different systems of measurement and representation, rather than differences in the underlying reality: Both signals show the time. The discrete points mark where the continuous analog signal is sampled. The frequency with which they occur is called the "sampling rate."

The higher the sampling rate, the closer the digital data will resemble the continuous analog data; the lower the sampling rate, the more information about the analog data will be lost. There are three important implications of a higher sampling rate when digitizing television pictures. The first is that a greater overall volume of information will result. The second is that the sampled signal will have greater fidelity to the original signal. Finally, these two benefits will be offset by a need for more resources to process, store, and correct errors.

The selection of an appropriate sampling rate is absolutely critical to the resulting quality of digitized television images. For this reason, the CCIR worked for more than two years to arrive at a family of sampling rates that would produce digitized pictures at predetermined levels of quality. Fortunately, the CCIR was also able to find a rate that translated both American NTSC television (525 lines/60 fields) and European PAL/SECAM television (625 lines/50 fields) into a digital format sufficiently similar to allow nearly-identical equipment for the two signals.

Engineers came up with a clever way to make the two systems compatible by finding multiples of both signals. The luminance signal is sampled at 13.5 MHz and the chrominance signals at 6.75 MHz. Explaining the CCIR standard requires a technical discussion beyond this chapter. However, this summary will explain the 4:2:2 standard, a term often encountered in the marketing brochures of digital equipment.

The television signal is composed of black and white information, "luminance," or brightness, called the Y component, and two "chrominance," or color, measures, called the U and V components. The 4:2:2 sampling standard means that for every four luminance samples, two color samples of each of the two chrominance elements will be sampled. To say it another way, 4:2:2 stands for a datastream composed of 4 luminance samples, 2 (U) color samples, and 2 (V) color samples. In a sense, 4:2:2 stands for 4(Y):2(U):2(V).

If all the components were sampled equally, the standard would be 4:4:4, meaning 4 (Y) luminance samples, 4 (U) color samples, and 4 (V) color samples. Similarly, 4:1:1 means 4 (Y) luminance samples to 1 (U) color sample, and 1 (V) color sample, and so on.[14]

One method to reduce the information processing loads is to adopt a lower level of sampling. Many systems start with 4:1:1, 4:1:0, or even 3:1:0, although the lower data rate severely compromises the color quality of the resulting picture.

Quantizing: The second step to convert analog to digital data

Sampling is the first step in converting continuous analog data to discrete digital data, whether it is time, temperature, music, or a television signal. The second step is "quantizing," or determining the number of measurement levels. Like sampling, decisions made about quantizing will determine the quality of the picture.

For example, suppose a television signal is sampled 858 times per line, using a 4:2:2 format. This sets the ratio of the luminance component to the chrominance components, but leaves open the number of levels along which each sample could be rated. The luminance could have 100 intervals between black and white, or 50, or 4, depending on how coarsely or finely the data is quantized.

The highest possible quality of image, either computer-generated or camera-captured, can be described as 8:8:8 RGB, which stands for Red, Green, and Blue, quantized at 24 bits per pixel (bpp). Twenty-four bits is the number of descriptors available to represent information about the sampled unit. "Bits per pixel," or "bpp," is the standard term used to specify the amount of data provided each picture element (pixel) of the television image.

Twenty-four bits per pixel is more data than necessary for today's television systems because it is more information than the human visual system (HVS) can actually perceive. The color palettes of high-end computer graphics systems offer 24 bpp still-image generation. However, professional television processors, which must manipulate moving pictures in real-time, quantize each sample at 10 bits, or even 8 bits for less-stringent applications. As the processing capacity of computers increases in the next few years, the depth of data will increase for television images, although it probably isn't necessary for home viewing.

Coding: The last step in digitizing the television picture

Samples of the analog signal have been taken and quantized at some level. Now it is time to "encode" the data into a standardized structure so that it can be processed, transported, transmitted, stored, retrieved, and ultimately transformed into a picture that the human visual system can perceive.

The digital data that represents the television picture could be encoded into an infinite number of forms. Although it would be possible to represent 10 bits per pixel using the ten Arabic numerals with which we are so familiar, engineers have instead chosen to use binary code to make digital television compatible with other digital data structures.

Binary means two; binary code is based on two symbols. Although the symbols could be "y" and "n," for yes and no, or "a" and "b," or any other two arbitrary markings, by convention binary code is composed of 0s and 1s. The important feature of 0 and 1 (or any other two-symbol set) is that one symbol represents "on" and the other "off," which translate quickly and easily into the presence or absence of electrical current. Computers use this binary code to represent all the numbers, alphabet letters, symbols, and images they process. The computer's ability to perform millions of on-off computations per second gives them extraordinary power and usefulness.

If the only role of binary numbers, or binary digits, were to reside inside computers, understood only by engineers and programmers, there would be no need for most communications industry professionals to know about them. However, binary digits have entered the public discourse through information theory and they are referred to so constantly it is rare to enter a present-day discussion of television delivery without reference to them.

A Bit of Information Theory

The contraction for "binary digit" is "bit." In information processing, a bit is a measure of information, just as an inch is a measure of space and a minute is a measure of time. Eight bits are grouped together to form a "binary word," called a "byte" (pronounced bite).

The term "bit" came into common parlance when Claude E. Shannon and Warren Weaver published "A Mathematical Theory of Information."[15] Shannon and Weaver codified the knowledge about transmitting messages by telegraphy and telephony into a unified theory. They began by identifying the parts of a communication system: source, message, channel, noise, and receiver. They then proposed an original definition of information as "data that reduces uncertainty." The basic unit of information theory, the smallest measure of information, is the "bit."

Information is not merely data. Any array of numbers or letters can be data, but it doesn't become information unless some uncertainty somewhere is reduced. For example, consider the following situation. Sixteen people named Kim Smith work in a huge, worldwide corporation. Two other co-workers are gossiping in the cafeteria when one of them mentions Kim. They quickly realize that there may be some confusion. They might discover which Kim they are talking about in the following way:

Is Kim blonde (= 0) or brunette (= 1)?	0 <1 bit>
Is Kim male (= 0) or female (= 1)?	1 <1 bit>
Is Kim over 30 (= 0) or under 30 (= 1)?	1 <1 bit>
Does Kim work in Accounting (= 0) or Sales (= 1)?	1 <1 bit>

Bit by bit, it takes a total of 4 bits of information for the two conversationalists to establish that out of the 16 possibilities, the actual subject of their confidences is the blonde Ms. Kim Smith, 28, who works in the Sales Department. (It takes 4 questions with 2 answers each to isolate the right Kim Smith; the formula to find which of the 16 answers is correct is $2_4 = 16$.)

Information on the move

People in the communications industries are concerned about distributing communications products, so they must consider the rate at which information can be sent. In digital systems, this inevitably brings up a vocabulary that bewilders those unaccustomed to the terms. At the many advanced television expos, the rapid enunciation of these numbers is a rite of passage, so anyone who wants to be listened to must take the time to master them. (It is not unusual for people to get them wrong, and they often go unchallenged!) In spite of the exotic size of these numbers the basic concepts are easily understood.

When a bunch of bits are transmitted over the air or transported over a cable system, they are collectively called a "bitstream." The speed of the information flow is expressed as "bits per second" (bps), just like the speed of a car is expressed in miles per hour (mph).

Since the amount of data in a television picture is so huge, the number of bits in question number in the millions, even billions. The prefixes we used earlier for the large number of cycles of radio waves will also be useful here: kilobits and kilobytes, to express thousands; megabits and megabytes to express millions; gigabits and gigabytes for billions; and terabits and terabytes for trillions of the little uncertainty reducers.

This nomenclature is used often enough that there are abbreviations for these terms. Capitalization matters; the "k" for kilo and the "b" for bit are lower case. The letters for mega, giga, tera, and byte (M, G, T, and B) are all capitalized. Thus, twenty thousand bits is "20 kb";

twenty thousand bytes is "20 kB." Twenty million bits is "20 Mb"; twenty million bytes is "20 MB," and so forth:

Information	Term	Flow Rate
20 kb	"kilo"	20 kbps (20,000 bits per second)
20 kB		20 kBps (20,000 bytes per second, 160,000 bits per second)
20 Mb	"mega"	20 Mbps (20,000,000 bits per second)
20 MB		20 MBps (20,000,000 bytes per second, 160,000,000 bits per second)
20 Gb	"giga"	20 Gbps (20,000,000,000 bits per second)
20 GB		20 GBps (20,000,000,000 bytes per second, 160,000,000,000 bits per second)
20 Tb	"tera"	20 Tbps (20,000,000,000,000 bits per second)
20 TB		20 TBps (20,000,000,000,000 bytes per second, 160,000,000,000,000 bits per second)

The knowledgeable reader is by now aware that the term "bit" has been used in two ways. First, a bit describes the smallest unit of information that reduces uncertainty. Second, a bit is an integer, a 0 or a 1, a mere placeholder in the dataflow of a message, whether or not the bit itself actually reduces uncertainty.

In their literature and in meetings and conventions, communications industry professionals sometimes describe the capacity of systems in terms of the amount of information they convey. They often use the term "pipe" as an analogy for channels or cables. For example, the capacity of a water pipe that spews out 1,000 gallons per minute would be 60 times larger than one that allowed 1,000 gallons to flow out per hour, and even larger than one that put out a trickle of 1,000 gallons per day.

Similarly, the capacity of a system (or a channel) can be measured in terms of how many bits per second can flow through it. (Think of it as the hydraulic theory of communication—the pipe is the channel and the bitstream is the liquid flowing through it!) Figure 2.6 is a graphic representation of these ideas.

Actually, the issues of how much and how fast information could flow through a channel have been concerns since the beginning of electronic communication. In 1924, Harry Nyquist, the researcher of telegraphy mentioned earlier, first determined the relationship of channel size to the carrying capacity as a ratio. The simple formula is called the

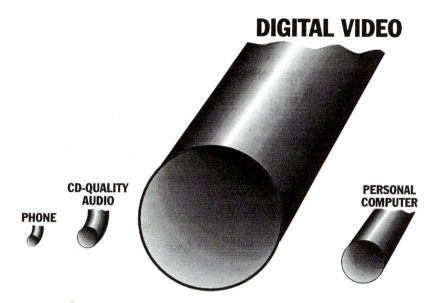

DIGITAL VIDEO

CD-QUALITY
AUDIO

PHONE

PERSONAL
COMPUTER

Figure 2.6 Different media require conduits that have sufficient bandwidth to transport them. Source: Videomaker Magazine.

"Nyquist limit": Channel capacity must be twice the bitrate, two to one, or 2:1.

Later, Shannon (the same man who developed information theory) refined this proposition, making the ratio more flexible. The "Shannon limit" says that theoretically channel capacity may be twice the bitrate, but in practice the ratio is modified by the type of encoding, the amount of accompanying noise, and the amount of power used to transmit the signal.[16]

In the last chapter, the definition for bandwidth was the size of a channel in terms of the number of frequencies on the electromagnetic spectrum that are occupied by the signal. This is the meaning engineers have in mind when they observe that the TV signal uses an enormous 6 MHz of bandwidth per channel.

In advanced communication systems, bandwidth is also stated in terms of the amount of information a channel will permit to flow through it in a given period of time. For example, at a convention seminar attendees might hear: "You can run 10 megabits per second through that pipe."

It stands to reason that the huge amount of information in a television picture, which must flow in the proper order and in real time for the viewer to receive a coherent picture, necessarily entails a rapid bitrate and a large-bandwidth channel—big, big pipes. Such big pipes, or conduits, are called "broadband" channels and wired systems that are big enough to carry high-quality, full-motion video are called "broadband networks." This label differentiates them from "narrow-band" channels, like dinky 3 kHz copper telephone wires.

Discussions of advanced television frequently include a reference to "pipes versus pictures" or "conduits versus content." One reason the relationship between the channel capacity and the programming is stated in such an adversarial manner is because, at present, conduits constitute a limit on content. For example, many consumers complain because their cable system doesn't carry the Golf Channel or the Comedy Central. This problem arises because the conduit has a limited capacity and cable operators cannot carry all the channels their subscribers would like to have.

If we start with Nyquist's 1924 formulation as a rough rule of thumb (channel capacity must equal at least twice the bitrate), we can chart the improvements in encoding and the handling of the problem of noise. Today, telephone companies routinely send 28 kilobits per second (28 kb/s) down 3 kHz telephone lines. In other words, they have achieved a bitrate that is more than 9 times the size of the channel, an extremely important accomplishment when we consider the issues of advanced television systems.

Although it is a great improvement, a 9:1 ratio still isn't enough capacity to permit a digitized TV picture to be transmitted over a 6 MHz channel. Further, for a variety of technical reasons, it is more difficult to achieve the 9:1 ratio with images than with sound. However, suppose for a moment that it were possible. A digitized NTSC picture equals 216 megabits per second (Mb/s), which still requires an 18.6 MHz bandwidth channel (216/9 = 24)—four times the 6 MHz per channel allotted by the FCC.

Turn this problem around the other way: The 6 MHz television channel could theoretically handle up to 54 megabits per second ($6 \times 9 = 54$). Too bad—that's about one-quarter of the 216 Mb/s needed to transmit a good-quality digital TV picture. (In fact, a 6 MHz channel will only carry between 3 and 4 megabits per second.)

The problem is seemingly intractable when considering the HDTV picture, as Woo Paik and Jerry Heller had to do. Its information-laden picture requires a bitrate of 1 gigabit per second (Gb/s), which would require at least a 250 MHz bandwidth channel!

Recall that the HDTV picture has 1,080 active lines and 1,920 pixels on each line, for a total of about 2.1 million pixels. If each pixel requires 8 bits to represent it, then every field of HDTV involves about 16.6 million bits of information. There are two fields to a frame and 30 frames to a second, so one second of HDTV video contains about 1 billion bits (1 Gb) of information!

Do it by the numbers:

Number of pixels:	1,080 lines × 1,920 pixels = 2,073,600 pixels
Number of bits in the HDTV frame:	2,073,600 pixels × 8 bits = 16,588,800 bits × 2 fields = 33,177,600 bits/frame
Number of bits per second:	33,177,600 bits/frame × 30 seconds = 995,328,000 bits/second, nearly 1 billion bits per second (1 Gb/s)

There is only one solution to this problem: Reduce the information in the television signal through a technique called compression. Even though digitizing the signal initially creates more data, it is impossible to achieve the needed reduction with an analog signal. Digital processing is the only way.

Digitizing the Television Picture

As we have seen, the number of pixels in the NTSC picture is variable, depending on how many times each line is sampled. The highest quality broadcast images have about 450,450 pixels to represent them, resulting from the product of lines times 858 samples per line. (When a lower quality image is acceptable, as is often the case, a lower sampling rate will result in fewer pixels.)

For each pixel, it takes 8 bits to represent the color so there are about 3.6 megabits (Mb) of information per field, and 2 fields per frame, so there are about 7.2 Mb in a frame of digitized NTSC. This converts to the bitrate by multiplying it times 30 frames per second:

Number of pixels in NTSC video:	525 lines × 858 samples = 450,450
Number of megabits per field in NTSC video:	450,450 × 8 = 3,603,600
Number of megabits per frame in NTSC video:	3,603,600 × 2 = 7,207,200

Bitrate for NTSC video: $7{,}207{,}200 \times 30 = 216{,}276{,}000$ or about 216 megabits per second (Mb/s).

The video must be transmitted or transported at that real-time rate, 216 megabits per second, for a coherent, real-time television picture to arrive at a receiver. In fact, these numbers describe some of the first digital video ever transmitted when images were sent back from Mars by the Mariner lander and from Jupiter and its moons by the Voyager spacecraft.[17]

Compressing the Digitized Television Picture

The first step towards solving the problem of the information overhead required by HDTV was to digitize the analog signal, enabling the application of compression techniques. Compression does not mean making the television picture smaller; rather, it refers to more efficient encoding of the signal to represent the same picture with fewer bits of information. It is the information, not the picture, that is compressed!

Compression schemes and processes are divided into two types: "lossless" and "lossy." Lossless techniques rearrange the data for ease or compactness in further processing; however, no data is actually dropped or lost. By contrast, lossy processes discard information that can never be retrieved. Sometimes the missing data can be reconstituted from the surrounding data, but often it is lost forever. Whether or not the viewer can perceive the discarded data depends on which data is lost and how extensive the loss is.[18]

The name for the equipment that compresses the picture and then re-constitutes it is called a "codec," standing for compressor-decompressor. Some codecs operate mainly with dedicated hardware, while other compression schemes use off-the-shelf, general-purpose equipment and use software to carry out the compression.

Types of Compression: Four Functions, Four Standards

There are four methods of compressing images that are endorsed by one of the standards-setting bodies.[19] All the officially-sanctioned schemes are in common usage. They are frequently referred to in the literature, in lectures, and in conversations at the many seminars and conventions about new technical developments in the broadcast, cable, telephone, and computer industries.

These four standards are "H.261" (or "P*64"), "JPEG," "MPEG-1," and "MPEG-2," each designed to address a different media context. All use both dedicated hardware and software for compression, which is faster than software-only solutions—but they are also more expensive.

P*64 or H.261: Compression standard for video over the telephone

Many business people, families, and friends would like to see their conversational partner. The CCITT (Consultative Committee for International Telegraphy and Telephony) developed a compression standard for video-telephone applications to send images over existing copper telephone wires.

The standard has two names, H.261 (pronounced "h dot two-six-one") and P*64 (pronounced "p by sixty-four"). This is a pragmatic standard that is generally understood to provide enough detail to tell if the smile on the face of the other person is sincere. H.261 provides a low-resolution, color image of a person's face or upper body, talking and gesturing against a minimally-changing or out-of-focus background.

JPEG: The standard for still images

The International Standards Organization established a standard for still images called JPEG (pronounced jay-peg), primarily for applications using digitized images in a computer environment. JPEG stands for Joint Photographic Experts' Group, the title of the technical committee that developed the standard.

JPEG is based on a technique called the "discrete cosine transform," discussed later in this chapter. JPEG provides high-resolution images, including photographs, computer-enhanced photos, computer generated images, and specialized medical images.

In 1993, several computer-based editing systems began using a version of JPEG compression (called motion-JPEG) to edit moving images because it affords a high level of compression on individual frames. Each frame is whole and complete, so motion-JPEG allows frame-accurate editing. (As we will see further in this chapter, compression techniques for motion video preserve only fragmentary information on some frames, so these schemes will not work for editing systems.)

MPEG-1: A standard for CD-ROM

The compression standard for moving images on CD-ROM is called MPEG-1 (pronounced em-peg one). MPEG stands for Moving Pictures Experts' Group, a committee composed of technical representatives from film companies, video manufacturers, computer companies, and CD-ROM developers and makers. Like JPEG, MPEG-1 compression uses the discrete cosine transform technique. MPEG-1 allows CD-ROMs to provide highly compressed moving pictures of only moderate quality at a data rate of 1.5 kilobits per second (kb/s).

Many companies abandoned the MPEG-1 standard in favor of proprietary compression algorithms, claiming that the process of setting the standard had become overly politicized. Critics of the final standard claim that representatives of the member companies jockeyed for a formula that would offer them a competitive advantage, rather than promoting the most technically superior process.[20] As a result, detractors say that the compromised programming code is too clumsy and complicated for efficient compression.

However, the Hughes/RCA satellite-delivered, direct-to-home (DTH) digital television signal, launched in the fall of 1994, used MPEG-1 for the compression needed to bring consumers 150 channels. Hughes wanted the higher quality promised by MPEG-2, but waiting for it would have delayed their 1994 launch by an entire year.[21] Although the digital picture itself was relatively free of artifacts, some consumers complained about the quality of the decompressed video, saying it looked blocky and lacked detail. (Later reports absolved the compression technique, claiming that the poor picture quality was really the result of a very narrowband satellite uplink.)

MPEG-2: A standard for broadcast-quality video

MPEG-2 is the standard developed for professional-quality moving images. As of late 1994, the discrete cosine transform-based algorithm was operating in several test facilities but was available in only a few products, usually in addition to another, proprietary compression system, such as in products released by General Instrument Corporation. However, in late 1995, Compaq and Apple announced they were working to put MPEG-2 processing in their multimedia computers to bring broadcast-quality video to the home desktop.

Using MPEG-2, a codec can compress broadcast-quality video at a rate of nearly 10 megabits per second (Mb/s). C-Cube Microsystems has developed a chip set that will allow real-time compression at this high data rate and AT&T has demonstrated an MPEG-2 decoder chip. MPEG-2 is the standard of choice for television transmission, cable transport, satellite-delivered television, and other applications requiring very high-quality moving images. Figure 2.7 shows the high-tech environment where MPEG conversion takes place.

The problem with MPEG-2 is that one way it compresses is to discard data that is redundant in adjacent frames, saving only the pixels that are different from the previous frame. It's impossible to edit from this compressed data because it takes too much processing power to reconstitute this much missing data, while decompressing a realtime bitstream at the same time. Recognizing this problem, research on an MPEG-2+ standard is underway that would allow editing of broadcast-quality compressed video.

The MPEG group is also investigating a standard for high-quality video telephony that will be called MPEG-4. (MPEG-3 was intended for HDTV but was dropped when processing power made it possible for the high definition picture to be compressed in MPEG-2.)

Figure 2.7 An MPEG-2 digital video processing and editing bay. Source: Optibase.

Proprietary standards: Compression techniques with no official sanction

There are several proprietary compression schemes that, while not blessed by official standards-setting bodies, are sufficiently well-known and promising to deserve mention. These techniques are vector quantization, wavelet, and fractal compression.

Vector quantization is a software-only compression technique developed by Scientific-Atlanta. It is relatively inexpensive but significantly slower than combined hardware/software codecs. VQ is also severely asymmetrical, which means it takes much longer to encode the video than it does to decode it, especially for high-quality images. Depending on the specific equipment used, compressing the video can take as little as ten minutes to encode one minute of video—or as much as an hour or two.

A "vector" is an engineering term for a straight line. In VQ compression, a line is drawn through the color spectrum of the television signal from black to white. Each color is represented at a place along the line. VQ allows very high compression by eliminating most of the color and luminance information from the data stream.

Unfortunately, the range of color is lost when it is replaced by a much smaller number of points along the vector. For example, although there may be 100 possible variations of a medium to dark blue, all will be shown as a single dark blue when VQ is used. As a result, although it is computationally simple, offers a high degree of compression, and is inexpensive, it doesn't deliver very satisfactory performance.

Wavelet compression was made possible through a discovery by Bell Labs' mathematician Ingrid Daubechies. The technique deconstructs and transforms the waves that carry television signals into wavelets (meaning "small waves"). It provides efficient procedures for compressing the information in waves of all types and yet preserves the important features of the wave itself. This preservation endows the procedure with higher quality than other compression schemes.

A leader in wavelet compression is Aware, Inc., based in Cambridge, MA. The technique is the basis for several innovative codecs, including the low-cost, software-only Captain Crunch codec by Media Vision and the ImMIX VideoCube. Both objective and subjective tests indicate wavelet-based compression may be superior to other compression algorithms.

Fractal compression is a new technique based on the similarity of the infinite detail of natural images, independent of the level of observation. For example, a coastline has a jagged outline, whether viewed from outerspace, a high hill, on the beach, or staring at sand particles. Stephen Mandelbrot and others have used simple mathematical formulas to produce images of extraordinary complexity and detail.

A company in Norcross, Georgia, Iterated Systems, does the most advanced work in fractal compression. As of late 1994, the technique is most successful with still images. The company's 1995 demonstration of compressed moving images at COMDEX showed that there is much work to be done before fractal compression can compete with the alternative techniques.[22] Figure 2.8 compares the fractal technique to other schemes for still image compression.

Probably the most speculative work is being conducted by researchers at Bell Labs, Yale University, and the University of California at San Diego. They are working with a company, HNC Software, to apply neural network theory to video compression. A neural network generates a cloud-like map of points (each of which represents thousands of bits) from billions of bits of data. The image is then regenerated from the "cloud."

Compression: A Complex, Multi-Step Process

Table 2.1 lists all the steps involved in compressing a television signal.[23] Broadly outlined, those steps are: 1) "perceptual coding"; 2) "spatial compression"; 3) "temporal compression."

Perceptual coding

Perceptual coding means exploiting the properties of the human visual system (HVS) to eliminate unnecessary data. Exploiting properties of human perception has long guided efforts to compress the enormous bandwidth of the television signal.

For example, people are much more sensitive to differences in brightness levels than they are to color information. In 1953, engineers on the National Television Standards Committee used this characteristic of perception when they converted the original black and white signal to the current color signal by coding it with more luminance (brightness) information than color information.[24] Today's 4:2:2 sampling standard works on the same principle: The luminance data is sampled at twice the rate as the color data.

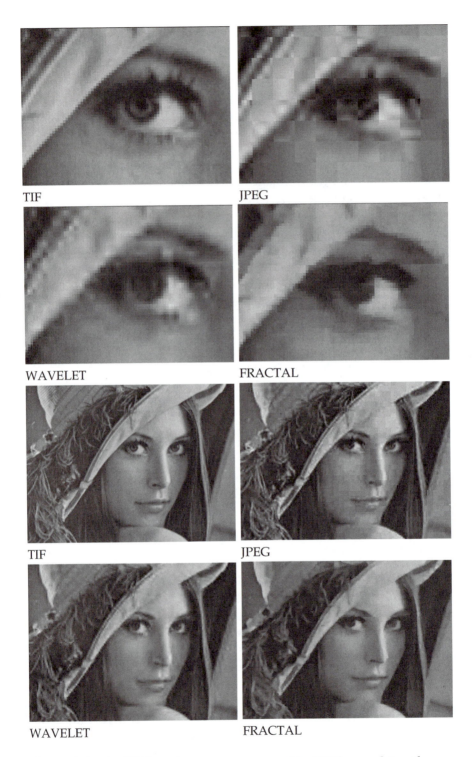

Figure 2.8 The TIF Lena image as compressed by JPEG, wavelet, and fractal compression techniques.

Table 2.1 Preprocessing and Compressing the Television Signal

Compression Technique	Amount of Compression	Description
Perceptual coding	Not counted	Altering the signal so that it only provides the information viewers need to see a good image.
PREPROCESSING		
Filtering	None	Removes noise from signal
Color space conversion	33%	Reduces 3 variables, red, blue, and green, to 2 variables
Digitizing	25%	Sampling, quantizing, and coding the analog signal
Decimating (Scaling)	75%	Subsample digitized signal and discard remaining information
INTRAFRAME COMPRESSION		
Data transformation	None	Rearranges data for further compression
Quantizing	60%	Assign values to rearranged data
Compaction encoding	31%	Rewrite data efficiently, using Huffman variable length coding, or arithmetic encoding.
INTERFRAME COMPRESSION		
Predictive coding	None	Codec looks at key frame, or I-frame, which is divided into 16 areas and amount of change in each area is estimated. Based on estimate, codec selects new I-frame or changes frame.
Decimation	50%	Compressed data is re-sampled and about 50% of the data is discarded.

Another important aspect of perception for compression is that research on vision pathways shows that there are two parallel but distinct channels to the brain from the retina of the eye. Experiments to monitor the activity of single cells in the eye's retina indicate that the first channel, the sustained system, processes the details of a stationary scene. The second channel, the transitory system, detects motion and rapid changes in the scene and updates the information quickly and continually.[25]

Based on an understanding of the HVS, researchers Robert and Karen Glenn developed a television transmission system that reduces the data in the television signal by taking advantages of these findings. In the Glenn system, which is compatible with current broadcast technology, the detailed non-moving portions of a picture are sent less often than the fast-moving portions. A further refinement is that while motion must be updated rapidly, it doesn't have to be presented in as much detail as the non-moving parts of the picture.

Removing redundant data

Spatial compression is also called "within frame" or "intraframe" compression. All these terms refer to techniques that remove redundant information from within a given frame of video. For example, suppose a TV picture shows a woman on a horse with mountains and sky in the background. The sky is mostly blue. Instead of transmitting each blue pixel, it is much more succinct to write: "Hey, computer! Pixels 1–110,898 are blue with white patches" (written in digital, of course).

There's another form of data redundancy in moving video: Duplication in time. Techniques employed to identify and remove the information that doesn't change are labeled as "interframe," "within frame," or "spatial" compression—all interchangeable terms. Let's go back to the woman on the horse. We see her in Frame 1. We see her in Frame 2 and there is almost no difference in the sky or the mountains. Only the woman and the horse will have changed within themselves and in relation to the background. The same goes for Frame 3.

One interframe compression technique takes advantage of the similarity between frames by keeping only the pixels that are different, the "difference signal." A sequence of a woman riding the horse, a plane landing, or a boat moving through the water are all amenable to compression through motion estimation and compensation and predictive interpolation. These prediction algorithms use powerful processing capabilities to anticipate how a scene will change based on the information that is already known.

Implementing Digital Television

The benefits of digital television are great, including better quality TV images, compression-related reductions in bandwidth costs for transmission and transport, and compatibility with computer and telephone networks. These advantages led John Hendricks, CEO, Discovery Networks, to remark at the 1993 National Cable Television Association convention in San Francisco that digital television will sweep the country in the next decade. He noted the difficulties with standards, cost, and consumer confusion, but saw them as problems to be solved, rather than barriers.[26]

Hendricks' predictions proved correct. Just a year later, the first major use of digital television emerged. In 1994, Hughes began their Ku-band direct broadcast satellite service, transmitting television signals near the 12 gigahertz frequency range. At launch time, the programming was encoded into MPEG-1, transmitted over the satellite, received by the consumers' 18" dish, and decoded by the settop box. In 1995, the compression was upgraded to MPEG-2.

Technical barriers to implementation

A few disadvantages to digital television still exist. There are problems with the digital signal itself. Developing and setting standards has proved problematic. Finally, the high cost of replacing the installed base of analog equipment for production, transmission, transport, and reception is a formidable barrier to implementation.

The problems associated with picture quality, aside from those caused by compression, include pixellation, contouring, and aliasing.[27] "Pixellation" means that the viewer can see the pixels, revealing the tiny squares that actually make up the images. The problem can result from inadequate sampling, too few levels of quantization, or coarse encoding.

"Contouring" results when adjacent areas of the picture are quantized at the same level within each region, and the two are different from one another. In an analog picture, the range of values allows tiny differences within an area; with digital conversion, these small differences are quantized at the same value giving the entire area a similar appearance. When two adjacent areas have two different but internally identical areas, the line between them may mark an unsightly contour.

"Aliasing" is similar to pixellation in that the images acquire a squared-off look. In this case, however, it is not caused by coarse pix-

Table 2.2 Layers of Digital Formats in a Cable System

System component	Competing digital formats
Programming	MPEG-1, MPEG-2, Motion JPEG, DigiCipher, and other proprietary formats
Network communication	TCP/IP, Asynchronous Transfer Mode (ATM), SONET
File servers	C, C++, object-oriented formats
Operating System	Microware, Microsoft
Settop box	Oracle, Sybase, Informix, COTS

els. Rather, aliasing results from too-high a sampling rate introducing unnecessary, spurious data.

Solutions to the problems with the digital signal exist, especially with increases in the processing power of chip sets for codecs. However, the difficulties of developing and setting standards are much less amenable to solution because they require intricate cooperation in the face of stiff competition.

Implementing digital television in an entire cable system is a major challenge. Engineers must address different formats for eight layers: analog-to-digital conversion, compression, encryption, multiplexing, transport, reception, and hardware. Dr. Sadie Decker, Vice President, Telecommunications, Inc., demonstrates the software challenges, as shown in Table 2.2. Although there is currently no definitive solution, Dr. Decker argues that the guiding principles must be simplicity, reliability, and maintainability.

The economics of digital television

A bitter joke swept the 1993 National Cable Television Association: If a pollster asked viewers what they wanted from television, the two answers they wouldn't hear would be "better picture quality" and "more channels to choose from." The point of the joke is that while there is little demand pull for DTV, there is a rapidly evolving supply push for it.

It is primarily the short supply of over-the-air bandwidth that is driving companies that deliver programming via terrestrial broadcasting and satellite to invest in compressed DTV. In April 1994, Hughes doubled its C- and Ku-band rates, a threat to direct-to-home services, such as PrimeStar, USSB, and Echostar that compete with multi-channel cable offerings.

Broadcasters believe compressed DTV creates an opportunity for them to increase the value of their current 6 MHz of bandwidth, because they could air five channels in the same frequencies where they now air one channel. For example, a local station could air their regular programming on one channel, broadcast a network-supplied all-news channel with local news inserts, offer a sports channel, and provide two time-delayed channels for convenience programming.

By contrast, cable operators are only slowly beginning to accept that there may be advantages from DTV. Until the arrival of direct-to-home satellite services (DTH), many system operators believed they had ample bandwidth. The 1993 Cable Regulation Act and the 1994 FCC-ordered rate rollback discouraged them from adding new basic channels. Moreover, DTV will require cablers to change every settop box, an enormous expense. However, competition from DTH, which offers the better digital picture and multiple premium and pay-per-view channels, are powerful competitive features.

Necessity drove direct-to-home services to adopt DTV first. Broadcasters are the next group most likely to implement it, and, as we have seen, they are investing heavily in digital equipment for their internal operations. Smaller cable operators may not adopt DTV for some period of time, depending on how the competition between cable operators and direct-to-home services plays out. If DTH provides strong competition, based on picture quality, it will force cable operators, and perhaps broadcasters as well, to make the necessary investment to implement DTV sooner, rather than later.

Another consideration is HDTV. Since DTH services are already digital, they are able to provide HDTV pictures with little modification to their hardware. They may be able to gain a competitive advantage by offering HDTV programming which the other delivery services cannot.

Prognosticating the future is difficult and rarely accurate. However, it does seem far more likely that DTV will be implemented within the next decade than HDTV will, since the economics are so much more favorable to digital conversion. However, even in the most favorable climate, progress is slow. Market researcher Robert Aston of Market Vision (Santa Cruz, CA) estimates that based on revenues from both business and consumer entertainment sources, in 1994, 2% of television material was digital; by 1995, that number had risen to only 5%.

Many see the adoption of digital standards as the stepping stone to HDTV sometime later in the 21st century. Moreover, DTV is essential if TV is to become interactive. The next chapter will examine the potential for interactive television.

Notes

1. Pamela Samuelson, "Digital media and the law," *Communications of the ACM*, 34:10 (October 1991):23–29.

2. T. Tucker, "From the boardroom to the desktop," *Teleconnect* (September 1993):50–53.

3. Amy Harmon, "A digital visionary scans the info horizon," *Los Angeles Times* (June 1, 1994):D6, D8. Also, Jonathan Seybold, personal interview by author, October 1994.

4. William J. Mitchell, "When is seeing believing," *Scientific American* (February 1994):68–73.

5. Samuelson, op. cit.

6. Information about these products came from product brochures distributed by the companies involved.

7. Christine Burnish, "HBO studio productions' digital dreams come true," *AV Communications* (June 1989):26–29.

8. Amy Harmon, "Invasion of the film computers," *Los Angeles Times* (August 15, 1993):A1, 22-23.

9. Charles Barish, "Superman's now super digital," *Videography* (October 1993):30-32, 101-102.

10. Bruce Stockler, "Water, water everywhere," *Millimeter* (October 1994):43–44.

11. Chris McConnell, "ABC Television on a digital spending spree," *Broadcasting & Cable* (July 18, 1994):61.

12. Craig Kuhl, "Test track, race track or fast track?" *Convergence* (March 1995):34–39.

13. "Virtual libraries get a boost from Feds," *Chronicle of Higher Education* (October 5, 1994):A26.

14. A. G. Uyttendaele, Digital Video/Video Compression, draft of contribution to ISOG from NANBA Technical Committee, available from A. G. Uyttendaele, Cap Cities/ABC, Inc., 77th W. 66th Street, New York, NY 10023-6298.

15. Claude E. Shannon and Warren Weaver, *A Mathematical Theory of Information* (Urbana: University of Illinois Press, 1949).

16. John R. Pierce, *An Introduction to Information Theory: Symbols, Signals & Noise,* 2nd ed. (New York: Dover Publications, Inc., 1980).

17. John Pierce, op. cit.

18. Arch C. Luther, *Digital Video in the PC Environment* (New York: McGraw-Hill, 1991).

19. Joan Van Tassel, "Compressed Video: Today and Tomorrow," *Videomaker* 12 (April 1994):73–77.

20. Personal communication from a highly placed member of the MPEG-1 committee who requested anonymity.

21. S. Scully, "PrimeStar buys compression for $250 million," *Broadcasting & Cable* 123, no. 32 (August 4, 1993):49.

22. Bob Doyle, telephone interview with author, October 1995.

23. This discussion of compression relies heavily on the excellent articles by Bob Doyle, "How Codecs Work," *NewMedia* 3, no. 3 (March 1994) and Bob Doyle, "Crunch Time for Digital Video," *NewMedia* 3, no. 3, (March 1994):43–50.

24. E. B. Crutchfield, ed., *Engineering Handbook,* 7th ed. (Washington, D.C.: National Association of Broadcasters, 1985).

25. The information about this research and the work of Robert and Karen Glenn came from: Earl Lane, "The Next Generation of TV," *Newsday* (April 5, 1988): Discovery section, 6.

26. S. Merrill Weiss, "Looking back at how far we've come," *TV Technology* (September 1994):33, 36, 67.

27. Luther, op. cit.

3

Interactive Television: Two-Way TV

Introduction

A child plays a video game using a Nintendo player that's hooked up to his television set. On her computer screen, an art director in New York can see a director in Los Angeles, her own image, and a mutually-shared "whiteboard" where each can see their own and the other's notes. As they make notes on the shared graphic, they are talking about their mutual project—all part of the Silicon Graphics Studio for professionals.

A student puts Compton's Encyclopedia on CD-ROM into his computer and plays the listing under Romania, listening to music and watching a short video clip of the capitol city, Bucharest. A financial analyst has CNBC playing in a small window of her computer screen. An Orlando engineer orders a pizza over the telephone. A fan of the Oakland Raiders pushes a button to watch the tight end in a close-up shot, in place of the wide angle shot aired by the FOX network. Finally, a family in Mt. Pleasant, Illinois use their remote to order the episode of ABC's 20/20 News they missed last Friday night.

In spite of the disparate technologies and activities these examples involve, every one of them has been called "interactive television" (IATV) by observers and organizations in the past two years. The range and number of IATV applications, as well as the complexity of the systems needed to deliver some of them, have made the design of IATV programming a slow process.

Dimensions of Interactivity

It is difficult to bring conceptual clarity to the subject of interactivity. To begin with a definition, "inter" is a prefix that means between or among. "Interact" means to act mutually or to engage in reciprocal acts. There must be at least two entities for interaction to take place. In addition, the interaction takes place within some context: Historical, social, political, economic, architectural, and so forth.

Reciprocal does not mean identical. You bow, I curtsey; she argues vociferously, he assents quickly. In the same way, while interaction involves reciprocal actions, they may vary greatly depending on the situation, the interactors, and the technology of communication.

There are many other ways to consider interactions:

Is response rapid or slow?

Is the overall length of each response long, short, or varied?

Does one party have control, is control shared, or is it passed from one interactor to the other?

Do responses relate meaningfully to initiations?

Is the mode of interaction verbal, textual, visual, or acoustic?

What is the purpose of the interaction and is it related to the content?

What is the result of the interaction?

Compare a minimally interactive medium, broadcast television, and a highly interactive medium, the telephone. A viewer watches a TV program, writes to the network, and receives a form letter in return. The interaction is slow, lengthy on the part of the network, and short on the part of the viewer. Control rests almost entirely with the network; the viewer can only change the channel or refrain from writing. The viewer may be responsive to the network, but the network is more responsive to advertisers and viewers in the aggregate than to the individual viewer. The style is audio-visual on the part of the network and textual on the part of the viewer. The purpose of the interaction by the network is to entertain and inform; for the viewer, it is to be entertained and informed.

By contrast, the telephone allows rapid interactions of variable lengths, control is accessible to both parties, and they are able to be highly responsive to one another. The style is audio or minimally graphic (considering black and white faxes), and the purpose is connection and exchange.

The above examples both involve communication exchanges where technology links the communicators. When a technology is interposed between two interactors it may also impose constraints on the interaction.

One of the main characteristics of technology that influences people's communication is its channel capacity. For example, the technology may offer different actors a greater or lesser ability to communicate within the interaction. When the participants have the same channel capacity for expression, the technology, or channel, is called "symmetrical." Symmetrical bandwidth gives both people an equally high, medium, or low capacity channel; a telephone is a good example of a channel with symmetrical bandwidth.

The opposite condition is "asymmetrical bandwidth," where one communication partner uses a channel that permits more information than the channel the other partner is using. Television is a good example of a channel with asymmetrical bandwidth.

Another characteristic of technologies is whether the interactors are networked together or stranded on a reception device. For example, the telephone is a networked reception device; television sets and radios are stranded devices. The computer can be either networked or stranded (standalone) and users move back and forth between the two conditions with a few keystrokes. When it's stranded, it's called "offline"; when it's networked, it's "online."

Tables 3.1 and 3.2 organize the interactive capacities of communication technologies along the dimensions of bi-directional channel capacity and degree of connectedness.

One point to note here is that interaction is tied to the concept of "feedback," another word for response, which is carried in the return channel. The influential early model of communication presented in Shannon and Weaver's Information Theory was a linear, one-way view of the process: SOURCE —> MESSAGE —> CHANNEL —> NOISE —> RECEIVER, often referred to as the SMCR model.[1] In this formulation, both interactors were treated as alternating sources and receivers, and the model worked well for analyzing the adequacy of telephone technology. However, note that there is no overt provision for feedback in the model.

When researchers began studying real interactions and messages, it became clear that the SMCR model was inadequate because the lat-

Table 3.1 Technologies with Symmetrical Communication Channels

Symmetrical Bandwith	Network	Stranded (Standalone)
Source: Low Return: Low	Phone, fax, online chat; multi-user domains; local area networks; search online text databases; text e-mail; some videoconferencing	PC computing: text or math; searching a text database
Source: Medium Return: Medium	Silicon Graphics' Studio, some local area networks; some video-conferencing	Some nonlinear editing systems like Apple QuickTime; PC computing: graphics
Source: High Return: High	Face-to-face conver-sation, Broadband networks, such as Hi-Ovis, Japanese test site	Specialized high-resolution applications, like medical imaging

est message is a response to the previous message.[2] Thus, messages are not independent of one another—earlier messages affect later messages. In other words, some messages serve as feedback to the previous message. Not all messages are direct feedback; people do change the topic into entirely new venues of conversation. Here is how feedback occurs in a typical greeting sequence:

Message 1: Hi!	Source to 2
Message 2: Hi!	Feedback to 1, source to 3
Message 3: How are you?	Feedback to 2, source to 4
Message 4: Fine. You?	Feedback to 3, source to 5
Message 5: Good. How's your family?	Feedback to 4, source to 6

Feedback is central to any framework for sequential interaction.[3] We will return to this important concept later in this chapter. For now, we will look at the technologies of interaction.

Table 3.2 Technologies with Asymmetrical Communication Channels

Asymmetrical Bandwith	Network	Stranded (Standalone)
Source: Low Return: None	Teletext	Audio CD; radio; datacasting; books
Source: Medium Return: Low	CD-ROM over a network; CompuServe CD; Interaxx shopping system with CD-ROM over TV and phone return path	Video games; PC with CD-ROM
Source High Return: None	Current cable TV	Current broadcast TV; VCRs
Source: High Return: Low	Video on demand; near video on demand; ACTV; EON system; Interactive Network; Pay-per view over cable	Avid nonlinear edit system; Search graphics database
Source: High Return: Medium	Some comm'l broadband test sites	Matrox editing system: videotape in and out of computer processor
Source: High Return: High	Experimental test sites: Starbright Pediatric Network	

Symmetrical, networked interactive channels

These channels are dominated by communication activities. Face-to-face interactions, phone calls, fax, e-mail, BBS chat lines, online multi-user domains, local area networks, and the Silicon Graphics' Studio broadband network—every item listed in this category is some form of person-to-person interaction.

One symmetrical activity—face-to-face interaction—is the most common model for all communication, the very standard against which communication and interaction are judged. Of course, it is the channel that is symmetrical, not the participants, so individual differences may result in a very asymmetrical exchange, as between a Marine recruit and a drill sergeant!

However, the face-to-face model is pervasive and many refer to it when examining communication through technologies, such as the telephone, computer, radio, etc. In fact, the ability of a communication device to mimic face-to-face communication is the accepted measure of the "intelligence" of a machine. This model was presented by Alan Turing, who during World War II helped develop the prototype computers that were used to decode secret German radio traffic. In 1947, Turing wrote an influential paper, "Computing machinery and intelligence," that presented the "Turing test": The ultimate test for a computer would be for a human being interacting with a computer to be unable to tell that the machine wasn't another human being.[4]

By allowing people to see and hear one another in realtime, broadband networks recreate many of the characteristics of face-to-face communication.[5] These similarities might include: 1) the use of spoken language; 2) the ability to converse in real time; 3) the ability to see the conversational partner.

An interesting advanced implementation of a broadband symmetrical network is the Silicon Graphics' (SG) Studio network, used primarily by professional creative people in the entertainment industry, such as producers, directors, and editors.[6] Both parties use powerful computers, linked by special telephone wires that allow the rapid flow of large amounts of information.

The SG Studio permits people in different locations to work together on images—film, video, or graphics. Both people see themselves and the other in small "thumbnail" windows along the left side of their screens. In the middle of their screens each sees the "virtual projector," and views the footage they both need to work on. They then "cut" a few frames at a time and "paste" them on a shared "whiteboard." Both parties can draw on the whiteboard and see what the other has drawn. At the same time, the link allows the interactors to carry on a voice telephone conversation. Figure 3.1 shows a person using the InPerson video conferencing software the SGI system uses.

Another example of a symmetrical, networked channel is the joint venture by Worlds, Inc. and Landmark Entertainment.[7] In 1995, they opened the Interactive World's Fair in cyberspace, using a hybrid technology. Users play a CD-ROM through their computer, then connect through their "modem" to an online server. (Modem stands for modulator/demodulator. The device changes the 0s and 1s of computer data into signals compatible with telephone networks.) Once the

Figure 3.1 Indy workstation showing video conferencing. Image courtesy of Silicon Graphics, Inc.

user connects with the IA World's Fair, they visit any one of several pavilions with hundreds, even thousands, of other computer users.

The technology works by sending only changes to the video through the narrowband telephone system instead of the video itself; multimedia computers take information from the CD-ROM and quickly add, delete, or redraw the visual images to reflect the visitor's position in the Fair and the presence of others who the user contacts and "chats" with.

The use of symmetrical, networked channels for interpersonal communication is not surprising. It makes people "equal in the eyes of the technology," so to speak—each party is equally constrained or empowered. Inequalities may result from the individuals involved or the situation but not from the technology itself.

Symmetrical, stranded interactive channels

This kind of channel is used primarily for information gathering, processing, and retrieval. Accessing statistics from a database on a disk drive or a CD-ROM are examples of this type of interaction. In a sense, e-mail fits here since people access, read, and compose their e-mail messages on their own terminals some time after it has already been sent.

The high cost of networking can be prohibitive. For this reason, medium and high bandwidth applications are carried out on a single computer whenever possible. Increasingly, libraries subscribe to services that send them new CD-ROMs with updated information and then put the CD-ROM on a local network to save the cost of connecting with an expensive service, like Mead Data Central (which offers Nexis, a collection of magazine and newspaper articles and Lexis, a legal database).

Asymmetrical, networked interactive channels

Asymmetrical, networked channels permit interactions in which one party requires larger channel capacity than the other communication partner. One example might be video on demand systems, where the downstream flow must be large enough for broadcast quality video, whereas the upstream flow need only be a selection number of a few digits.

Similarly, interactive TV shopping requires at least color still photos to move downstream to the consumer; moving video would probably be more interesting and involving to the viewers. The upstream information flow from the buyer is merely textual and numerical data: the item and its size, color, quantity, a name, a credit card number, and an address.

Asymmetrical networked channels tend to be used for entertainment and transactions that require images. Shopping, video on demand, and the ability to order movies and TV shows over cable with a remote control device fall into this category. Such services are available to subscribers of advanced television test sites located in various communities around the U.S.

The current architecture of most cable systems doesn't allow for interaction because there is no provision for upstream communication from the viewer to the cable headend.[8] In order for consumers to have even narrowband communication back to the headend where a cable system's programming originates, operators must install new settop boxes and upstream amplifiers to deliver the viewers' messages.

In the meantime, most cable systems that offer pay-per-view service use the telephone as the upstream communication system: Viewers dial into a computer which sends a pulse to their settop box, opening the pay-per-view channel in the customer's home. The ability of the computer to make changes in the consumer's settop box is called "addressability," which means the operator can provide or deny access to any of the channels on the cable system.

One service, ACTV, uses current cable systems to offer a kind of interactive television in the sense that it allows viewers to alter the actual picture they see, rather than manipulating an overlay or using a playalong system that doesn't affect the TV images.[9] ACTV either uses 3 or 4 cable channels or "multiplexes" them on one channel. (Multiplexing is a coding technique that loads more than one signal on a single channel.) The user chooses different camera angles on a sporting event or different versions of a program. They are actually moving between one channel and another.

ACTV has developed some ingenious methods for personalizing their programming. For example, a children's show host asks, "Are you a boy or a girl?" If the child is a girl and presses "1," she is assigned to one channel, while boys are assigned to a second channel. Throughout the program, the host can then refer to the "good little [girl] or [boy], messages seemingly directed to the watching child.

Broadcast television shopping uses the telephone as the interactive return loop, connecting the buyer to an operator who processes the transaction. One force driving the development of interactive shopping is that the entire transaction can be computerized, shifting the processing costs of input and verification to the buyer.

Asymmetrical, stranded interactive channels

The premier models of this kind of channel are the familiar broadcast media: radio and television. Radio is a narrowband medium with no return loop or upstream communication capacity. Television is a broadband medium with no return loop.

A number of interactive programming services have started up, even though the optimal systems to realize interactivity have not yet been built. These services have found a variety of ways to get around the stranded nature of the reception devices (radios, televisions, computers) until they become networked. Taking their cue from cable operators who implemented pay-per-view and using the telephone for upstream messages, they adopt other media to let their customers communicate with them.

The Interactive Network (IN) is an interesting early example.[10] Broadcast television provides the images downstream to viewers. IN subscribers pay $15 per month and buy a special laptop device for $199 that receives simulcast FM radio signals and displays them on a small screen, shown in Figure 3.2. The viewers touch buttons to respond. IN provides playalong games with sports events and game shows. At the conclusion of the program, the viewer plugs their phone cord into the laptop device. It automatically calls a central computer, downloads the viewer's choices, and rates the sender against all other players for prizes.

EON is another early developer of interactive programming that also uses radio signals.[11] EON employs FM radio to deliver information that complements broadcast television programming, allowing viewers to playalong with game shows and sporting contests, to buy from a shopping channel, to predict outcomes of mystery shows, and to participate in opinion polling. EON technology will be embedded in a cable box or a video gaming box. Radio signals are also used for a wireless return loop. As users watch television, they reply to questions or place orders with a remote control device that sends a radio signal to a central computer for forwarding and processing.

Figure 3.2 The Interactive Network laptop device. Source: Interactive Network.

Proto-Interactive TV: Video Games and CD-ROMs

Currently, the largest market for interactive video is programming for stranded devices, such as video game players for TVs and audio-visual CD-ROMs for personal computers.[12] This market is interesting, not only because of its size, value, and rapid growth, but also because it is the laboratory for the development of interactive television.[13]

The long period required to build broadband networks means that program producers can't create products for the mass market, which would pay for the high cost of development. They can, how-

Figure 3.3 CD-ROMs are one of the most popular forms of packaged interactive entertainment.

Figure 3.4 One of the most popular video games, *Doom*, by Atari. Source: Atari.

ever, make products for video games and CD-ROMs which teach them what consumers want, how they use interactive products, and how much they will pay for them.

The key interactive property of video games is speed. Few PCs can match dedicated players for rapid feedback—critical to a game.[14] Thus, CD-ROMs for PCs tend to be informational or exploratory, which allow for longer response times without evoking user frustration. However, this situation is changing. PC processing is becoming powerful enough to match the players for speed, so in the next 2 to 3 years, price (multimedia computers are about $2,000, players cost only about $200-$300) will be more important than performance.

Although stranded devices could incorporate more interactive features than they currently do, such as keeping a record of user moves and characteristics and using them to determine the direction of the story or game, little use has been made of the ability of computers to "learn" about their human users. Future interactive programs will probably incorporate more dimensions of interactivity that will make them better and more complex.

We have now considered some important interactive technologies. Now we will examine human-to-human and human-to-machine interaction, types of technology-mediated interaction, and the programming that provides interactive experiences.

Feedback, Equilibrium, and Interaction Types

Earlier we stressed the importance of feedback in interaction and communication. Researchers of human-machine interaction have looked at systems and observed that there are two types, depending on the feedback they allow: cybernetic systems and homeostatic systems.[15]

Cybernetic systems have a goal, often incorporated within the technology itself, which defines success or failure. They use feedback to guide the system's responses toward an established norm or outcome. The outcome is a fixed point. A good example of a cybernetic system is a thermostat. Set it for 72 degrees: When the temperature drops below 72, the furnace goes on; when the temperature moves above 72, the furnace shuts off.

Homeostatic systems allow a "balance" between two inputs. The balance may fluctuate within some range that is determined by the interaction itself, sometimes established outside the technology. Unlike cybernetic systems, the equilibrium of homeostatic systems does not settle at a single fixed point; rather it can rest within some range, perhaps even a large one. The role of feedback is to guide the system towards this balance.

Often biological systems are homeostatic. Consider how the amount of energy a person expends walking a mile changes over time. When the exercise first starts, it may be quite difficult; as the person gets in shape, it gets easier—the equilibrium between effort and accomplishment is changed. Another example of a homeostatic system is a relationship between two people. Both members provide input. The relationship fluctuates between intimacy and distance, openness and exclusivity, and other emotional poles determined by the participants.

These are the characteristics of systems. Now we turn to characteristics of individual users. Research in interpersonal communication finds that individuals interact for two overarching reasons: affiliation and dominance.[16] A summary of the dimensions of systems and humans is presented in Table 3.3.

Affiliation refers to closeness and intimacy in a relationship. When people seek to affiliate, they establish commonalities and links between one another. Although differences in social status and personal characteristics exist, there is a movement towards equality of participation and commitment on the part of those in the relationship.

Dominance refers to the distance and differences between people in relationship. The dominant person seeks to control the other, based on inequalities in power and status. Parents dominate their young

Table 3.3 Dimensions of Systems and Individuals

Characteristics	*Cybernetic Dimension*	*Homeostatic Dimension*
Interaction type	Transaction and simulation	Correspondence and comparison
Outcome	Deterministic	Emergent
Motivation	Achieve a goal	Engage in a search
Success	External	Internal
Activity	Task	Social
Reward	Sense of accomplishment	Sense of identity
	Affiliative Orientation	*Control Orientation*
Interaction type	Transactions and correspondence	Simulation and comparison
Typical activity	Exchange	Evaluation
Basis of interaction	Interpersonal similarities	Interpersonal differences
Outcome	Agreement, coordination, coorientation	Acceptance, rejection, ordering
Conversational markers	Mutual deference, expressions of liking and caring	Instructions, orders, question & answers, competition

Table 3.4 Typology of Interactions

	System Characteristics	
	Cybernetic	*Homeostatic*
Individual Motives to Interact:		
Affiliation	Transaction	Correspondence
Dominance	Simulation	Comparison

children; many feminists believe men attempt to dominate women; supervisors control their staff.

When the interactional characteristics of systems and individuals are considered together, a typology of interaction types results, as shown in Table 3.4.

Transactions

Transactions are among the most simple, predictable, and common of all interactions. Their appeal is simple: Interactors get something they didn't have before. Transactions fulfill the human need to have and to hold. Through them, people get together to buy, sell, trade, and exchange; they eat, work, earn, and learn.

There are many examples of transaction-based interactive television applications, although most use the telephone as the response loop, until two-way networks are deployed. For example, the InTouch system in Portland, Oregon, offers interactive shopping by embedding special signals with the over-the-air television signal. If a consumer sees something advertised on television that they want to buy, they press buttons on a remote. The remote is connected to a modem inside the settop box which dials up on a dedicated telephone line and sends the ordering information to a computer.[17]

Another shopping service is by Audio Services, Inc. The company plans to use an empty cable channel to send still photographs to consumers. When they decide to buy, they press buttons on the remote and their message is communicated with a modem through either the telephone or cable network. Finally, Interaxx is distributing still photographs on a CD-ROM connected to the television set, with responses also made by telephone.

Pay-per-view (PPV) and video-on-demand (VOD) are also examples of transactions. With pay-per-view, a telephone call to a computer signals an instruction to the subscriber's settop box. The settop box permits the consumer to access the PPV channel. VOD is like a video jukebox. The viewer uses a remote control to request a movie and the programming service delivers it from a central storage facility.

All these instances of transactions have a four-step process in common: The consumer makes a request and a supplier fulfills it; the supplier bills for the product and the consumer pays for it. Pictorial information is important to transactions both as a sales tool and as a product. Successful marketing requires the buyer be able to examine the merchandise and sellers want the highest possible quality images. In VOD and near-video-on-demand (NVOD) applications, the images are actually the product and consumers have been conditioned to expect excellent broadcast quality pictures.

The problems experienced by transaction services are those of trade: consumer satisfaction and protection (including pornography), misrepresentation, and fraud.[18] Providers need to make sure they have policies in place to deal with each of these issues before launching the service. Failure to address consumer concerns risks potentially severe legal and public relations problems.

Simulation

Simulation interactions are those where the person enters a pre-designed program or software environment and manipulates it according to their own goals and wishes. Examples of such interactions are games and computer-aided instruction and design.

Simulation interactions range from simple command-perform sequences to complex choices in detailed fantasy environments. However, their interactivity increases as the designed environment has a more and more real, or "virtual," appearance. For example, players of a game where they are virtual fighter pilots will prefer one that conveys a rich, elaborate environment and realistic action and speed. Similarly, computer-aided design is more valuable when the designer can incorporate as many features as possible of the constructed object.

This type of interaction appeals to the need to win, master, and achieve. The feedback is designed to tell the user if and how far they have moved from their goal. For example, in a game, you score points—or you don't. In computer-aided instruction, a test tells you whether you have learned the material. If you pass, you continue; if you fail, you repeat the material. In most applications, the more sensitive and rapid the feedback, the more satisfying the service will be.

Pictorial information brings authenticity and materiality to simulation interactions, such as games and computer-aided design. To be convincing, they require verisimilitude, the appearance of being true or real. Subscribers need this fidelity to reality so they can involve themselves in the programming environment.

The IATV systems under construction or being tested now do include programming services that provide simulation interactions. The Time Warner system in Orlando, Florida, now provides Jaguar 64-bit video games on demand and plans to add a video game service where several players in different households can compete in the same graphical environment.[19] In several cable systems, the Sega Channel downloads Sega games to individual players and the company is exploring multi-user game playing as well. Some cable systems are also providing their subscribers with high-speed, broadband connection to the Internet and other online services.[20] Giving people such fast access will encourage the rapid development of graphically-rich 3-D environments that simulate physical environments.

Providers and users of simulation services encounter problems of excess: addiction and the distribution of sexual and violent content. Addiction is a psychological and social disorder that holds significance for system managers because subscribers may use services beyond their means to pay for them.[21] Aside from the personal tragedy that any kind of addiction inevitably engenders, it is especially problematic for service providers because it places them and their customers in an adversarial relationship that can sour future relations and generate negative publicity.

Sexual and violent content also pose problems for providers of services that feature realistic simulated environments. Violence in interactive media has already become an issue with respect to some of the most popular video games and CD-ROM titles, such as *Mortal Kombat* and *Doom*. In 1993 negative news reports of excessive violence in CD-ROM games surfaced. The intense level of concern led to the formation of three ratings services, the vendor-created Recreational Software Advisory Council (RSAC), the developer-oriented Interactive Digital Software Association (IDSA), and the Software Publishing Association (SPA).[22] Indeed, the potential for addictive involvement, combined with realistic violence, is cause for some concern. Questions about these effects, particularly on young people, will continue to be raised about interactive media, perhaps even more vociferously than they have been about television because simulation is so realistic.

An important feature of the debate is that interactive simulation is much more influential than passive television viewing because participants make decisions and take action. Critics of the interactive content note that it is bad enough to watch a jungle fighter slash an enemy's throat; it is even worse to put the knife in the virtual hands of a ten-year old and have the child slashing a realistically simulated throat.

Correspondence

Correspondence has two dictionary meanings. The first is "to send messages to someone over time," as in an active correspondence between pen pals. The second is "to move towards increased similarity," to become more like the other person in the relationship.

Correspondence is about social contact. In the context of IATV, it means communication between the subscribers. They will talk, teach, collaborate, play, confront one another, and engage in conflict. A powerful reward participants gain from this contact is a sense of personal identity—they find themselves in their interactions with others. For this reason, some observers believe pictorial person-to-person communication may be the avenue by which subscribers are induced to sign up and pay for interactive services. (This inducement is often referred to by the trendy term "killer app," standing for killer application. It means a popular service that will fund the cost of the whole system.)

The goals of correspondence interactions are determined by both of the participants, emergent from the interaction itself and adapted to their personal needs. This flexible outcome has no fixed goal, unlike the transaction and simulation types of interactions which have goals like winning a game, retrieving information from a database, buying a compact disc, or renewing a library card.

The role of pictorial information in correspondence is to establish and enhance intimacy. However, there are questions about whether the images serve that purpose. In the Hi-Ovis experiment in Japan, people didn't like being on-camera as it required them to arrange their appearance for a performance.[23] The lack of consumer acceptance of AT&T's Picture Phone also casts doubts on the viability of visual communication.[24]

However, it is clear there are some applications where images are welcome. In business dealings, being able to see the person you are negotiating with is important enough for companies to spend thousands of dollars on travel for a single meeting. Further, while people may not want picture-phones for ordering the Thanksgiving Day turkey from the butcher, they are eager to see their loved ones who may live far away.

In other applications, such as routine transactions, play, or exploratory communication it is possible for subscribers to have their own graphic representations. These software icons are called "avatars." Some networking programs allow a subscriber to create a personal avatar offline in a drawing program, then import it into a message. Once the avatar is sent to the other participants, only changes to the avatar are sent: smiling, frowning, crying, jumping up and down, bowing, etc.

One of the most interesting applications of correspondence interaction comes from Starbright, a nonprofit group that delivers high-tech entertainment to sick children and their caregivers. Starbright began by providing bedside carts with video games. However, one of the biggest problems sick kids face is isolation. Headed by producer Steven Spielberg, the group expanded to develop a worldwide online network connecting children, their parents, and their medical caregivers, as shown in Figure 3.5.

Figure 3.5 Film director Steven Spielberg at a Starbright focus group testing their network. Photographer: William Ericson.

The Starbright Network first launched in November, 1995, linking together five U.S. hospitals. The two-way broadband network is one of the most advanced of its kind, designed by Worlds, Inc. The software allows the children to design their own personal avatars to represent themselves in communication with other children located in any of the network locations. Together, the kids roam (or at least their personal avatars do) around sophisticated rendered environments, communicating with one another through voice and text.[25]

Interpersonal problems arise for services that provide correspondence interactions, such as harassment, electronic stalking, and domination by netbozos. (Netbozo is a derisive term used to describe individuals who engage in offensive behaviors such as personal attacks, digressions from established conversational threads, domination of discussions through overly-lengthy messages and in annoying communicational blitzes, etc.)

Uncontrolled, netbozos will drive sensitive, thoughtful, and efficiency-oriented subscribers from the network. For example, netbozos caused many Santa Monica public officials to stop logging into the nation's first governmentally-sponsored network, the Public Electronic Network (PEN).[26]

Harassment and stalking are also important issues for online services and there is no reason to think that the presence of images will lessen these problems. Online services have developed policies to protect their (mostly female) subscribers from others (mostly male), including message-tracking that prevents anonymous messaging and protects real-life identities and addresses protection. It is clear that any communication service that provides interpersonal contact must cope with annoying and potentially dangerous behaviors.

Comparison

Like correspondence, comparison interactions fill the need for identity.[27] People search the social landscape to understand who they are and how they fit in. They find answers through comparison interactions which allow them to register and compare their ideas, opinions, answers, votes, predictions, decisions, and traits with those of others. Subscribers choose populations, sub-groups, and other individuals with whom to compare themselves.

The language of comparison is unique—it is largely numeric and graphic. In order to compare even moderately large populations, it is necessary to collect, process, and display the information in numerical form. However, many people find numbers difficult to understand and visually uninteresting. Presenting the same information in bar and pie charts and graphs that incorporate color, geometric forms, and animated sequences brings comparison information to life.

Video brings credibility to comparisons. Advertising agencies tape some interviews for presentations to their clients to demonstrate the results from research. Seeing and hearing actual consumers talk about a product is infinitely more convincing than pages of numbers.

There are no examples of interactive services which are comprised purely of comparison interactions, although many services offer some opportunities for comparison within the framework of other types of interactions. The Cincinnati public opinion Qube system and local cable operators' electronic town halls with telephone call-ins are examples of programming that use comparison interactions.[28]

The problems with comparison are those of manipulation. Data distortion, fraud and misrepresentation, and a false view of public opinion due to biased, self-selected samples can all occur. For example, the temptation to influence public opinion in a high-speed communication environment, especially when commercial stakes are high, may prove irresible. System managers should prepare for attempts to manipulate comparisons by means fair and foul. Already, manufacturers have been hurt badly by highly contagious, net-generated public opinion. One familiar example is the controversy over the flaw in Intel's Pentium chip, where difficulties for the company were exacerbated by the rapid communication of comparison information among consumers over the Internet.[29]

Combined interaction types

Transaction, correspondence, simulation, and comparison: These are the pure interaction types. Table 3.5 shows the characteristics of the four types. However, combining them is possible and even desirable. AT&T tested a prototype interactive system using 140 employees and their families. They found that successful interactions included entertainment, information, transaction, and communication. In other words, it's fun, you learn something, you get something, and you relate it to someone else.

The AT&T sample isn't representative of consumers in general and must be interpreted with care. However, the findings do suggest that system design cannot rest on a single interaction type or style. One important point to consider is that when interaction types are

Table 3.5 Comparison Chart of the Four Interaction Types

Model Type	Activity	Need	Communication Style	Uses for High-Res Video	Problems
Transaction	Buy, sell, and trade	To possess	Moves from florid and descriptive to terse	Sales appeal	Trade problems Fraud & misrepresentation Consumer protection
Correspondence	Communicate with others: Talk, Teach, Coorient, Collaborate	Social contact and personal identity	Supplied by users	Intimacy	Interpersonal problems: Netbozos Harrassment
Simulation	Games Computer-aided instruction and design Information retrieval	To win, master, and achieve	Too varied to classify	Verisimilitude	Problems of excess: Addiction Violence
Comparison	Registration, to compare with others: Opinions, Traits, Votes, Predictions	Social identity	Information must be aggregated and processed: Numeric Charts Graphs	Credibility	Problems of manipulation: Public opinion Voting fraud Data distortion

combined, both the positive appeals and the complicating problems of each of the combined types will appear in new forms.

Interaction with Content

So far this discussion has focused on how people interact with technological systems. However, it is important to make a distinction between interaction with technology and interaction with programming, the material that the human users encounter in the course of the interaction.

An on-going research program at Stanford University by Clifford Nass and his associates examines how people treat computers that are programmed to behave like social actors.[30] For example, in several experiments, the content is designed to function as a teacher, worker, and/or evaluator. In most experiments, the computer's mode of expression is a human-like voice chip. Replicated across many studies, the fascinating finding is that when computers are programmed to act like people, humans treat them like people. The researchers base their conclusion on the following results:

1. People apply social norms to computers (*author's note: to computer content*);
2. People apply notions of "self" and "other" to computers (*author's note: to computer content*);
3. Voices are social actors and notions of "self" and "other" are applied to them;
4. Computers are gendered social actors and gender is a powerful cue (*author's note: computer content is gendered*);
5. People respond socially to the computer itself (*author's note: to the content of the computer*);
6. People do not see the computer as a medium for interaction with the programmer;
7. People respond differently to computers programmed with different personality characteristics.

Now let us take this idea a bit further. If people respond to interactively-delivered content as if it were another human being, reacting to personality differences and subtle relational cues, how closely can a person-programming relationship mimic a person-to-person relationship?

What would this person-programming relationship be like? If it were like a person-to-person relationship, research shows it would rest on 6 bases that would be reflected in the communication between the two entities: The conversation would include intimacy (affection, immediacy, receptivity, trust, and depth); similarity; relaxation (as opposed to composure); informality; equality; and a prominent social orientation.[31]

The participants would learn about one another through conversation. Each party would form expectancies from their previously-acquired knowledge about similar others, the other's mode of dress, accent, social class, knowledge level, and ethnicity.[32] Over time, the conversations would move from impersonal topics and surface subjects to beliefs, values, and the most deeply-held wishes and dreams.[33]

The structure of the conversation would include turn-taking, gaps, overlaps and interruptions, sequential observation, repair organization, and collaborative maintenance.[34] The content of the conversation would incorporate empirical information exchange, socio-emotional expression, and relational commentary. These utterances might take the form of observations, opinions, suggestions, agreement, compliments, invitations, jokes, complaints, arguments, emotional displays—the entire panoply of human communicative behavior.[35]

Research indicates that people have preferences for a certain speech tempo. So if they treat interactive content like social actors, they'll probably appreciate a virtual conversationalist that is more talkative, engages in a predictable amount of talk, allows shorter pauses, and coordinates their talk with their human partner.[36] People also like immediate language, expressions of certainty, and the use of relational pronouns, like "I," "we," and "us."[37]

The human and the programming would show both affection and conflict. They would express affection through privately understood language, special names, expressions of trust, and self-disclosure.[38] Most importantly, the partners would show they were "in sync" with one another by matching, complementing, reciprocating, and providing compensation for the other's conversational behaviors.[39]

Conflict between the person and the programming could escalate from mild difference to disagreement, dispute, battle, and war (but not, thank goodness, litigation!).[40] They would resolve confrontations by providing a remedy, legislating a solution, remediation, reaffirmation, or failing to resolve it at all—just like human couples.

How would the computer know which aspect of programming to present to the human user? When someone understands a person, they have a "mental model" or a "cognitive representation" of the perceived other. Now that the ability of computers to process large amounts of information has increased so dramatically, it is possible for a computer to have a mental model of a human interactor and to reconfigure dynamically that model as the relationship progresses.[41] The computer can learn the person's stated preferences and track that against their actual behaviors.[42] It can monitor cognitive styles, affective styles, weaknesses in user understanding, and on-going performance. It can learn about the person's interests, reasoning ability, and reading level.[43]

A computer could track the individual's conversational patterns and alter its own interactive content to correspond to its human partner. However, the computer has several more expressive modalities to use for its content than people do. In addition to having an animated or photographic representation that gives it the appearance of being a human and an audio chip that endows it with a human-like voice, the computer can use alphanumeric displays, diagrams, models, charts, sketches, and text to make its points.[44]

Oh, and one more thing. It shouldn't make things too easy! Surprise, spontaneity, challenge, and effort are important aspects of human behavior. The computer's programming must be able to be surprising and spontaneous. It needs to incorporate emergent behaviors that arise out of the interaction itself so that the interactive content is a source of novelty and pleasure for the human member of the team.[45]

Summary

IATV is a fascinating topic that reaches into human beings' deepest desires for personal connection and communication. It also meets many of people's most pragmatic needs, from shopping to getting the orienting information they need to survive.

Interaction also involves problems, such as universal access, privacy, misbehavior, consumer protection, copyright and intellectual property rights, and system security. Connections between people have consequences, both wonderful and terrible, that occur over wires just as they do in face-to-face relationships. As IATV systems come on-line, these issues will arise.

The benefits and dilemmas posed by interactive media are already upon us. As the sophistication of software design improves, the good, the bad, the ugly, the transcendent, the frightening, and the thrilling will all have their way with us. It is a prospect that is both daunting and exhilarating.

However, fully interactive systems won't come anytime soon. Mitch Kapor, developer of the popular computer program, Lotus 1-2-3, once described the American home's communication system as one that had "a multi-lane highway going in and a cowpath coming out." He meant that homes have multi-channel broadband programming services coming in on the cable wire and only a narrowband telephone wire going out. As we will see, it is not yet clear how long this asymmetrical imbalance will predominate.

Notes

1. Claude E. Shannon and Warren Weaver, *A Mathematical Theory of Information* (Urbana: University of Illinois Press, 1949).

2. Everett M. Rogers, *Communication Technology: The new media in society* (New York: The Free Press, 1986).

3. Kathy Kellermann, "Extrapolating Beyond: Processes of Uncertainty Reduction," *Communication Yearbook 16* (Newbury Park: Sage, 1992):503–514.

4. Alan M. Turing, "Computing machinery and intelligence," *Mind 59* (1953):11–35.

5. Some of these characteristics are covered in C. Nass and J. Steuer, "Agency & Ethopoeia: Computers as Social Actors," *Human Communication Research* 19:4 (June 1993):504–527. The ability to see others over a broadband system must be added to the list.

6. Joan Van Tassel, "Silicon Graphics online studio debuts," *Malibu Times* (October 18, 1994):A-11.

7. Joan Van Tassel, "The WWWorld's Fair," *WiReD* (August 1995):3.

8. Roger Brown, "The return band: Open for business?" *CED* (December 1994):40–43.

9. William Samuels, president, ACTV, personal interview with author, April 1994, Beverly Hills, CA.

10. D. C. Denison, "My excellent interactive adventure," *Boston Globe Magazine* (July 11, 1993).

11. Mark Berniker, "Eon mines for some IVDS gold," *Broadcasting & Cable* (August 8, 1994):29.

12. Jim Carlton, "Nintendo sues Samsung unit," *Wall Street Journal* (January 16, 1995):B4, B6.

13. Ken Locker, personal interview with author, February 1995, West Los Angeles, CA.

14. Ron Rice, in a note to the author, observes that the development of 6-speed CDs with acceleration may soon put the PC on a par with dedicated video game players.

15. I found this helpful article most impressive and it shaped my thinking about how to consider interactive systems: Stephen R. Acker, "Designing

communication systems for human systems: values and assumptions of 'socially open architecture'," *Communication Yearbook 12* (Newbury Park: Sage, 1988):498–532.

16. Michael Argyle and J. Dean, "Eye contact, distance and affiliation," *Sociometry* 28 (1965):289–304.

17. Jack Galmiche, president, ISI, parent company to InTouch, Portland, OR, telephone interview with author, March 1994.

18. Jeffrey Krauss, "Security issues and the NII," *CED* (September 1994):24. For a short but interesting view of the legislative concerns about security as it relates to consumer privacy, see Kate Gerwig, "Government grapples with privacy issue," *Interactive Age* (November 14, 1994):27, 32.

19. Jim Chiddix, telephone interview with author, March 1994, and in Time Warner Cable press release, "Time Warner Cable's Full Service Network to add Atari Jaguar 64-bit games," November 14, 1994.

20. See Chapter 10 for a full discussion of cable system trials of online access to subscribers.

21. Thomas Landauer, *The Trouble with Computers: Usefulness, Usability and Productivity* (Cambridge: MIT Press, 1995). For an amusing account, see Josh Quittner, "Johnny Manhattan meets the Furrymuckers," *WiReD* (March 1994):92–95, 138. Also, Suzanne Stefanac, "Sex and the New Media," *NewMedia* (April 1993):38–45.

22. The flavor of the discussion is captured in John C. Dvorak, "Blow their heads off!" *PC Magazine* 13, no. 10 (November 8, 1994):93.

23. William H. Dutton, Jay G. Blumler, and Kenneth L. Kraemer, eds., *Wired Cities: Shaping the Future of Communications* (Boston: G.K. Hall, 1987).

24. William H. Dutton, "Driving into the future of Communications? Check the rear view mirror," paper delivered at POTS to PANS: Social issues in the multimedia evolution from Plain Old Telephony Services to Pictures and Network Services, the BT Hintlesham Hall Symposium, Hintlesham, Suffolk, March 28–30, 1994.

25. Peter Samuelson, personal interview with author, March 1995, Santa Monica, CA, and Steven Spielberg, telephone interview with author, July 1995.

26. Joseph Schmitz, personal interview with author, July 1994, Santa Monica, CA.

27. Albert Bandura, "Self-efficacy mechanism in human agency," *American Psychologist* 37 (1982):122–147.

28. Carol Davidge, "America's talk-back television experiment: QUBE," in William Dutton, Jay Blumler, and Kenneth Kraemer, eds., *Wired Cities: Shaping the Future of Communications* (Boston: G.K. Hall, 1987):75–101.

29. Laurianne McLaughlin, "Pentium flaw: a wake-up call?" *PC World* 13, no. 3 (March 1995):50–51.

30. Clifford I. Nass and Jonathan Steuer, "Voices, Boxes, and Sources of Messages," *Human Communication Research* 19, no. 4 (June 1993):504–527; C. I. Nass et al., "Anthropocentrism and computers," *Behavior and Information Technology* 14:4 (April 1995):229–238; C. I. Nass et al., "Are respondents polite to computers? Social desirability and direct responses to computers," submitted to Public Opinion Quarterly (1995); Clifford Nass, Jonathan Steuer, and Ellen R. Tauber, "Computers are social actors," unpublished paper.

31. Judee K. Burgoon and J. L. Hale, "The fundamental topic of relational communication," *Communication Monographs* 51 (1984):193–214.

32. Kathy Kellermann, "Extrapolating Beyond: Processes of Uncertainty Reduction," *Communication Yearbook 16* (Newbury Park: Sage, 1992):503–514.

33. I. Altman and D. Taylor, *Social penetration: The development of interpersonal relationships* (New York: Holt, Rinehart, & Winston, 1973).

34. Jenny Mendelbaum, "Communication Phenomena as Solutions to Interactional Problems," *Communication Yearbook 13* (Newbury Park: Sage, 1989):255–267.

35. Jo Liska, "Dominance-seeking Language Strategies: Please Eat the Floor, Dogbreath, or I'll Rip Your Lungs Out, Okay?" *Communication Yearbook 15* (Newbury Park: Sage, 1991):427–456.

36. R. M. Warner, "Speaker, partner and observer evaluations of affect during social interaction as a function of interaction tempo," *Journal of Language and Social Psychology* 11:4 (1992):253–266.

37. D. J. Cegala, "A study of selected linguistic components of involvement in interaction," *Western Journal of Speech Communication* 53, no. 3, (Summer 1989):311–326.

38. J. K. Burgoon and J. L. Hale, op. cit., 56-57.

39. J. K. Burgoon et. al., "Adaptation in dyadic interaction: Defining and operationalizing patterns of reciprocity and compensation," *Communication Theory* 4 (1993):295–316.

40. William A. Donohue, "Interaction Goals in Negotiation: A Critique," *Communication Yearbook 13* (Newbury Park: Sage, 1989):417–427.

41. Acker, op. cit., 418–419.

42. Eric S. Fredin, "Interactive Communication Systems, Values and the Requirement of Self-Reflection," *Communication Yearbook 12* (Newbury Park: Sage, 1988):533–546.

43. Diana Gagnon, "Toward an Open Architecture and User-Centered Approach to Media Design," *Communication Yearbook 12* (Newbury Park: Sage, 1988):547–555.

44. Michael Shrage, *Shared Minds* (New York: Random House, 1990).

45. Fredin, op. cit., 541.

4

Connected, Switched Television: The Convergence of Television with Telecommunication

Introduction

Flash forward to the year 2015. The information superhighway is a reality: Schools, hospitals, most businesses, and the majority of U.S. households are connected to a network capable of carrying all manner of information: encrypted data, voice, graphics, and video. Other technologically advanced nations, including France, the U.K., Holland, Germany, Denmark, Japan, Singapore, and Australia are similarly wired.

Inexpensive videocams are everywhere. People can shop at local stores from their living rooms. They can hold two-way video conversations with friends, relatives, and business associates. They can access pictorial databases and download and save the images.

For all these applications to become a reality, there must be a digital broadband network where communication flows two ways, like the current telephone network—except that it would carry graphics and video as well as voice and data. Chapters 1, 2, and 3 covered innovations from the television and computer industries that are shaping advanced TV systems.

However, connectivity and switching are properties of the telephone system. When television acquires these two characteristics, it brings these two industries closer to "convergence." Convergence is the notion that the television, telecommunications, and computer industries are all becoming a single, seamless communication system.

Two decades ago, convergence was a theory—today, many believe it is becoming a reality. In a few short years, the previously separate communication fields have begun to merge, blur, and come together. Observers have concluded that this convergence is occurring at all levels: physical (systems and equipment), functional (processes), organizational, managerial, and individual. William Stallings provides an overview of the convergence of computers and communication:

> There is no fundamental difference between data processing (computers) and data communications (transmission and switching equipment).
>
> There are no fundamental differences among data, voice, and video communications.
>
> The lines between single-processor computer, multi-processor computer, local network, metropolitan network, and long-haul network have blurred.
>
> One effect of these trends has been a growing overlap of the computer and communications industries, from component fabrication to system integration. Another result is the development of integrated systems that transmit and process all types of data and information. Both the technology and the technical standards organizations are driving toward a single public system that integrates all communications and makes virtually all data and information sources around the world easily and uniformly accessible.[1]

Figure 4.1 is a conceptualization of the convergence of communications industries to a single information format.

This chapter begins with a description of the components of infrastructure. It then turns to its main topics: connectivity (the linking of all those individuals who wish to participate in the communication system) and switching capacity (the purposive routing of messages to and from specific stations and individuals).

The Communication Infrastructure: Hardware, Software, Firmware, and Wetware

"Infrastructure" means the underlying physical, technical, and human organization necessary to carry out some activity. Two methods for delivering television are over-the-air and over wires. Both systems are complex, composed of four types of parts: "hardware," "software," "firmware," and "wetware." Hardware refers to the actual physical components of an information processing system: The stuff—the boxes, connectors, and cables, composed of metals, plastics, silicon, and other materials.

Software originally meant the detailed instructions of a computer

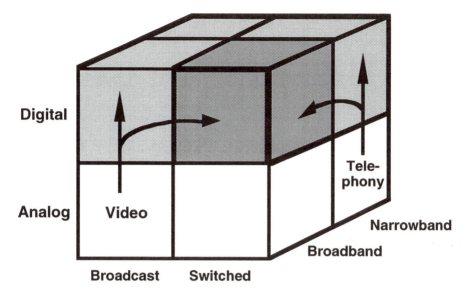

Figure 4.1 Industry convergence to switched digital format. Source: Carl Podlesney, Scientific Atlanta.

program that controls how the hardware operates. However, the usage of the term has changed over time so that today it refers to anything that the user sees on the screen. Thus, software includes Microsoft's *Windows* program, the *Myst* CD-ROM-based game, and a movie, such as *Natural Born Killers.*

Firmware has characteristics of both hardware and software. It is software that is permanently loaded into memory so that it is like part of the hardware; however, firmware can be re-programmed so it is flexible like software. "Wetware" is a literary term, coined from the cyberpunk futurist fictional genre. It is also a play on a Cold War KGB euphemism for murder, "wet work." Wetware refers to people—the human beings who use the information processing system.

Naturally, using "-ware" as a suffix has led to a proliferation of such words. For example, "bioware" describes technology that gets built into wetware—a pacemaker. A disgruntled employee might refer to her unpleasant supervisor as a "sicko piece of nastyware" and split pea soup might be called "glopware."

Although it is possible to say that each of these types of parts—hardware, software, firmware, and wetware—will play an important role in the evolving high-speed, image-capable, two-way communication infrastructure, no one is sure how any one element will finally become integrated into the whole. The character of the whole system is even less clear. In fact, there is a cottage industry of prognostication, which we will cover in detail in Chapter 11. However, one characteristic that advanced television systems will ultimately incorporate is connectivity.

Connectivity = Complexity

Familiarity breeds contempt, familiarity breeds . . . if nothing else, complication. Connectivity permits people to develop intimacy, closeness, and familiarity by linking them together. When people are close, they often form relationships. Indeed, for most of us, few aspects of our lives are more important than the interpersonal relationships we share with others.

By making it possible for people to develop relationships, connectivity leads to social complexity and multidimensional interdependency. This mouthful of syllables simply means that in socially complex networks, people depend on each other in many ways to fill a variety of wants and needs.

In his book, "Complexification," mathematician John Casti introduces complex systems with the story of Chang and Eng.[2] These Siamese twins first appeared in a report written by Mark Twain in "Those Extraordinary Twins."[3] Chang was a heavy drinker and his connected brother, Eng, was teetotaler. Notes Casti: "What makes them special even today is the fact that they were two humans linked together in a very unusual fashion, a connection that led to interesting consequences. This connective structure led to a system considerably more complicated than that representing the typical man or woman. So in trying to understand the complicated system 'Chang and Eng', it's essential to take this connectivity into account; Chang and Eng can't be understood by thinking of them as two disconnected individuals."

Connectivity leads to complexity by linking the members of a set to one another along one or more dimensions. The number of contacts that members of the set make is only one dimension along which members could be linked. They could be multidimensionally connected by type of relationship, type of communication content, depth of emotional bond, amount of liking, proximity, and many other variables, such as how often they lift a brew, shop, play video games, rob banks, make love, etc.

When people first connect, their relationship exists only as potential. However, over time, if individuals continue to communicate, they become linked in ways that depend on their characters and the nature of the experiences they share. Generally, the greater the number of nonequivalent dimensions that link people together, the greater the complexity of the relationship. When individuals are members of a group of people, the extent to which they are connected—the greater the number of dimensions that link them—the more complex their social system is.

Consequences of complexity

Complex systems differ from simple systems by the presence of frequent feedback. Compare a simple small pond with the huge, complex ocean. Throw a stone into the pond and ripples from its impact make small concentric waves for a few minutes. Then the motion stops. A physicist could probably describe this motion with just a few equations.

By contrast, the ocean is in such continuous motion that, even though they're there, it's difficult to even see the concentric waves. The motion of the ocean exists, at least partly, because there is constant interaction between the different wave motions. No number of equations could explain oceanic movement because the system is too complex. In other words, the way the ocean moves is unpredictable.

Complex social systems are something like complex biological systems; their complexity doesn't so much depend on size as on the sheer density of connection. Even in a very small social system, when people are multidimensionally connected there are opportunities for constant "feedback," which is itself fed back into the system. (In systems which have a goal, feedback is positive or negative information about the outcome of an action which allows the system to correct its course. Another kind of information is called "feedforward." Feedforward is anticipatory information about the likely outcome of a potential action. For example, suppose a teacher decides to cheat on his income tax and claim he drove 40,000 miles. Then he thinks about obtaining extra receipts, altering his mileage log, and perhaps mechanically changing his mileage indicator. The anticipatory thinking about the hassle of creating a false record and the consequences of getting caught—feedforward—is likely to result in a less exaggerated tax filing.

Complex social systems also anticipate the results of actions and consider the results of past actions, perhaps many times over. This "feeding back" of feedback is called "iteration." Densely connected social systems are characterized by iterative feedback so they grow,

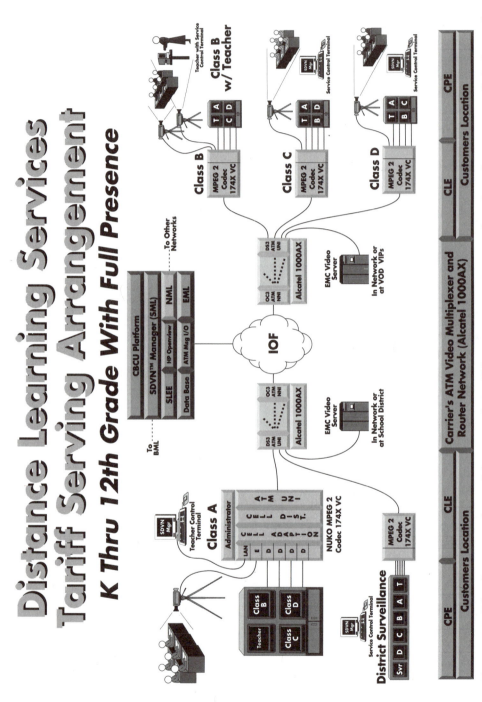

Figure 4.2 Schools are complex social systems that will become even more complex when classrooms are linked to each other as well as the outside world, like in this Alcatel project. Source: Alcatel.

change, develop, differentiate, and evolve, depending on the type and amount of iterative feedback. Although this process can be broadly understood, it is simply too complicated to allow detailed prediction.

Casti points out other consequences of complexity: hierarchy, increased number of possible outcomes, influence from indirectly connected set members, and unpredictable behavior. In describing hierarchy, he uses the term "cover set" to mean the highest level and "subset" to refer to the lower levels. Thus, in the hierarchy of the family, the cover set is "family," the parents are one subset, and the siblings are another subset.

Much of the unpredictability and richness of complex systems results from the vertical interactions between cover sets and subsets and from the horizontal relationships between subsets, says Casti. Everyone who has grown up in a close-knit family or been part of a highly coordinated team can attest to the validity of his observation.

A potential source of unpredictability is the indirect influence of people not actually in the set, but connected to those who are. This influence increases as the number of relationships rises and leads to an expansion of possible interactional options and outcomes for both individual members and the system as a whole. A further consequence of dense connection is that unexpected relationships may emerge from the rich fabric of associations, leading to nonintuitive and surprising outcomes.

Jonathan Gill, a former professor at M.I.T. who served as an advisor on high technology in the Clinton White House, says that another reason connectivity creates additional outcome possibilities is due to an increased "idea space."[4] The greater the number of people who can contribute their thoughts about a topic, the greater the range of ideas for potential action exists, enriching the entire social sphere.

Likewise, the number of alternatives for individuals is also increased because they are influenced by others to whom they are only indirectly related as well as by those with whom they are directly related. A mathematical technique called q-analysis demonstrates that a member of a set can be affected by another member who may be separated by several intervening members.

The abstract findings of q-analysis turn out to be surprisingly true: virtually anyone in the world is closer than you might think. One famous experiment tested the "Small World problem." The results demonstrated that an individual in one city can find someone they don't know in another city quite easily.[5] Subjects in Omaha and Boston were given a letter with the name of a person (who they did not know) written on the envelope and told to record the steps they took to locate the mailing address of that other person. On the average, it only took 5.2 steps to locate the missing address. For many, it took even less effort: 48% of the searchers were able to locate the unknown person with 3 calls. Imagine how easy contact will be when global networks are in place.

In addition to the influences of indirect links, there are other differences between complex and simple systems: limited divisibility, distributed decision-making, and multiple communication loops. Simple systems can be decomposed into their parts without doing much damage to the overall system because the linkages between them are so minimal. However, taking a complex system apart may well destroy the system. For example, it isn't possible to eliminate the nervous subsystem of a mammal without killing the animal because the nerves are richly connected and crucial to the operation of the overall organism. Similarly, a marriage is destroyed when one of the partner dies.

Simple systems typically have centralized decision-making, while complex systems operate with many such centers. The pattern of distributed decisions makes for stability in complex systems because any one choice will affect only a small arena. Political scientists have observed the instability in dictatorships and monarchies because of the weaknesses of centralized decision-making.

However, by far the most important difference between simple and complex systems is the fact that simple systems have few components which interact with one another very little, while complex systems have a large number of components with a high level of interaction, including many feedback and feedforward loops. The presence of so many communication paths may contribute to overall system collapse when internal stresses are present.

Except for distributed decision-making, all the above properties add to the likelihood that complex systems may become unbalanced. According to scientist Ilya Prigogine, when a system grows "far from equilibrium," the relationship between cause and effect may cease to be linear, so that small causes may engender large effects and the system becomes unpredictable.[6]

An example of social unpredictability caused by the presence of dense connection is the Pentium chip debacle experienced by the semiconductor manufacturer Intel, mentioned in the previous chapter. The company marketed the chip knowing there was a flaw in the math processing, but presumed it would almost never occur, or that if it did, there would be no serious consequences. When a newspaper printed a report of a professor's discovery of the error, discussion of the problematic Pentium raged in every nook and cranny of the Internet.

Initially, Intel said the flaw was unimportant and the company would do nothing. However, Intel's position collapsed in the face of rapid contagion of anger among computer users. Afterwards, observers noted that the extraordinary speed and reach of communication over the Internet forced the company to change its policy. In other words, a richly connected system of computer users increased the instability of Intel's operating environment.[7]

The connectivity of advanced networks has critical implications for everyone in our society. We are currently connected by the moderately narrowband telephone system, which has influenced virtually every aspect of social life. Some effects include the ability to meet and contact others, to exchange information (especially fact-based data), to act with great efficiency due to increased possibilities for timely coordination, and to maintain vibrant connections between otherwise isolated individuals.

Broadband Connection—"Bandwidth Equals Proximity"

Consultant Steve Rose coined this aphorism, which means that the more bandwidth a communication channel has, the more the people using it will find the experience is like being physically close to their partner. Rose's observation conceptually captures a decade-long effort by a British research team at the Communication Studies Group (CSG) to understand how using technology to communicate changes people's messages. The results of their research suggest predictions about how advanced broadband networks may affect our communication and to understand how we will use these systems.

The CSG research began with the most mundane of objectives: J. Short, E. Williams, and B. Christie were commissioned to investigate how Britain could reduce the cost of governmental communication.[8] The researchers handed office workers a simple questionnaire designed to find out how they rated the various communication technologies they used in the course of carrying out their everyday duties.

The employees assessed their responses to using different interaction modalities—face-to-face, video, audio, two-way, real-time teletype (now computer conferencing), and written documents. They evaluated these media as sociable/unsociable, sensitive/insensitive, personal/impersonal, and cold/warm. Finally, office workers assessed each medium for its perceived appropriateness and effectiveness for such tasks as giving and receiving information, idea creation, problem-solving, bargaining, persuading, conflict resolution, and so on.

Their findings suggest that people will probably like communicating over broadband networks. Analysis showed that people most preferred face-to-face communication, followed by video, audio, and written communication modalities. The higher a medium's rating, the more it was judged to be sociable, sensitive, personal, and warm; that is, the medium conveyed the "social presence" of the participants, including their verbal and nonverbal communication.

Social presence = social cues

The CSG researchers quickly recognized that they were onto something much more fundamental and important than just how the British government could save money (important as that was). They had discovered what it is about a communication technology that makes people like using it: It conveys social presence. But what exactly is social presence and how do participants perceive it? A CSG colleague, Derek Rutter, defined it more precisely as "the number of social cues that can be communicated across a given channel."[9]

In social science parlance, social cues include observable verbal, paralinguistic, and nonverbal behaviors. Translating the shoptalk into plain English, verbal behaviors are words, organized into phrases, sentences, and paragraphs. Paralinguistic behaviors are communicative sounds that accompany words, such as "um-hmm," "hunh," "yeah," "oooooh," throat-clearing, laughs, chuckles, giggles, gulps, sighs, and so forth. Nonverbal behaviors include facial expressions and body gestures.

The following dialogue is an example of the various components of a spoken message:

Sara: "Bring me that book, please."	Verbal
Sara directs her eyes to Gary, nods her head, points toward book on mantel, smiles	Nonverbal
Gary: "Um-hmm."	Paralinguistic
Gary nods in understanding, moves to get book.	Nonverbal
Gary: "No problem."	Verbal

The CSG scientists concluded that the more total cues a communication technology, or channel, carries, the more people will like it. However, there turned out to be another piece to the puzzle. It's not just total cues—it's the kind of cues. The answer to this final, clever knot took most of a decade to understand and required a reading of anthropologist Gregory Bateson.

Bateson was married to another famous anthropologist, Margaret Mead, and they were preeminent in the academic world. Bateson was a fascinating thinker in his own right and formulated a special way of thinking about the way people communicate to one another beyond the apparent content of the messages they exchange.

Bateson wrote that one level of a message is what people say, the "report" level: The weather is pleasant, I'm feeling tired, and so forth. The report level is characterized by cognitive or factual information.[10] The second level is called the "command" level. This level provides information about the relationship between sender and receiver, and how the receiver should interpret the sender's message. This relational data is sometimes called socio-emotional information.

What the CSG researchers came to understand is that there is a connection between the form of expression of a message and the information conveyed. Most factual data at the report level is communicated verbally. By contrast, at the command level people don't necessarily use words to convey socio-emotional information. Instead, they employ the sounds that accompany words, paralinguistic cues, and facial expressions and bodily gestures, nonverbal cues, to convey socio-emotional messages.

This finding confirmed earlier research. In 1971, research by Meherabian had indicated that people base their evaluation of a communicator's intent on tone of voice and facial expression, rather than words.[11] Later work by Burgoon confirmed that verbal cues are more important for factual, abstract, and persuasive messages, while nonverbal cues predominate in relational and emotional messages. The results also indicated that adults place more reliance on nonverbal than verbal cues to determine social meaning and that this reliance is greatest when there is a conflict between the two modes. Finally, people differ in the reliance they place on the two kinds of cues.[12]

Short, Williams, and Christie were the first to emphasize this relationship between types of cues and channel bandwidth. They concluded that communicators value channels that allow them to receive social cues. The greater the number of received cues the more accurately the receiver can decode the sender's intended message—if the message is socio-emotional in nature.

To put this conclusion another way: Face-to-face communication (and broadband channels) provide both visual and auditory cues that give people the socio-emotional information about how the speaker regards them and how they intend for the message to be taken, decoded from the available paralinguistic and nonverbal cues. This increased accuracy makes the receiver feel more secure in constructing an appropriate response to the message.

These findings can be summarized by the following dichotomies:

Report Level	*Command Level*
Cognitive	Emotional
Factual	Evaluative
Verbal	Nonverbal and paralinguistic
Abstract	Grounded

Not only does the number of cues a channel permits (its bandwidth) help receivers interpret messages aimed at them, it also profoundly affects the communication patterns of message senders. Further study showed the Communication Studies Group researchers that as channel capacity decreased, people became more aggressive in

Figure 4.3 Interactive TV like this Apple IATV box will link people and provide both textual and socio-emotional cues. Source: Apple Computers, Inc.

their language, even hostile. They theorized that communicators increase the level of these behaviors to compensate for the barriers to social presence imposed by narrowband channels. In other words, interactors became more strident when trying to express their messages across channels that restrict the number of cues they can send.

The highly rated face-to-face and video modalities that allow their partners to receive a complete repertoire of social cues are sometimes referred to as "rich" media, such as television and films. The telephone stands midway between visual media and textual media—it conveys verbal and paralinguistic cues, but filters out facial expressions and bodily gestures. Written documents severely restrict the cues available to participants, permitting only verbal and graphic information. Text-based computer communication permits the fewest cues, not even transmitting the identifying clues that can be gleaned from a letterhead, quality of the stationery, typeface, and format.

The work of the Communication Studies Group is supported by research into users' communication over computer networks. Kiesler, Siegel, and McGuire observed the tendency for online computer users to "flame," a writing style that employs intemperate, usually angry, language when addressing others. They argue that the "emoticons" (the smiley [:–)], frowney [:–( ], and winkey faces [;–)]) employed by computer users to indicate how they feel are a way that net-communicators overcome the limitations imposed by the narrow bandwidth of computer interaction.[13] (By the way, read these emoticons sideways to see their "faces"!)

The implications of this research for advanced broadband networks are many and they go well beyond the pragmatic situations— shopping and downloading files—that we considered at the beginning of this chapter. While no technological communication replaces face-to-face interaction, these advanced television systems will approach in-person experience more than any other channels we have adopted so far because they will permit the simultaneous exchange of socio-emotional information.

The new communication infrastructure will enable people to use broadband, interactive media for sensitive, delicate communications that are highly emotional. They will be able to assess the truthfulness and sincerity of their communication partners and that information will help them feel comfortable that they are responding appropriately.

In the workplace, technology has not yet—and may never—deliver on its promise to make telecommuting a reality. While conveying emotional information is often only somewhat important in organizational settings, there is tremendous emphasis on the communication of status. The broadband capability of advanced networks allows them to carry status data, which may make superiors more comfortable having off-site workers. They will be able to see their subordinates at work, even when they are home, and assess the degree of effort involved.[14] Extended, complex negotiations will also be possible because participants will believe they have a good "feel" for the situation and for how much they can trust the people on the other side.

Broadband connection will affect the "communities of interest" that have arisen on narrowband computer networks, as well. The current communities tend to be linked by shared interests and communication is dominated by the exchange of cognitive information. Communities in touch over richer media will also have the capacity for emotional bonding, which is likely to make them much more durable and resilient than today's computer interest groups.

As we will see, connection also has negative aspects. Nearly every book and article about today's television comments on how engrossing it is and how narcotizing it is on individuals, society, and culture.[15] People may be even more mesmerized by the immersive potential of the new systems. Already, even with mere text-based systems, universities are concerned about students who sign onto computer networks, stop attending classes, and neglect to get sufficient sleep or nutrition.

We have considered the effects of broadband connectivity: complexity, intimacy, and socio-emotional interaction at the personal, organizational, and social levels. However, the advanced interactive networks of the future will be switched as well as connected. And, as we shall see in the next section, switching is the complement of connection.

Connectivity = Affiliation, Switching = Social Control

While connectivity means linking people together, switching provides discretionary control over contacts, allowing people to target messages to some people and to exclude others. For example, when a person places a telephone call to a specific number the call is switched so that it goes to the number dialed and only that number. The technology of switching will be covered in Chapter 6. Here, we will consider the meaning and implications of switched communication.

While connectivity provides the means for people to form relationships and to get close to each other, switching allows senders to establish barriers and prevent communication from taking place. Yet without switching, connectivity would result in chaos. All messages would go to everyone in a cacophony of communication. In other words, switching is the complement of connectivity: providing the necessary social distance that makes connectivity functional and useful.

By allowing receivers to choose who (or what) they contact, broadband switching allows people to design their own media menu. They become their own programmers, tailoring the information they receive to their special wants and needs. Whether they like to watch mysteries, action-adventure, cartoons, comedy, music videos, political events—the new advanced networks will allow them to access stored video exactly to their taste.

Switching de-channelizes the media landscape. With television and radio, every channel is on the air all the time. (Choosing a channel is a way of activating a switch at the receiver end, while the senders just keep on sending.) Cable systems operate the same way: every channel the system offers, including every premium channel, actually enters the home. The settop box controls which channels household members can access, depending on the services the subscriber has paid for.

Fully switched, connected networks will gradually change this method of offering entertainment and information services. Rather than send all the channels at once, providers will only send the programming people want at that time. Instead of using the system's bandwidth to carry 60 or 70 channels of TV, switched systems will bring a handful of programs. The bandwidth can then be used for two-way messaging, allowing people to send video messages to another household or media library, as well as receive them from a centralized source.

Another way de-channelization will result from switched TV systems is that people will be able to access video material from sources of their choice. Libraries, museums, record companies, individual artists, program producers, musical groups, and performers of all kinds—they will be able to offer their video products directly to the consumer bypassing the intermediaries of programming services and packagers.

Just as switching will change the way individuals respond to the entertainment media environment, it will also alter organizational behavior and communication. While connectivity creates complexity which in turn promotes hierarchy, switching acts in the opposite way—it flattens hierarchies. It doesn't eliminate them, but it does reduce the number of layers.[16]

The flattening of hierarchy was first noticed in the 1970s, when organizational researchers predicted that computer networks (which are fully switched) would reduce the need for middle management. They observed that top level managers increasingly accessed shared databases to get up-to-the-minute information that had previously been provided to them by data-processing middle-level supervisors.[17] In the 1990s, the white collar layoffs of corporate downsizing were in part a response to the more horizontal organizational structure that switched computer networks made possible.

A final benefit of switched communications is coordination. Targeting receivers allows different messages to be sent to different groups so that each can engage in specialized, patterned action. From the narrowband telegraph (with adolescent delivery boys serving as switches!) to telephone and computer networks, the gains in coordination from switching cannot be overemphasized.

For example, grass roots organization by telephone is now a well-organized phenomenon as seen during every election cycle. Coordinated political action among computer network users is a new phenomenon, but appears to be equally effective. One documented example occurred with the nation's first government-supported computer network, the Santa Monica Public Electronic Network (PEN).[18] The system allowed residents to communicate with city departments and each other. One of the first acts of PENners was to use the system to successfully oppose the privatization of an ocean beach facility. Broadband, two-way switched networks will offer all these features, plus rich interaction.

Table 4.1 summarizes the effects of connectivity and switching on communication at different levels of social aggregation. Person-to-person refers to communication between individuals, while person-to-station interaction occurs when an individual contacts an institutional network node or retrieves information from an archive or database.

Organizational communication occurs between linked individuals who share some form of membership tie and, usually, a shared database that can be accessed by anyone in the group. Members can send messages to the group as a whole, to any combination of members, or to just one member of the group.

Broadcast mode is invoked when people communicate to groups held together only by their common interest, as contrasted to groups held together by an organizational tie. In broadcast mode, individuals send messages to an entire group at the same time, from a few people to millions. This mode also allows a person to send messages outside their immediate group to a wider audience.

Some of the effects noted in Table 4.1 have been demonstrated by research; others are speculative or based on anecdotal reports. Many

Table 4.1 Conceptualizing Connected, Switched Networks

Type of Connection	Person-to-Person	Person-to-Station	Organ'l Comm.	Broadcast Mode
Narrowband connection	Cognition-based relationships[19]	Data upload and download load efficiency[20]	Data exchange: process	Dispersed interest communities[21]
Rich, broadband connection	Emotion-based relationship[22]	Environmental mental immersion[23]	Trusted dispersed work groups[24]	Emotion-bonded disersed communities
Universal switching capability	Discretionary contact	Specialization and de-channelization[25]	Flatten social hierarchies[26]	Social action[27]

effects are quite fascinating and readers may want to follow up the cited works that will lead them to more detailed information about the effects and how and why they occur.

The preceding chapters have described the features of advanced television systems: broadband, digital, interactive, connected, switched delivery systems. The actual construction of this infrastructure is underway in many regions of the world. The next chapter will look at the status of television in the world and evaluate the movement towards a global system.

Notes

1. William Stallings, *Data and Computer Communications*, 4th ed., (New York: Macmillan Publishing Company, 1992):803.
2. John Casti, *Complexification* (New York: HarperCollins, 1994).
3. Mark Twain, *Pudd'nHead Wilson & those extraordinary twins* (New York: Harper, 1922).
4. Jonathan Gill, telephone interview with author, April 1994.
5. Jeffrey Travers and Stanley Milgram, "An experimental study of the 'Small World Problem'," *Sociometry* 32 (1969):425–443.
6. G. Nicolis and I. Prigogine, *Self-organization in non-equilibrium systems* (New York: Wiley, 1977).

7. Laurianne McLaughlin, "Pentium flaw: a wake-up call?" *PC World,* 13:3 (March 1995):50–51.

8. J. Short, E. Williams, and B. Christie, *The Social Psychology of Telecommunications* (New York: Wiley, 1976).

9. Derek Rutter, *Looking & Seeing: The role of visual communication in social interaction* (New York: Wiley, 1984).

10. Gregory Bateson, *Steps to an Ecology of Mind* (New York: Ballantine Books, 1972).

11. A. Meherabian, *Silent messages* (Belmont, CA: Wadsworth, 1971).

12. J. K. Burgoon, "Nonverbal Signals," *Handbook of interpersonal communication,* Eds. M. L. Knapp and G. R. Miller (Beverly Hills: Sage, 1985):344–390.

13. S. Kiesler, J. Siegel, and T. W. McGuire, "Social psychological aspects of computer-mediated communication," *American Psychologist,* 39:10 (1984):1123–1134.

14. Constance Perin, "Electronic social fields in bureaucracies," *Communications of the ACM,* 34:12 (December 1991):74–79.

15. Mary Hayes, "Working online, or wasting time?" *Information Week,* no. 525 (May 1, 1995):38.

16. N. M. Carter and J. B. Cullen, *Computerization of newspaper organizations: The impact of technology on organizational structuring* (Lanham: University Press of America, 1983).

17. S. R. Hiltz and M. Turoff, *The network nation: Human communication via computer* (Reading: Addison-Wesley, 1978).

18. Joan Van Tassel, "Santa Monica's Public Electronic Network," *WiReD* (March 1994). For a formal study, see Joseph Schmitz, Everett Rogers, Ken Phillips, and Donald Paschal, "The Public Electronic Network (PEN) and the homeless in Santa Monica," *Journal of Applied Communication Research,* 23:1 (February 1995):26–43.

19. Short, Williams, and B. Christie, op. cit.

20. J. R. Galbraith, "Designing the innovating organization" *Organizational Dynamics* 10 (1982):5–25.

21. S. R. Hiltz and M. Turoff, *The network nation: Human communication via computer* (Reading: Addison-Wesley, 1978).

22. Short, Williams, and Christie, op. cit.

23. Josh Quittner, "Johnny Manhattan meets the Furrymuckers," *WiReD* (March 1994):92–95, 138.

24. Perin, op. cit., 74.

25. Joshua Quittner, "500 TV channels? Make it 500 million," *Los Angeles Times* (June 29, 1995):D2, D12.

26. Carter and Cullen, op. cit.

27. Van Tassel, *WiReD,* op. cit.

Television Goes Global

5

The Globalization of Television

Much of known human history can be seen as a movement towards globalization. In the beginning, the mobility of tribes was limited to the distance they could travel on foot. Later, people domesticated animals that allowed groups to migrate to ever more faraway places. Finally, the invention of the wheel and sophisticated boats with rudders transported people thousands of miles from their homelands.

In the twentieth century, globalization has accelerated, as measured in: " . . . the increase in the numbers of international agencies and institutions, the increasing global forms of communication, the acceptance of unified global time, the development of global competitions and prizes, the development of standard notions of citizenship, rights and conception of humankind."[1] None of these activities would be possible without the control and coordination made possible by global communication structures.

The globalization of television is now penetrating not only countries with complex, state-of-the-art communication infrastructures, but also those that have been heretofore untouched. The figures are daunting: Worldwide, there are about 860 million homes with television sets. This section will examine the phenomenon in several different ways, including: 1) the technological development in various regions of the world which will become integrated into a global structure; 2) plans for a Global Information Infrastructure; 3) global distribution of programming; and 4) barriers to global television.

A world of difference

Enormous inequalities exist between different countries in the world, so that international television has come much sooner to some areas

121

than to others. The countries that have the most advanced communication infrastructures are Singapore, France, Germany, and Japan, followed by the United Kingdom and the United States.

The comparison of these media-rich societies with the rest of the world is stark. Europe and North America constitute about 15% of the global population, yet they own 65% of all the TVs and radios, read 50% of the newspapers, and publish 67% of the world's books.[2] The industrialized nations have 6 times the number of radios per capita (2,017 radios per 1,000 people) as the underdeveloped nations, and 9 times as many TVs (798 TVs per 1,000 people).

Contrast this communications wealth with Africa, the continent with the least developed information infrastructures. In 1990, Africa had 37 TVs and 172 radios per 1,000 and published 1% of the world's newspapers. In contrast to the industrialized nation's 513 book titles per million persons, underdeveloped nations publish 55 titles per million persons. Comparisons of the penetration of telephones are worse. In the developed nations, there are 246 telephones per thousand people, 1.1 in China, 0.7 in India, and 0.7 in Nigeria.

The Global Information Infrastructure

Television can be delivered over-the-air or through some kind of hardwired infrastructure. Over-the-air systems include Very High Frequency (VHF) and Ultra High Frequency (UHF) broadcast television; satellite delivery from direct broadcast satellite (DBS), TV receive-only (TVRO), a direct-to-home (DTH), or direct satellite service (DSS); wireless cable via multi-channel multi-point distribution service (MMDS); local multi-channel distribution service (LMDS); and low power TV (LPTV). Chapter 7, the Wireless World, will discuss the characteristics of wireless systems in detail.

Wired systems include the familiar cable television systems; video over circuit-switched telephone networks; video over packet-switched computer networks; and fiber optic networks owned by power companies and other organizations. The architectures and characteristics of wired systems will be covered in Chapter 6, the Wired World.

For now, we will consider these broadband technologies as potential pieces that will ultimately be united in an interconnected, switched global communication infrastructure. However, only one of these technologies is currently a fully operational, globally-effective method for television delivery: satellites.

The first step toward a global broadband network is agreement on global standards that allow the transfer and exchange of data, voice, graphics, and video across the universe of wired and wireless systems. Standardization is essential for every conceivable function of interactive broadband communication, such as interactive TV, access to computer networks, telemedicine, distributed work groups and collaborators, and videoconferencing. We will cover this important issue further in Chapter 5, A Matter of Standards.

The notion of a nationwide, interconnected, broadband two-way network, capable of carrying voice, data, and video, is often referred to as the Information Superhighway or the National Information Infrastructure (NII). When this idea is considered in a worldwide context, it is referred to as the "Global Information Infrastructure," or GII.

Efforts to create a GII are underway. In 1995, at their Ministerial Conference in Brussels, Belgium, February 25 and 26, the G7 nations (the U.S., Canada, Japan, Germany, Britain, France, and Italy) agreed to work together to develop the GII.[3] The group approved 11 pilot projects to start the effort. Italy's business leader, Carlo de Benedetti, Chairman of Olivetti SpA, spearheaded the movement within Europe despite opposition from some nations, especially France. Benedetti urged the Europeans to dismantle communication trade barriers, ensure competition, agree on technical standards, and develop intellectual property rules.[4]

The G7 nations set an objective to interconnect the major high-speed (34 to 155 megabits per second) facilities of all the G7 countries by 1997. Their optimistic time frame called for initial tests in June, 1995, initial links by December, 1995, additional links by June, 1996, and deployment and final testing through mid-1997.

The 11 approved projects include:

Create and make accessible an inventory of national and international GII projects;

Set standards for global network interoperability;

Provide cross-cultural education and training;

Compile digitized electronic libraries;

Create electronic museums and galleries;

Increase linkage and integration of databases of information about the global environment and natural resources management;

Enhance management of emergency situations;

Explore global healthcare applications;

Place government information online;

Create information exchange for small and medium business enterprises;

Integrate maritime activities.[5]

The U.S. National Telecommunications and Information Administration of the U.S. Department of Commerce has proposed five essential principles of telecommunications that should apply to the GII: A flexible regulatory framework, competition, open access, private investment, and universal service. In his discussion of these points, Graf notes that a successful rollout of the GII must accommodate economic realities.[6] For example, there have to be customers with sufficient buying capacity and suppliers of services must be able to deliver them at a cost that can be afforded by those customers.

In spite of the hopeful timetable for the GII, it is just in an embryonic stage, a glimmer on the horizon. By contrast, the movement towards global television is occurring rapidly. The next section will describe today's evolving system in various regions of the world.

Moving towards global television: The U.S.

The U.S. has a sophisticated infrastructure of television delivery on both the local and national levels that allows a virtually universal distribution of programming. Over the air, the broadcast television system delivers free service to 98% of the population through local network affiliate and independent stations. Only satellite-delivered programming is national in scope, beamed to about 3 million owners of large C-Band satellite dishes (TVRO service), transmitted at 4 GHz and 6 GHz, and nearly 1 million users of the smaller Ku-Band dishes (DBS service), using spectrum between 17 GHz and 31 GHz.[7]

U.S. entrance into a global television system is inhibited by the characteristics of its two wired systems: cable television networks and the national telephone network. The cable system is fragmented, even though it is ubiquitous. More than 11,000 local cable networks pass 95% of the U.S. population. About 64% of the households that cable passes actually subscribe to it.[8] Constructed over the past 30 years, most systems are now in their third developmental generation, offering 50 or more channels and pay-per-view capability. Currently, many operators are upgrading their systems to allow narrowband upstream communication for limited interactive capacity, but the local systems remain difficult to interconnect into a national structure.

The technology of the U.S. telephone network lags well behind that of Europe, Japan, and Singapore. The 15-year long deployment of Integrated Digital Services Network over copper wire has been so slow to arrive that the technology has been surpassed by fiber optic cable. It is uncertain if ISDN will ever be implemented because telcos will install the more advanced fiber technology. (Scoffing at the tortoise-like speed of implementation, cynics say ISDN stands for 'It Still Does Nothing').

Moving towards global television: Canada and Latin America

Canadian television is more like U.S. TV than that of any other country in the world. Like the U.S., Canada has a strong system of local television stations and, since most of its 10 million TV households are located within 50 miles of the U.S. border, most receive American television programming as well.[9]

Canada is one of the most cabled nations in the world, with penetration rates of 65%, rising to 80% in some urban areas.[10] There is also significant competition between telcos and cablecos to deliver television. The Canadian Radio-television and Telecommunications Commission held hearings in 1995 to decide the best way to regulate the communication infrastructure.[11]

Currently, the cable industry faces no limitations on the services they can provide, while telephone companies are restricted from delivering video services. The Canadian regulators have the same rationale for the disparity of regulation as U.S. regulators: The telephone companies earn vastly more money. Canadian telephone earnings totaled U.S. $15.3 billion while cable revenues amounted to U.S. $1.6 billion.

Canadian cablers are fed programming from the U.S. via wire as well as satellite. Vyvx, a private television carrier, was first to bring broadcast-quality, fiber-optic television distribution to both the U.S. and Canada. In addition, British Telecommunications will provide direct transoceanic connectivity between U.K. broadcasters and Canadian distributors via fiber optic cable. Through BT's teleport to France, Canada will also be linked to continental Europe.[12]

South of the U.S., the communication infrastructure is far less sophisticated. In Mexico, television did not diffuse to the entire country until the mid-1980s. It is still controlled by a single family-owned company, Televisa, which is closely allied with the reigning government. A few urban areas have cable systems in the wealthier areas, including Mexico City, Guadalajara, and Acapulco. In 1994, the Mexican company, TV Por Cable Nacional (40% owned by U.S. company Falcon Cable), had plans to expand their systems in northern Mexico. However, the devaluation of the peso in early 1995 forced the organization to put the plans on hold, as the costs of programming soared due to the devaluation of the currency.[13]

Recently, however, new technologies have emerged. Mexico now has five wireless cable operators, with 2 other systems in the offing. The service has proved especially competitive in populous Mexico City.[14]

Three DTH (direct-to-home) services will launch in 1996, including Hughes DIRECTV, which plans to beam 144 television channels (72 in Spanish and 72 in Portuguese for Brazil) and 60 music channels. The service will be called Galaxy and will cover the Caribbean, Central, and South America as well as Mexico.

Other DTH services are planning to deliver programming to Latin America, attempting to reach between 9 and 13 million multichannel households in Latin America, out of the total of 77 million households.[15] Galavision, a partnership of PanAmSat and Mexico's Televisa, plans to launch in 1996. Panamericana Cable Communications started a DTH service called Inkari in November, 1995.[16] Nahuel-Sat S.A. in Argentina has been delivering television to Argentina, Chile, and southern Brazil since 1994.

Half of all Latin America's cable homes are in Argentina, nearly 5 million. Cablevision is the biggest cable operator, located in Buenos Aires. The largest U.S. cable operator, Telecommunications, Inc. (TCI), plans to acquire 80% of Buenos Aires' Cablevision de Argentina, which currently has about 450,000 subscribers in that city.

Chile is developing a sophisticated wired network that will deliver television, and Telefonos de Chile SA will build a multimedia interactive network to 105,000 homes in the capital city of Santiago.[17] In 1995, SBC (Southwestern Bell) invested $316.6 million in VTR Inversiones which provides cable TV service to Chilean households.[18] And in Valparaiso, Chile, 36,000 homes out of a total of 286,000 homes are passed by Cablevision, 50% owned by U.S.-based United International Holdings, Inc.[19] In Brazil, cable is just beginning. UIH is partners with

Net Sao Paulo (a tiny system with only 70 current subscribers) to pass 1,500 homes.

Moving towards global television: Europe

The European media market is a huge one, amounting to nearly 31% of global activity.[20] In May 31, 1994, the European Union working group, presided over by Commissioner Bangemann, drafted a report that urged the union to recognize that the globalization of communication is inevitable and that Europe's private industry must lead the way. The report urged the Europeans to move quickly to avoid being overwhelmed by American and Japanese competitors.

This move towards privatization is new for Europe. Since the nineteenth century, telephone services have been provided by government telecommunications departments, referred to as PTTs, or public telegraph and telephone agencies. In 1994, the Commission of the European Communities recommended that PTTs' monopolies end in January, 1998.

In just the past decade, European nations have licensed privately-owned and operated channels to offer commercial programming, many via satellite. These new services have proved so popular that some of the continent's public broadcasting systems have lost 47% of their audience.[21] Privatization is well along in Italy, the U.K., and Germany, while France and Spain have changed to a more limited degree.

The extent of Europe's cable infrastructure varies widely from country to country. In Germany, cable passes about 63% of the population; in the Benelux and Scandinavian nations, that figure rises to between 70% and 95% and, in the U.K., it's only about 22%. Approximately 24% of French households are passed by cable and neither direct-to-home television via satellite nor cable have enjoyed much success. In 1994, individual and community antennas connected only 1 million homes out of 21 million TV households.[22] About 14% of Spanish homes are passed, and Italy and Portugal have no cable at all.[23]

In the United Kingdom, the deregulation of both the television and telephone industries has spurred massive investment by U.S. companies. Comcast, Cox Cable, Falcon Cable, NYNEX CableComms, Insight Communications, Jones Intercable, TCI, U.S. Cable Corporation, SBC (Southwestern Bell), and US West have created a patchwork of partnerships with one another and with U.K. companies to build cable systems that pass almost 3 million U.K. television households.[24]

In Germany, the national telephone company will lose its monopoly status in 1998, so competition will heat up in that country. RWE, a power company, plans to offer broadband wired service for computer users, as do two partnerships, one between British Telecommunications and a German industrial group, Viag Ag, and the other between Northern Telecom and a subsidiary of Daimler Benz.[25]

European countries are also experimenting with the most advanced television technology. Companies in both the U.K. and Belgium are conducting interactive television trials. In Sweden, TV-4, a privately-owned station that plans to go on the air in early 1996, is building a state-of-the-art tapeless station. TV-4 will build a server-based fiber optic network, sending digital, compressed video throughout the station.[26]

The countries of Eastern Europe must catch up to develop television infrastructures like those of their Western European neighbors. The government of Poland allotted a broadcasting license to Polsat, a satellite company, which had been broadcasting into the country from the Netherlands in 1992. Polsat competes with two state-owned services and fifty percent of its programming must be produced locally. French pay TV giant, Canal Plus, also plans to beam in an encrypted channel for Poland viewers. In Russia, the government is beginning to issue licenses for over-the-air broadcast stations with national, regional, and local coverage. NTV is the first commercial station that features news free of government control.[27]

Romania is the most-cabled Eastern European nation—50% of households are passed by cable, but only 6% of the country's TV households actually receive it. In Slovenia, cable passes 27% of the homes, while 80% subscribe. Twenty-three percent of homes in the Czech Republic could get cable, and 12% do. And in Hungary, 1 million customers, located in major markets, get some kind of cable service. Kabelkom, the region's first and largest cable MSO that launched in March 1991, entered a 50/50 partnership with a joint venture between U.S. giant Time Warner and United Communications International.[28]

Moving towards global television: Pacific Rim nations

The Pacific Rim is currently the world's fastest-growing television market. The economic growth of the area, combined with the potential audience size, have made these countries a source of revenue for hardware manufacturers and programming suppliers. The information in Table 5.1 makes clear how important this emerging market will become in the future.

There are four ways Pacific Rim countries receive television: Local broadcast transmitters, cable, wireless satellite (microwave), and

Table 5.1 The Television Audience in Selected Pacific Rim Nations[29]

Country (Pop./mil)	Per cap. Income($)	#TVHH (in mil.)	Cable Pen. (%HH passed)	% Satellite Penetration
Australia (44.5)	13,480	13.4	not avail. (began 1994)	.06
China (1100)	380	140/sets	21.4	.71
Hong Kong (5.8)	15,380	1.7	3.57	18
Japan		40	20	20
India (911.6)	310	40.3	24.5	5
Indonesia (199.7)	670	11	–0–	9
Malaysia (19.5)	2,790	2.7	–0–	–0– (banned)
New Zealand (3.5)	12,060	1.8	–0–	–0–
Philippines (68.7)	770	6.8	3.7	–0–
Singapore (2.9)	15,750	10	3.7	–0– (banned)
S. Korea (44.5)	6,790	13.4	(begins '95)	3.1
Taiwan (21.1)	10,000	5.2	50	3
Thailand (59.4)	1,840	12.2	.86	.16

direct-to-home satellite services. The level of economic development, roughly approximated by per capita income, determines which type of delivery prevails.

The per capita income of Pacific Rim countries can be divided into three categories. Table 5.1 indicates that Japan, Singapore, Hong Kong, Australia, New Zealand, and Taiwan are all at $10,000 or above; South Korea, Malaysia, and Thailand average between $7,000 and about $2,000; and the Philippines, Indonesia, China, and India average less than $1,000.

The private development of cable systems requires a population with a relatively high disposable income. Modern fiber optic cable systems are being franchised in Taiwan, Singapore, South Korea, and China. China is an exception to the general rule of per capita income determining delivery mode because the government is paying to build modern cable facilities.

In countries where per capita income is low, cable systems are too expensive, costing an average of $800 per household to build. However, wireless cable (MMDS), which costs approximately $300 per household may be possible. Direct-to-home programming via satellite is also expensive, largely due to expenses incurred for transponder time. However, reception costs are shifted to the receiver, who must buy their own downlink dish, so satellite-delivered TV is largely restricted to the wealthier citizens.

As a result of the economic realities, audiences in poorer countries are more likely to receive TV programming via broadcast channels or MMDS rather than by cable or satellite. One example is Vietnam, where religious broadcaster Pat Robertson is negotiating with the government to invest $3 million to provide programming over MMDS to more than 10 million Vietnamese viewers.[30]

Satellite-delivered television, both for the direct-to-home and retransmission (broadcast stations and cable headends) markets, is booming all over Asia. PalapaB2P brings television to parts of Asia, Indonesia, and part of the Philippines. Using AsiaSat-1, Rupert Murdoch's highly successful Star-TV has been able to reach an estimated 220 million people in Asia, India, and the Middle East. Even tiny Thailand receives programming from two transponders, since June 1995.

To accommodate the increasing demand, many new satellites are going up. In 1994, Intelsat lifted a high-powered 704 satellite over the Indian Ocean and AsiaSat-2 launched in 1995, expanding Star-TV's offerings to 100 channels of programming. Apstar-1A, PanAmSat-4, RimSat G1 and G-1, Shiva 1, and Express 6 also launched and began transmitting in 1995.

In wealthy Hong Kong, Wharf Cable is constructing a 500,000 home system, the region's first large capacity fiber-coax system. Cable will also come to Australia in the next few years, as the government has approved licenses for 9 new cable companies, who can offer 180 channels of programming.[31] Via satellite, Galaxy plans to offer direct-to-home digital channels that will reach most of the country's 6 million households.

Nearby, New Zealand has been slow to receive multichannel television services. The first cable system came online in 1993 in Wellington and a test system now passes 600 homes. Satellite delivery is the most likely pathway for global television to reach this island nation.

Japan, another high income market, receives programming from its national service, NHK. NHK broadcasts two terrestrial-based channels and two satellite channels. In addition, there are 5 private commercial terrestrial networks, all of which have regional affiliates.

Cable television is coming to Japan, too. Currently, cable passes about 6% of the population. The Ministry of Post and Telecommunications has already deregulated cable and will lift restrictions from satellite-delivered TV in 1996.[32] Time-Warner has partnered with U.S. West, Toshiba, and Itochu Corporation to build a $400 million cable network. And TCI joined Sumitomo to put up $500 million to construct a cable network for 775,000 households in Tokyo, invest in other cable systems, and to develop programming for the Japanese market.[33]

Competition in the direct-to-home satellite market is coming as well, with private interests taking on the NHK publicly-supported service. In 1993, Japan Satellite Broadcasting (JSB) began offering programming, capitalized at more than $300 million.[34] In 1997, Space Communications Corporation of Tokyo plans to put up a satellite with 24 transponders to deliver TV and business communications throughout Asia and Hawaii.

South Korea is one area that is growing enough economically to support cable. In March, 1995, the country's 5 public stations stopped their programming over cable and M.Net in Seoul began broadcasting 20 commercial cable TV channels. The system launched with 230,000 subscribers, but South Korea's 48 cable operators plan to roll out across the country as quickly as demand will allow.[35]

In spite of their classification as countries with low per capital income, China and India must be considered significant markets. Regardless of the overall per capita income, the populations of both countries are so enormous, each has a substantial number of people who possess the means to buy TV programming services.

Shanghai is the center for advanced Chinese television. More than 1 million of Shanghainese receive cable television, and, in 1992, the city allowed Oriental Television to broadcast, the only nongovernment TV station in the country. Responsible for its own financial well-being, it quickly captured 42% of the city's audience for its more popular programming. In 1993, Oriental TV actually offered live coverage of the Academy Awards and the Super Bowl![36]

Cable passes 20% of China's households, owned and operated by provincial and municipal governments. Chinese citizens are prohibited from owning satellite dishes, so direct-to-home delivery isn't a factor at this time.[37]

Indian audiences currently receive television from a government-owned and operated service, Doordarshan. In 1995, Indian television underwent major change. Doordarshan TV was granted autonomy and private broadcasters were allowed to both send and receive programming via satellite. India also passed new regulations over its chaotic cable industry that has more than 60,000 operators typically offering only 6 or 8 channels. The government approved a franchise by U.S.-cable company Falcon International and partner, the Hindustan Times Ltd., to acquire 100,000 cable subscribers in New Delhi. They hope to expand eventually to 1 million subs in India.[38]

The regions least-served by international television are the Middle East and Africa. In the Middle East, this situation exists largely because governments enforce religious values that prohibit their citizens from seeing much of the programming that is viewed in the rest of the world. In Africa, the lack of television service is caused by poverty, with the exception of South Africa.

However, recent developments will change Africa's isolation. AT&T plans to circle the continent with a high-capacity submarine fiber-optic cable. "Africa One" will first link coastal populations to global information networks, then connect regional communication centers throughout Africa.[39] Satellite delivery is also coming, through the 41-member RASCOM (Regional African Satellite Communications Organization and Comsat World Systems) which manages the IntelSat satellites. The consortium will use digital technology to reduce costs to bring television to about 40 million TV households in Africa. Finally, NTL will send television signals to Mauritius, an island off the coast of South Africa, for retransmission over local broadcast facilities.[40]

As we have seen, the infrastructure for global television varies greatly by region and it will take many years for a unified system to come into being. By contrast, programming is already a global enterprise.

The global television market in programming

Both television and movies are premier American products. Entertainment has been a strong U.S. export since the 1920s, when American-made movies achieved worldwide popularity. This trend was continued by U.S. television programs, whose licensing in other countries has given the nation's television industry a strong global dimension since the 1960s.

Within the U.S., ambitions for a worldwide television market emerged during the Kennedy administration's New Frontier policy when "U.S. policymakers first envisioned a global television system linked by satellite technology. Their utopian discourse suggested that, in the face of Third World 'unrest' and growing Soviet competition, television would play an important role in promoting an 'imagined community' of citizens throughout the Free World."[41]

Today, entertainment is the second largest export of the American economy, bringing in $8 billion in overseas sales. In 1995, the U.S. sold $4.5 billion a year of TV and movie product sales to Europe, while the U.S. buys merely $300 million from Europe annually. Eighty percent of Europeans watch U.S. films; 2% of Americans watch European films.[42]

The forces propelling the globalization of television programming are mostly economic, rather than political or ideological. One such force is the change in the financing of television programs in the U.S. Prior to 1978, TV networks owned outright the programs they aired. In that year, Congress passed the "financial interest and syndication rule" that prohibited networks from such ownership.

At first, the nets financed programs in their entirety. However, as the network audience declined under the onslaught of cable viewing and VCR penetration, networks gradually reduced the amount of funding they would provide to a production company. Production companies, receiving only about ⅔ of the money needed to produce an episode, turned to distributors and syndicators to make up the difference, a process known as "deficit financing." In recent years, producers have also turned to international co-production arrangements, increasing the number of programs designed to appeal to a multinational audience.

Another factor encouraging a global TV program market is the privatizing of television around the world, leading to the creation of many additional television channels. To fill the increased demand, there are now six international markets for the purchase of television programs: NATPE (National Association of Television Programming Executives), Monte Carlo, MIP-TV (Marche International des Programmes de Television), MIPCOM (Marche International des Filmes et des Programmes Pour la TV, la Video, le Cable et le Satellite), May Screenings, and MIP Asia.[43]

NATPE is held in the U.S. in January of each year. Primarily a market for domestic syndication where original and off-network reruns are marketed to over-the-air network affiliate stations, independent stations, and cable networks, NATPE has become a popular market for Latin American and Canadian customers.

For European buyers, the early-February Monte Carlo market is probably the most important venue, where producers and syndicators introduce mid-season shows to the European market. MIP-TV, a March market, used to be where Europeans took their first look at U.S. television pilots for the fall. Lately this market has declined in importance as U.S. networks have moved the announcements for their fall schedules to June. The new timing has made the New York May Screenings more important and more than 400 buyers from around the world descend on the Big Apple to look at new programs from producers.

MIPCOM and MIP-TV were established by U.S. program producers and distributors as European sales venues for their products. MIPCOM is a mid-October market, while MIP-TV, held in April program, is geared to U.S. network-style programs. MIP-TV is especially useful because the U.S. fall television season has shaken out the successful programs, allowing buyers to predict their popularity, consistency, and quality. MIP Asia is a new market, held for the first time in Hong Kong in late November of 1994. It caters to the new television stations and satellite delivery services in the Far East.

Feature films are important as TV programming as well as for theater fare. Foreign television sales now generate as much profit as a film's domestic box office, so increasingly film producers who come to Europe for the Cannes Film Festival in May attend the MIP-TV market in April on their way to Cannes.[44]

Trends in the international TV program market

The last decade has seen an enormous expansion in the number of channels worldwide. This larger market has resulted in greater demand for programming, of which the U.S. has been the primary beneficiary. American programming composed 28% of that shown on Italy's Italia 1 and about 8% of Germany's RTL and Pro 7.[45]

There are now three centers in the U.S. for creating international programming. Traditionally, New York and Hollywood have marketed programming, first to Europe, then to Asia. In the past few years, Miami has also become a production center for programming aimed at the growing Latin American television market.[46]

The one-hour action-adventure series is an enormously popular genre world-wide. In 1994, one estimate indicated that international buyers wanted 20 to 30% more of the one-hour product than was available.[47] It is a format that has proven difficult to copy and no other country has been able to duplicate U.S. success in this area. Drama is universal and audiences in societies as disparate as Israel, Russia, and Egypt enjoy "Dallas" and "Kojak."

By contrast, comedy does not travel well, on the whole remaining a local phenomenon. One exception is cartoons. A generation of European thirty-somethings smile when they hear "yabba-dabba-doo," from "The Flintstones."

In the last few years, the market for individual programs has been eclipsed by the export of entire channels, many of which began as U.S. cable channels. MTV and CNN stand out as especially successful and many more are scheduled to follow.[48]

MTV reaches 231 million households in over 100 countries, including Europe, Russia, South America, and Asia. Viacom, the parent company, is also readying Nickelodeon and the more traditional music channel, VH-1, for the international market. Following its debut in the United Kingdom and Germany, VH-1 will launch in Latin America in 1996 as a 24-hour channel beamed from Miami. Turner Broadcasting sends CNN's international channels to 24 million viewers, accounting for about 14% of the company's total 1994 revenue.[49]

An extraordinarily popular channel format is TV shopping. From Canada and the Caribbean, to China and Japan, the United Kingdom and Israel, there are hundreds of millions of viewers buying merchandise they see on television. The business success of this kind of marketing depends on a pre-existing credit card economy and a delivery infrastructure for rapid, accurate fulfillment, however, limiting the countries where these ventures can be profitable.

Examples of channels for export abound. ESPN plans to launch a 24-hour satellite sports channel in Brazil and the Country Music Channel is now beamed to Latin America and Thailand. The Discovery Channel will also be available to Latin American viewers. HBO is a favorite of the Czech audience, while students in Thailand and Taiwan can attend Jones Intercable's Mind Extension University: the Asian Campus.

CNN continues to spawn competition. Spanish cable homes will soon receive "Telenoticias," an all-news channel prepared by Reuters TV. The Associated Press, which has long supplied print news and photos to customers around the world, plans to enter the global video business, making footage from anywhere available everywhere. NBC plans to export its success with CNBC to ANBC—a 24-hour Asian business channel that will go to about 24 million households in 15 countries.

Global multimedia production While the computers for multimedia come mainly from the U.S., the software is an international effort. Major multimedia production is underway in the United Kingdom, France, Germany, Italy, Japan, and India. Developers are creating programming for PC, Mac, and Philips' CD-I platforms, as well as for dedicated game players. The most important global commercial markets for multimedia product are COMDEX (U.S.), E3 (Electronic Entertainment Expo-U.S.), and MILIA (Cannes, France).[50]

Barriers to the globalization of television

In spite of the enormous growth of world communication, there are factors which could prevent, or at least slow down, a global television system. These factors include: infrastructure disparity; censorship and other restrictions by regional and national governing bodies; and violations of intellectual property rights.

As discussed earlier, there are large differences in the communications infrastructure between the regions of the world. The disparity not only limits the economic development of these areas, it also slows the globalization of television and other communication modalities, such as telephone and communication networks.[51] The solution to this problem is difficult, given the high cost of constructing sophisticated communication infrastructures.

The satellite distribution of television programming to elites anywhere in the world already occurs. However, the vision of a truly global television system, especially one that is interactive, connected, and switched cannot depend solely on wired networks. At an approximate cost of $800 per household connection, the world would need to invest about $700 billion dollars. Hopes for a global TV system rest with some form of wireless infrastructure and its lower implementation costs.

The absence of communication infrastructure is not the only barrier to the globalization of television. The need for international communication to support economic, political, and cultural activities motivates modern nation-states toward allowing their citizens and institutions to engage in transborder data flows. However, too much openness can have serious negative consequences that may result in political instability, the devaluation of local culture, religion, and customs, the consolidation of the yearnings of ethnic and linguistic groups that transcend national boundaries, and the importation of images that encourage violence and rebellion.

Some governments restrict communication to protect against this political and social disruption. In the Middle East, programs are edited to eliminate references to sex and between-gender interactions. In Indonesia and Malaysia, citizens are forbidden to have satellite dishes. Singapore does not allow residents to tie into the Internet. Historically, most countries have controlled the broadcast media very stringently, guarding access to television and radio very closely. It is no accident that a successful coup is marked by the capture of the governmental palace, army headquarters, and the national television and radio facilities.

Other countries are more concerned about their potential loss of cultural identity. Canada and France are examples of nations that have recurrent concerns about a flood of international information, particularly from the U.S. One strategy they have adopted is to place quotas on the amount of foreign programming television distributors may show.

Led by the French, in 1989 the European Union's 15 member states agreed to restrict foreign programming to 50% of non-news and sports programming, which had to be of European origin, where practicable. In February, 1994, the French successfully blocked movie and television programming from being included in the free trade provisions of the General Agreement on Trade and Tariffs (GATT).

In February, 1995, the French were able to influence the EU Executive Commission to close the loophole created by the words "where practicable." The group placed less stringent restrictions on thematic channels, such as cartoon and all-movie channels, which must have 25% of their programming coming from European production companies. Video-on-demand and other new technology services were exempted altogether from quotas.[52]

The French position is controversial, even within the European Union. For example, Germany and the U.K. are far less concerned. The London "Economist" magazine noted that 85% of Europe's film directors are over 50 years old, and the editorial page of the Munich Sueddeutsche Zietung observed: "Prime time on European television still belongs to Hollywood and the wee hours will belong to a lonely, doubtless very 'cultural' European movie. As it circulates through the sprockets, it will discharge only one function: to meet the quota."[53]

An extension of this issue appears to garner more widespread support throughout Europe: The next battleground will arise over the computer networks and the plans for a Global Information Infrastructure. Once again, France is playing a dominant role in limiting U.S. influence. EU Commissioner Edith Cresson, former French prime minister, said the 15-member EU "is as culturally and industrially threatened by education and training software of Nintendo and Microsoft as it is by U.S. television series."[54]

Another barrier to global television is the lack of a global legal system. Enforcing intellectual property rights against theft has surfaced as a severe problem of world trade, as exemplified by the conflict between the U.S. and China.

From 1993 to 1995, U.S. software makers became increasingly concerned as more and more bootlegged CD-ROMs surfaced, loaded with expensive computer programs and interactive games. The U.S. government and the Hong Kong Customs and Excise Department identified 29 factories inside mainland China manufacturing unauthorized software on CD-ROM.

By 1994, the problem had become immense: In 1994 alone, such piracy cost U.S. software makers $1 billion. The Copyright Subcommittee of the U.S. House of Representatives issued a study in February, 1995, showing the importance of core copyright industries to the U.S. "Core copyright industries include movies, books, sound recordings and computer software. They contribute more to the U.S. economy and employ more workers than any single manufacturing sector . . . the value added to the economy is $226.5 billion yearly," noted the subcommittee study.[55]

Despite the barriers to a wholly integrated global television system, the above discussion makes it clear that the world is well on the way to global distribution of programming. The development of an integrated infrastructure will take a much longer time, but meanwhile, satellite distribution between continents to local microwave or local television transmission facilities ensures that people around the world have access to both locally and internationally-produced television.

Some scholars believe that the media are a source of culture, in addition to interpersonal communication from family, friends, neighbors, and teachers. This opinion raises the question of the influence international programming might have on the development of culture around the world.

Does Global Television Presage a Global Culture?

A focus on the world as a single unit is a major idea of the twentieth century that permeates intellectual thought, historical action, and commercial development. Beginning in the 1980s, ubiquitous international communication, reliable transportation, international commercial outsourcing, the loss of jobs, and the massive movement of entire populations have driven home to many people that the geographical basis of culture may be changing.

For example, the 12 to 15 people sharing a meal in a bed-and-breakfast in Scotland may well have more in common with any one of them than they do with their immediate neighbors. This phenomenon occurs because the bed-and-breakfasters are likely to be a German computer programmer, a British professor, an American documentary filmmaker, an Australian engineer, a French statistician and an Italian architect—all members of the "Information Society," who read the same books, see the same movies, listen to the same music, and will exchange Internet e-mail addresses at the end of the meal.[56]

However, it is unlikely there will be a global culture that is like the geographically-situated culture we take for granted in our everyday lives. If one were to develop, it could take hundreds, perhaps thousands of years to evolve. If we broaden our definition of culture, then "it may be possible to point to trans-societal cultural processes which take a variety of forms . . . processes which sustain the flow of goods, people, information, knowledge and images which give rise to communication processes which gain some autonomy on a global level. Hence there may be emerging sets of 'third culture,' which themselves are conduits for all sorts of diverse cultural flows which cannot be merely understood as the product of bilateral exchanges between nation-states."[57]

This formulation of culture holds that global communication may lead people to joint memberships in overlapping spheres of cultural influence: They are born into their first culture, ethnic culture; they are raised in their second culture, civic culture; and they work and play in their third culture, the global communication culture.

Some postmodernist theorists concern themselves with these questions and there is a significant research tradition that explores the dimensions of an emerging global culture. Since it is a growing and changing phenomenon, there is little agreement among scholars about the important variables, concepts, and processes which might be involved.

Other scholars believe that the new communication technologies are destroying the Global Village recently created by the mass media. "While the new media communication system promises decentralization, individualism, and a break from passive viewing, it will end the global village as a dominant environment and the tribal culture which it has helped to recreate in the modern world. That culture bespeaks a commonality, a unity which derives from powerful bonding forces: a shared vision, shared values, a shared sense of security and comfort, much as America provided the many ethnic groups a new home in the New World and coalesced into the melting pot which was our societal core."[58] In this view, the structure of the technology itself is the model for its effects; that is, the proliferation of digital delivery and distribution technologies, away from the centralized, unified analog systems creates a society that loses those centralized, unified features.

At present, we cannot draw solid conclusions about the evolution and effects of a connected, switched global system of television because it is still too far in the future to make accurate predictions. However, we do know that a precondition of such a system is agreement upon a set of standards to guide the equipment, processes, and datastreams that will comprise a global infrastructure. The next chapter will consider the problem of global standards.

Notes

1. Mike Featherstone, "An Introduction," *Global Culture: Nationalism, globalization and modernity,* ed. Mike Featherstone (London: SAGE Publications, 1990):6.

2. Peter Golding, "The communications paradox: inequality at the national and international levels," *Media Development* 41 (April 1994).

3. "G-7 nations hop on info superhighway," *Los Angeles Times* (February 27, 1995):D5.

4. "Business leaders schedule meeting on G7 issues," *Telecommunication Reports,* 6:7 (March 17, 1995):6

5. "G7 countries agree to 11 'GII testbed networks'," *NextNet,* 4:5 (March 13, 1995):13.

6. James E. Graf, "Global Information Infrastructure: First Principles," *Telecommunications* (May 1994):72–73.

7. As of April 24, 1995, these figures were provided by DBS Digest, Pueblo, Colorado, P.O. Box 11036, Pueblo, CO, 81001.

8. *Facts at a Glance: International Cable* (Washington D.C.: National Cable Television Association, Spring 1995):1.

9. Marc Doyle, *The Future of Television* (Lincolnwood, IL: NTC Business Books, 1993).

10. *Facts at a Glance,* op. cit., 1.

11. Joanne Ingrassia, "Canada's infohighway war: Cable, telcos fight to enter each other's businesses," *Electronic Media* (March 20, 1994):23, 52.

12. Services of Vyvx and BT taken from public relations literature from each respective company.

13. Thomas P. Southwick, "Down Mexico Way," *Cable World* (April 10, 1995):8.

14. Annette K. Hugh, "MVS Multivision pushes ahead despite peso's plunge," *Wireless International,* 2:5 (April 1995):6–7.

15. John Shackelford, "Miami: Launch pad for Latin America," *International Cable* (April 1995):20–28.

16. Wayne Walley, "Latin TV sets plans for DBS," *Electronic Media* (March 20, 1995):40, 44.

17. "C-Cor to provide line extenders for Chilean multimedia project," *Wireless International* (April 1995):20.

18. "International Scene," *Radio Communication Report,* 14:5 (March 13, 1995):26.

19. *Facts at a glance,* op. cit., 21.

20. Sylviane Farnoux-Toporgoff, "The European Union postures the information society," *New Telecom Quarterly* (3rd Quarter, 1994):3–6.

21. Fred Hift, "EBU adapting to compete with commercial TV," *Electronic Media* (April 10, 1995):72–73.

22. "Looking for that little extra," *Cable and Satellite Europe* (November 1994):68–70.

23. *Facts at a glance,* op. cit., 1.

24. Ibid.

25. "European utilities eye telephone business," *Wall Street Journal* (December 2, 1994):B4.

26. National TeleConsultants, "TV4. Sweden Goes Server Route from Get-Go," *Broadcasting & Cable* (April 3, 1995):S-2.

27. C. Stratimirovich and B. Mallalieu, "Polsat license bid giant-killer," *Hollywood Reporter* (February 1, 1994):1–5.

28. "Magyar Chiefs," *Cable and Satellite Europe* (November 1994):38, 40.

29. Data for this table was found in *TBI Television Broadcasting International,* 1995 (Washington, D.C.: National Association of Broadcasters, 1995).

30. David Tobenkin, "Robertson close to deal on Vietnam cable," *Broadcasting & Cable* (January 1995).

31. "Hitch-hikers' guide to Asia's cable and satellite markets," *Cable and Satellite Europe* (November 1994):22–28.

32. "Japan's Cable Growth," *Forbes* (November 6, 1995):44.

33. Rich Brown, "TCI invests in $500 million Japanese venture," *Broadcasting & Cable* (May 30, 1994):28.

34. Doyle, op.cit.

35. Associated Press, "Cable arrives to revolutionize South Korea for better or worse," *Los Angeles Times* (March 2, 1995):D4.

36. Associated Press, "Chinese being given greater remote control," *Los Angeles Times* (April 17, 1995):D5.

37. Yu-li Liu, "The growth of cable television in China," *Telecommunications Policy,* 18:3 (1994):216–228.

38. Indrajit Lahiri, "Star struck in India," *World Broadcast News* (April 1995):86–88.

39. "AT&T's Africa cable," *Telecommunication Reports,* 6:7 (July 1995):15.

40. Information taken from Comsat World Systems press release, "Africa and Middle East Regions forecasted as next growth areas for international broadcasting." April 20, 1995, 6560 Rock Spring Drive, Bethesda, MD, 20817.

41. Michael Curtin, "Beyond the Vast Wasteland: The policy discourse of global television and the politics of American empire," *Journal of Broadcasting and Electronic Media* (Spring 1993):127.

42. Tyler Marshall, "Hollywood faces a new fight with Europe on quota issue," *Los Angeles Times* (February 10, 1995):A2.

43. Morrie Gelman, "Monte Carlo: Special Issue" *The Hollywood Reporter* (February 1, 1994):S3–S16.

44. Louise McElvogue, "Cannes now a prime-time player," *Los Angeles Times* (October 13, 1995):D4, D5.

45. Meredith Amdur, "Adventure a strong U.S. export," *Broadcasting & Cable* (August 29, 1994):39.

46. Shackleford, op.cit.

47. Michael Freeman, "It's the hour of the hour," *Media Week,* 4:23 (February 1, 1995):18.

48. Charles Haddad, "Turner's epic expansion overseas," *International Cable* (April 1995):30–41.

49. Ibid.

50. Jeffrey Young, "World Class," *The Hollywood Reporter Special Issue on International Interactive* (January 11, 1994):S8, S22.

51. H. Mowlana and Laurie J. Wilson, *The Passing of Modernity: Communication and the Transformation of Society* (New York: Longman, 1990):43–75.

52. Tyler Marshall, "EU panel urges tighter TV import quotas," *Los Angeles Times* (February 9, 1995):D-1.

53. Ibid.

54. Associated Press, "Europeans seek quotas to blunt U.S. dominance of cyberspace," *Los Angeles Times* (February 23, 1995):D-8.

55. Brooks Bolick, "Piracy tab $1 bil in 94; data 'cannot be ignored'," *The Hollywood Reporter* (February 17–19, 1995):1, 35.

56. K. Gergen, *The Saturated Self* (New York: Harper Collins, 1991).

57. Mike Featherstone, "Global Culture," op. cit., 1.

58. Richard Weinman, "Anytime, anywhere communication," *New Telecom Quarterly* (4th Quarter, 1994):18–22.

6

A Matter of Standards

Standards: Pros and Cons

The arguments for and against setting universal standards rest on interoperability, technological innovation, and economic development. The acceptance of standards allows products from different manufacturers to work together, fosters the design of new products and services, and can even result in the creation of entire industries and product categories.[1]

Interoperability among discrete pieces of equipment that compose an entire system, as shown in Figure 6.1, is the first benefit of standards. It ensures that light bulbs plug into sockets, that batteries fit into flashlights and drum-beating rabbits, and that videocassettes fit into VCRs. Interoperability is a powerful incentive, as any international traveler who has carried the assortment of plugs and adapters necessary to shave or dry hair can attest.

It's important to manufacturers, as well. Standards reduce the risks of research and development because producers can depend on relatively stable conditions for their products. With the assurance that they can recoup their investment, they are willing to devote time and money to developing new products that enable specific applications. Consistent interfaces also allow companies to turn their attention to cost reduction, increasing the efficiency of production and the quality of the core product while decreasing installation and maintenance costs.

The overall market may gain as well. Standards reduce barriers to entry to the market, introducing a competitive environment. Competition usually acts to increase innovation. Thus, standards encourage market differentiation because producers introduce additional features for each product, offered at price points that reflect their desir-

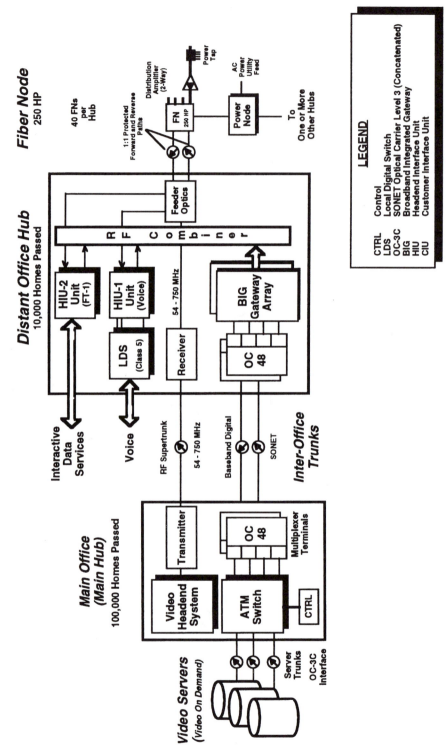

Figure 6.1 A complex integrated multiservice distribution system that must have interoperable components. Source: Carl Podlesney, Scientific-Atlanta.

ability. At the same time, interchangeable products reduce consumer confusion because buyers understand price differences based on features, but become bewildered by incompatible products.

For example, in 1948, CBS produced vinyl records that ran at 33rpm, while RCA introduced the 44rpm standard. As a result, the record industry experienced a slump that lasted 4 years.[2] Similarly, 15 separate videotape formats all failed until VHS won the standards war; only then did the video cassette industry prosper. One explanation for the relatively slow adoption of computers might be that in the past 15 years, there have been 200 different computer and video game formats—all incompatible.

When a company is developing the internal design of a product, standards are desirable; but they are absolutely necessary for the interconnection of a system when multiple suppliers are involved. Without standards, it is impossible to provide a smooth upgrade path for improving the product over time.[3]

The absence of standards may prevent a market from developing at all. Leaders from several sectors have cited a lack of accepted standards to explain sluggish performance of a sector. "No standards is slowing the growth of the multimedia industry: developers have 20 platforms and platform conversion costs 50% of the going-to-market cost," said Stewart Bonn, Senior Vice President of the Compact Disc Group of Electronic Arts.[4] In the same vein, according to the banking industry, "Lack of standards makes home banking impossible."[5]

Setting standards and having access to accepted standards are beyond merely desirable; they are necessary merely to participate in international commerce. Out of the total U.S. exports to Europe of $83 billion, $48 billion are subject to some kind of European Community (EC) standards.[6]

However, there are some negative aspects to setting standards. The most serious is that a large installed base, built upon accepted standards, can act to stifle innovation. The costs of building a consensus to change the standard, designing new products, replacing manufacturing equipment, and, to consumers, replacing old products, can be enormous. The difficulty of introducing high definition television, as described in Chapter 1, is an excellent example of just how tough it is to change the standards of a product with an enormous installed base.

Internationally, standards can be (and have been) used to erect trade barriers. The three major regional trading blocks—the U.S., Europe, and Japan—all accuse the others of such behavior. A U.S. government publication states: "If the GATT cannot sustain an international economic order based on free trade principles, standards will be used, increasingly, as nontariff trade barriers and also as part of national, or regional, industrial policies. This is now happening both in Europe and Japan."[7]

Finally, there are costs associated with the standards-setting process itself. A decade ago, producers sent an employee to sit on a committee, assuming the work could be done in the worker's spare time. The pace of technological change, the increasing complexity of the work, the larger number of stake holders, and the tremendous stakes have all added to the investment of time and money needed to reach consensus on standards.

The Standards-Setting Process

Standards-setting bodies move through stages in the process of reaching agreement. Steps common to both public and private efforts are:[8]

1. *Place item(s) on the agenda:* The organization that wants to put standards in place writes a proposal, outlining the current state of the technology and the scope of the proposed standard;

2. *Development:* A working group is appointed and a subset of the group prepares a draft of proposed technical specifications for circulation to all members. This stage may take 1 or more years, depending on the complexity of the technology and the proposed standard;

3. *Stakeholder review:* After the working group members have made their recommendations and comments, a revised draft is prepared for interested stakeholders to make comments;

4. *Iteration:* The comments from the stakeholder review go back to the working party and the process starts over until the standard can no longer be improved;

5. *Consensus:* Many standards, even formal ones, are voluntary, with market forces providing the primary compliance mechanism. Key players whose participation fulfills a critical piece of an end-to-end solution must be convinced, cajoled, or pressured into accepting the standard. If it proves impossible to reach agreement, the industry may be doomed to a period of

"duelling standards," which may depress the overall growth of the market;

6. *Testing and validation:* All the products and processes covered by the standard are tested to make sure they work as claimed. Sometimes disagreements can be resolved by a technical test of solutions, such as occurred between the HDTV over-the-air transmission systems by General Instrument (QAM-64) and Zenith (VSB-16).

7. *Publication:* This final step allows stakeholders to design products that will interoperate, conform to import rules, and meet legal requirements for safety. In recent years, there have been calls for releasing specifications electronically.[9]

These processes are common to many different standards-setting bodies, both public and private. When it comes to the television of the future, there is no shortage of such organizations—at least 28 different groups are involved. A plethora of officially-recognized agencies, consortia, trade associations, individual companies, and non-profit organizations that address every part of the system from end-to-end designs to chip sets in the settop box, have entered the fray. The next section will describe the most influential groups among them.

International Standards-Setting

The formal international standards-setting body is the International Telecommunications Union, a specialized body of the United Nations headquartered in Geneva, Switzerland. However, discontent with the ITU has spurred the creation of private organizations to develop standards that are more sensitive to market needs.

The embryonic ITU came out of the first instance of formal international standards involving communication: the telegraph. In 1849, European countries, including Germany and Austria, signed an agreement to interconnect telegraph lines. They set technical standards for the equipment, agreed on the codes, and devised a method of charging for telegraphy services.

That agreement grew into today's International Telecommunications Union, although the organization didn't actually adopt that name until 1875. By 1885, the agency had expanded to include the telephone. A parallel organization, the Radiotelegraph Union, coordinated international standards for communication by radio. In 1932, the

ITU reorganized to take its present form of an administrative staff, supported by two parallel committees, the CCIR (International Radio Consultative Committee) and the CCITT (International Telegraph and Telephone Consultative Committee).[10]

The membership of the ITU is composed of about 175 member countries. The organization handles both upstream and downstream activities in standards-setting activities. The upstream process is the way information comes into the ITU from member countries and recognized national standards-setting bodies. The downstream process is how the ITU publishes and disseminates its "recommendations," the term the agency uses to mean standards.[11]

The upstream process is handled by 15 work groups. Every four years, the ITU holds a Plenipotentiary Conference, attended by representatives from all the member nations. Recommendations are proposed by the ITU work groups or come up from national work groups. Agreement on standards occurs by a consensus adoption in the conference. Unresolved issues go back to the work groups, along with new questions. Their conclusions about these assignments will be considered at the next Plenipotentiary Conference.

This ponderous structure and leisurely pace contrasts sharply with the volatile communications industries, one of several factors that undermines the ITU's preeminence in setting standards. Many also criticize the organization's tilt towards its historical constituency, national telephone and telegraph utilities. Still another problem has been the difficulty of obtaining global consensus in an era when the role of standards in building and controlling markets is widely recognized. The economic importance of standards has led to the creation of regional standards organizations and greater activity in the private sector.[12]

The ITU has been moving to correct these perceived deficiencies. In 1992, the agency voted to put several reforms into effect in 1994. These changes included restructuring the CCITT and CCIR into standardization sectors allowing a wider spectrum of participants (other than national governments), adopting a more market-driven process, and providing more timely access to information about standards, such as electronic dissemination.[13]

Another agency, the International Standards Organization (ISO), promotes worldwide cooperation in intellectual, scientific, technological, and economic activity. In that role, it has defined the Open Systems Interconnection (OSI) standards for computer networks. OSI is the only internationally-approved protocol for computer networks, providing a layered approach to encoding messages. It defines standards for 7 layers:

LAYER 7: *Application*—Provides access to the OSI environment for users and also provides distributed information services;

LAYER 6: *Presentation*—Provides independence of the application processes from differences in data representation and syntax;

LAYER 5: *Session*—Provides a controlled structure for communication between applications. Establishes, manages, and terminates sessions between cooperating applications;

LAYER 4: *Transport*—Provides reliable, transparent transport of data between end points. Provides end-to-end error recovery and flow control;

LAYER 3: *Network*—Provides independence of the upper layers from the data transmission and switch technologies used to connect systems. Responsible for establishing, maintaining, and terminating connections;

LAYER 2: *Data link*—Provides for the reliable transfer of information across the physical link. Sends blocks of data with synchronization, error control, and flow control;

LAYER 1: *Physical*—Concerned with transmission of unstructured bitstream over a physical medium. Deals with mechanical, electrical, functional, and procedural characteristics to utilize the physical medium.[14]

Regional Standards-Setting Organizations: ETSI, TTC, and ATIS-T1

In addition to the creation of informal standards-setting groups, the perceived deficiencies of the ITU has also led to the establishment of other formal standards organizations serving regional trade zones. Table 6.1 shows regional organizations.

Private international standards-setting bodies: DAVIC and MMCF

Dissatisfaction with the responsiveness, speed, and efficiency of the ITU also led to the formation of private groups. DAVIC is an acronym for the Digital Audio/Visual Council, an ad hoc organization which acts as an overall coordinating agency to identify conflicts and gaps in the many standards needed for the emerging television system. An ad

Table 6.1 Regional Standards-Setting Organizations

Region	Process
ETSI-European Telecommunication Standards Institute/Europe, established in 1988	Expedite standard-setting, expand membership to vendors and users. Published standards for public networks, WANs, PBX, radio for portable networking, and others.
TTC-Telecommunications Technology Committee/Japan	Private, voluntary group of telcos, equipment mfgrs., and foreign cos. Dominated by NTT. Works by consensus or, if that fails, vote.
APT-Asian Pacific Telecommunity	Has 23 member nations that helps explain and apply standards.
AIC-Asian ISDN Council	Membership includes ASEAN countries, to advance ISDN networks.
APEC-Asia Pacific Economic Cooporation	Supplies compatibility information from members to international bodies.
ATIS-Alliance for Telecommunications Industry Solutions, T1 Committee/U.S.[30]	2,000 members from hundreds of mostly telcos. They work on standards and make recommendations to U.S. and int'l standards organizations.
ITSC-Inter-regional Telecommunications Standards Conference/international, 1992. (Became GSC-Global Standards Collaboration Group.)	Members are regional standards bodies, including ETSI, the T1 Committee, the TTC, Canada, Australia, and Korea. Private organization to reduce trade barriers caused by stds. and discuss tech at pre-standardization stage.

hoc organization is one that is assembled for a specific purpose and ceases to exist when it has finished its work. DAVIC's goal is to hold open discussions and to conduct interoperability tests and experiments, leading to a global consensus among experts on digital compression, software, hardware, networks, satellites, consumer electronics, and research laboratories.

For its operation, DAVIC used the model of the Moving Picture Experts Group, which developed standards for the digital compression schemes, MPEG-1 and MPEG-2, as discussed in Chapter 2. According to Bob Luff, a member of the management committee:

"DAVIC is a body of technical representatives from more than 100 companies worldwide. They share a common goal of assisting in the success of emerging digital audio/visual applications and services by the timely availability of internationally agreed-upon specifications of open interfaces and protocols that will maximize content creation and hardware interoperability across service providers, networks, and countries."[15] The standards would cover many delivery systems—fiber optic cable, coaxial cable, twisted pair copper wire, and satellite—and a broad range of applications.

Formed in October of 1994 by Leonardo Chiariglione (who also headed MPEG), DAVIC identified 17 critical interface points and began discussions to ensure compatibility. While otherwise retaining its mission, in March, 1995, the group decided to defer to another organization, the ATMForum, to develop standards for "asynchronous transfer mode," a protocol for fiber optic networks. (The ATM data structure is a potentially contentious issue because satellites don't need to use it in their delivery systems. For that reason, program providers who distribute their product over satellite don't want to have to encode their images into the ATM data structure.)[16]

Another informal organization is the Multimedia Communications Forum (MMCF), an international consortium of vendors whose 100+ members include AT&T, DeteBerkom Gmbh, Ericsson, IBM, Motorola, Siemens, Northern Telecom, NYNEX, Cabletron, U.S. Sprint, and Compression Labs. The group's goals are to facilitate open exchange, define, and recommend architecture, standards, and quality of service levels for cross-platform functionality. MMCF provides an up-to-date, centralized source of information on market-ready products, end-user applications, and multimedia requirements.[17]

The MMCF's End-user Application Committee examines desktop videoconferencing with collaboration, video playback, multimedia messaging, database access, networked kiosks, distance learning, interactive services, games, shopping, and sports entertainment. MMCF meets quarterly and maintains links with the other major standards-setting organizations. They plan to offer demonstrations and publish recommendations in order to accelerate market acceptance and vendor interoperability.[18]

U.S. Standards-Setting Organizations

The U.S. standards-setting scene can only be described as a splendid anarchy. A host of governmental agencies and private trade associations, professional organizations, general membership groups, third-party certifiers, and ad hoc consortia establish their agendas and do their utmost to make them the custom, if not the law, of the land. It is clear that standards-setting has become an important and inescapable marketing tool.

The government sets approximately half the standards in the U.S. through its regulatory agencies. Most deal with safety and other consumer issues. The National Institute of Standards and Technology (NIST), under the Department of Commerce, acts as the nation's central laboratory for developing and disseminating measurement standards. However, the U.S. has no comprehensive, governmentally-set standards policy and if NIST attempted to put one in place the agency's efforts would be roundly rejected by American business interests.

An important agency for the emerging television system is the Federal Communications Commission. As we have seen in Chapter 1, television industry players requested the agency to establish and monitor the standardization for HDTV. While the FCC has no authority over digital protocols, compression schemes, or switching options, it does exert influence in other areas.

The FCC has had a historical commitment to universal service with respect to telecommunication networks and that predisposition may carry over to broadband networks. Universal service is the notion that everyone ought to be able to have telephone service and that monopoly telephone companies have a responsibility to provide service categories to include all people in society. The FCC also regulates telephone and cable systems and can influence how they conduct their businesses through that power.

Another agency, the American National Standards Institute (ANSI) is a private, self-designated national coordinating board for standards. ANSI's central role as a third-party certifier has given the group a quasi-official status and it is a member of the International Standards Organization (ISO) and the International Electrotechnical Commission (IEC). ANSI receives contribution money from private companies and raises 28% of its income through the controversial practice of selling standards information and providing standards-related services. Some critics believe that the current system favors the strongest players in the market and would prefer to see standards disseminated freely.

The other private bodies providing standards for the evolving television system include every conceivably related professional group. In addition, associations and consortia are organized around every manufactured piece of the emerging advanced communication system. Key professional groups include the Institute of Electrical and Electronics Engineering (IEEE), the Internet Engineering Task Force (IETF), and the Video Electronics Standards Association (VESA). Private trade associations abound, like the Electronic Industry Association (EIA), the Infrared Data Association (IRDA), the International Multimedia Teleconferencing Consortium (IMTC), Personal Conferencing Work Group, the ATMForum, ATM25, and the International Multimedia Association (IMA).

Private standards-setting groups for equipment

There are competing designs (and, of course, standards) for virtually every piece of equipment that goes into broadband networks. Table 6.2 identifies these bodies and the standards they are involved in setting.

Barriers to Global Standards

Over the last decade, changes in the context of setting standards have made the process much more difficult. In the past, standards were set to formalize requirements for existing technology. Today, standards are often set which exceed state of the art.[19] HDTV is an instance of this change, where the demands for progressively-scanned/60 frame-per-second monitors cannot be manufactured.

In addition, company personnel used to be able to work on technical committees in their spare time, relatively free from oversight. Now the investments for research and development are so large and the stakes so enormous, senior managements pressure technical representatives to push the company design at all costs. Finally, the increasingly complex network of professional groups, trade and industry associations, consortia, alliances, vendors, suppliers, and end-users, with their divergent perspectives, makes reaching agreement ever more arduous. Even when agreement is obtained, compliance is voluntary and enforcement is impossible.

The emergence of yet another layer of actors, the regional trading blocks and their associated standards-setting bodies, the European Telecommunication Standards Institute (ETSI), the Japanese Telecommunications Technology Committee (TTC), and the U.S. T1 Committee, brings even more powerful forces into the fray. Yet, in some ways this decentralization may be exactly what is needed. The emerging

Table 6.2 Some Private Standards-Setting Bodies for Network Equipment

Group/standard	Purpose
ATMForum	Standards for asynchronous transfer mode, including optical switch and data structure for SONET networks.
Int'l Standards Org./ (OSI)	Open Systems Interconnection protocol is a platform to develop standards for computer networks.
IBM/SNA	Systems Network Architecture was set by IBM for its computer networks.
Ethernet, SNMP, TCP/IP	Computer network standards and protocols.
CD-ROM	Standards are identified by color, called "books." Red Book = audio CDs Yellow Book = 650MB multimedia CD-ROM Orange Book = recordable CD-ROMs, CD-R White Book = Philip's CD-I
Sony-Philips/DVD and Toshiba-Warner/DVD	Developed competing standard for digital videodisc. Single standard now agreed upon.
UC-Berkeley/RAID storage	Redundant Array of Inexpensive Disks has 5 levels of standards, depending on how disks are configured.
VESA-Video Electronics Standards Assoc. has group VOST (VESA Open Set Top)	Trying to set standards for settop boxes: open standards, consistent interface, uniform hardware standard.

communications infrastructure is far too complex, intricate, and ambitious to be codified by any one entity, even an international one like the ITU. There are multiple trade-offs to be made between performance, features, costs, and uses that depend on the individual, the society, and the culture where the technology is introduced.

It's true the pace of the process is slower than it might otherwise be if there were a standards committee with czar-like powers. But the dispersion and iteration of decision-making may ultimately evoke wide agreement, avoid error, and incorporate superior technology than a faster, seemingly efficient process.

Market conditions can also retard standards. While competition usually spurs innovation, it can also stifle it by allowing the larger players to manipulate standards and to erect barriers to market entry by smaller players. Likewise, a massive installed base of an inferior product can hold back superior solutions, the problem of "excess inertia."[20] An example of excess inertia is the NTSC 525-line TV standard. The Europeans adopted the 625-line standard, which yields a clearer, more detailed picture, but the installed base of NTSC TV sets in the U.S. discouraged adoption of the higher standard.

A too-rapid pace of technological change, "excess momentum," can also make standards-setting quite difficult. The history of innovation in communications technology is replete with sudden breakthroughs that replace existing products long before anyone expects, rendering standards obsolete. The commercial replacement of the ½" Beta format by VHS is an example where the officially-recognized standard faltered before a dynamic competitor. There are also examples of incorrect perceptions of how people will use products and consumer reinvention of products that make standards ineffective or irrelevant.[21]

The Future of Global Standards

Ironically, it appears that the more competition there is, the more cooperation will be required! This strange conclusion occurs because of the property of communication systems that the more people who are hooked up to it, the greater the benefits to members and the greater the profits to the system provider. In economic parlance, this is called "increasing returns to adoption" or "network externalities."[22]

Another paradox is at work here. As the benefits increase, cooperation in the midst of competition becomes imperative. This situation suggests that the process of standards-setting will require additional formalization in order to ensure perceived fairness by all (or at least a substantial majority) of the participants. Many technical people believe that the process has become overly politicized and that powerful companies block designs that are not theirs, the "Not Invented Here" syndrome. Formalization will also entail greater efforts to validate designs and verify their efficacy. Finally, greater coordination between groups and greater participation of Third World representatives needs to occur to establish a truly global communication system.

A final irony is that as the process of standards-setting becomes more formalized, it is likely to codify standards that have already won acceptance in the marketplace. The speed of technological change and the increase in inexpensive electronic communications, which let user groups find out about and purchase new products, rapidly outstrip the ability of technical groups to muster the political agreement necessary to set standards.

No matter the pace of change, however, actual systems must be built with existing products to precise specifications. If we are to have a global communication system, they must interoperate, by meeting the challenge of standardization. In the next chapters, we will look at the wired and wireless building blocks of the emerging system.

Notes

1. Frank Schwartz, "Set-top standards," *Electronic Design* 42, no. 9 (September 19, 1994):151.

2. Timothy Somheil, "Real life in a box," *Appliance* 50, no. 8 (August 1993):41.

3. Steve Rose, interview by author, Malibu, CA, May 13, 1995.

4. Somheil, op. cit., 41.

5. John Dickinson, "Financial crash on the digital highway: lack of home banking standards," *Computer Shopper* 14, no. 4 (April 1994):68.

6. U.S. Congress, Office of Technology Assessment, *Global standards: building blocks for the future,* TCT-512 (Washington, D.C.: U.S. Government Printing Office, March 1992):8.

7. Ibid. 84.

8. Bob Horton, "Standardization and the challenge of global consensus," *Pacific Telecommunications Review* (September 1993):16–22. The stages I use here were developed by Dr. Bob Horton. I have modified them somewhat to include Step 6, testing and validation.

9. Peter R. Wilson, "Standards: past tense and future perfect?" *IEEE Computer Graphics & Applications* (January 1991):47.

10. Raymond Akwule, *Global Telecommunications: The technology, administration, and policies* (Boston, MA: Focal Press, 1992):43–53.

11. Bob Horton, op. cit.

12. S. M. Besen and J. Farrell, "The role of the ITU in standardization," *Telecommunications Policy* (April 1991):311–321.

13. Bob Horton, op. cit., 21.

14. William Stallings, *Data and Computer Communications,* 4th ed. (New York: Macmillan, 1992):803.

15. Bob Luff, "Why take an interest in DAVIC?" *Communications Technology* (April 1995):18–19.

16. Lindsey Kelly, "Group delaying interactive goal," *Electronic Media* (March 20, 1995):22, 36.

17. "Multimedia Communications Forum establishes workgroup to establish MIB," *OSINetter Newsletter* 9 (January 1994).

18. M. Abel, R. Bell, C. Perey, and W. Zakowski, "The MMCF transport services interface," *New Telecom Quarterly* (4th Quarter, 1994):38–44.

19. Peter R. Wilson, op. cit., 3.

20. Bruce M. Owen and Steven S. Wildman, *Video Economics*, (Cambridge: Harvard University Press, 1992):270–271.

21. Mark Berniker, "Microware creates de facto operating system for interactive TV," *Broadcasting & Cable* 124:30 (July 25, 1995):30.

22. Bruce M. Owen and Steven S. Wildman, op. cit., 265–268.

The Wired World 7

Introduction

It is difficult to convey the sense of purpose, mission, and almost romantic dedication that so many people bring to the giant job of building the complex infrastructure required for advanced television systems. They attend numerous standards-setting meetings, conferences, and conventions, questioning, debating, pushing the perspective of their industry, company, camp, or individual opinion. Twenty-five year old whizzes jump into the verbal fray, arguing vehemently with grizzled veterans over how to best build these advanced networks. It's exciting, passionate, and engrossing. Participants have a sense of making a new world, or at least bringing to fruition the century-and-a-half dream of communicating at a distance.

Distance communication, beyond beating drums and signaling with flags, begins in the 19th century with sending messages over wires. The telegraph, the telephone, and the lightbulb were all outgrowths of scientists and inventors understanding more about the new technology of electricity. The same issues that were involved in developing the earliest systems recur again and again as ever more sophisticated solutions are devised.

Spectrum, transmitters, receivers, radio waves, electrical pulses, light energy, relays, repeaters, wires, and switches—these were the components of the first wired systems and they still constitute the majority of the components of today's systems. Tremendous advances have occurred with each of these elements, so that they are changed beyond all recognition from their first generation predecessors. Some twentieth-century media, such as television and satellites, have brought entirely new technologies with unique components and configurations.

Still, the problems addressed by earlier developers are strikingly similar to the difficulties encountered today. Cost, capacity, configuration, switching, consumer issues, economic benefits, and public policy—these considerations recur again and again in the creation of wired infrastructures.

From Network Television to Networked Television

Fifteen years ago, viewers watched three commercial broadcast networks and one public broadcast network. In the largest markets there were a handful of independent broadcast stations—all transmitted over the airwaves. Today, more than 90% of U.S. households are passed by cable. Sixty-two percent of all U.S. television households can watch more than 35 channels of television and many choose among more than 50 channels, including broadcast stations, basic cable networks, premium channels, and digital music channels—all received over cable.

Cable networks are just one kind of wired system that will deliver television in the future. Broadband telephone, computer, and perhaps even electrical power networks capable of carrying full-motion video are all potential vehicles for tomorrow's television. However different in other respects these industries and organizations may be, wired infrastructures have their own complexities, regardless of who attempts to build them. The issues include functionality, components, design, type of switching, the location of intelligence in the system, configuration, and capacity. The concern about the cost of building wired systems will be covered in Chapter 11.

Functionality: What the Emerging Advanced Television System Needs to Do

To define functionality—what the system should do—one begins at the desired end point. Ultimately, the configuration of the emerging communication infrastructure must carry high definition, digital, compressed images, over an interactive, universally connected, switched, global network.

> **Function 1:** *High bandwidth*—The system must deliver high quality images.

Function 2: *Digital*—The system must be able to accept, process, package, and deliver digital signals. For the present, the system must also be able to pass through analog programming and to convert analog signals to digital signals and vice versa.

Function 3: *Compression capacity*—The system must compress and decompress digital signals.

Function 4: *Interactivity*—The system must allow for symmetrical, two-way broadband communication.

Function 5: *Connectivity*—The system must provide for near-universal access.

Function 6: *Switching*—The system must be fully switched, allowing message-routing from any point to any point.

Function 7: *Global compatibility*—Ultimately, the system must be world-wide in scope.

Interactive Systems: "They're much more complex than anyone thought!" —Jim Chiddix, Time Warner

Jim Chiddix ought to know. He's been on the forefront of deploying wired interactive television systems for entertainment giant, Time Warner, for the past 10 years. Chiddix' work has grown from the video-on-demand/near-video-on-demand system in Queens, New York, in the mid-1980s, to the Full Service Network (FSN) in Orlando, Florida.

The FSN in Orlando is an interactive television system that provides an analog tier of service (broadcast television and cable networks) and a digital tier of services. The digital tier includes video-on-demand, home shopping, library access, and a variety of information services. Ultimately, it will offer video telephony, mobile telephony, and networked game playing. The FSN's interactivity is severely asymmetrical, providing much more downstream bandwidth for programming to enter subscriber households than upstream bandwidth coming from the home.

Nevertheless, the FSN architecture has proved to be remarkably difficult to build for both Time Warner and the other corporations that are constructing advanced TV systems. Many companies have found they need special expertise to organize all the work that must be done and the hundreds of vendors it takes to put it all together. That expertise is provided by a "system integrator." Few organizations are capable of providing integrator services; perhaps IBM, AT&T, Andersen Consulting, Broadband Technologies, and only a few others can successfully take on this work.

The telephone company BellSouth Communications is currently operating a broadband network in the state of North Carolina that links several hospitals and a large number of schools. According to Peter Brackett, Research Manager, Advanced Data Networks, Science and Technology Organization for BellSouth, the telephone company built its own in-house unit for integration management because the management felt no one else would be able to do the job. "No vender was ready for prime time and everybody had blood on them. There were incompatibilities between switches, transport and gateways," said Brackett.[1]

There is a downside to using a system integrator. Companies may be reluctant to outsource this function, feeling that they are "giving away their soul," or at least the soul of the business. System builders worry about becoming overly dependent on an integration company.

One solution is for conduit providers to give more responsibility to vendors, especially those providing major components. Vendors such as Northern Telecom, Siemens, and Fujitsu recognize that their customers need help with these complex broadband networks. At the equipment level, integration would be facilitated by widely-accepted standards.

No matter who puts the system together, it is a thorny problem. While the components and processes are well-understood in theory, their practical implementation into a single system is a challenge. Just consider the amount of equipment that is needed.

The Components of an Interactive Television System

The majority of the components are located in the transceiver node (headend or central office), where television signals come in and are stored or immediately retransmitted; the rest of the pieces of equipment are distributed throughout the network, at nodes or reception sites. Table 7.1 describes the components that make up an interactive television system, where they are located, and what they do.

Table 7.1 Components in an Interactive Television System

Component	Location	Function
Satellite dishes: C-band, Ku-band	Headend	Signal reception;
Microwave, FM, VHF, & UHF antennas	Headend	Signal reception;
Fiber cable node	Headend	Signal reception;
Router, bridge, or gateway	Headend	Connect signals from one network to other networks, such as telephone nets to cable or computer networks;
A-D, D-A converters	Headend, settop	Convert signals from analog to digital or vice versa;
Descrambler	Headend, settop	Specialized processor that can descramble incoming and upstream signals;
Laser diode	Headend, terminal node	Converts electrical pulses into on-off light (optical) signals;
Modulator	Headend, terminal node	Varies some characteristic of a signal as the information to be transmitted varies so signal carries information;
Processor	Headend, settop	Converts bitstream into ATM format;
Multiplexer, or MUX/demultiplexer	Headend, terminal node	Loads multiple streams of information onto the same channel, or transport medium; disassembles multiplexed bitstream;
Fiber optic cable	Network transport	A transport medium, made of silicon, that carries light signals;
Photodiode	Headend, term. node	Converts optical (light) on-off signals to electrical pulses;

Table 7.1 *Continued*

Component	Location	Function
Switch	Headend	A device which opens and closes in order to complete or break a circuit. Also used to route data streams. Can be human, mechanical, electronic, or optical;
Combiners and splitters	Headend node	Bundle separate signals together and separate bundled signals;
Coaxial cable	Network transport	A conducting wire that carries great quantities of data (about 1 gigahertz) at very high speeds (about 200 megabits per second);
Amplifiers	Network	Amplify signals, every ½ mile or so;
Tap	Receiver	A passive box (no electronics) which splits signal from coax and reduces its amplitude for delivery to reception site;
Gateway	Receiver site	The junction box where a cable enters a reception site, a household, or business;
Settop box	Receiver	Essentially a computer: retrieves downstream signals and decompresses them. Addresses, scrambles, compresses, modulates, and sends signals upstream;
Software	Throughout	Operating system software (OSS); Application software, Server software; Operation, administration and management (also called OA&M, or OSS-Operational Support System) software; Transaction billing software.

Many of the components are produced by different manufacturers with sub-components contracted to many more companies. It is easy to see why standards, whether voluntary or enforced, are so essential for all this equipment to interoperate.

Given the list of off-the-shelf components, it is worth asking the question: what isn't here yet? While the following equipment is understood in theory, some requires substantial development for practical implementation. Here are products that still need work:

- A system for powering telephones for cable systems;
- Low-cost terminal equipment for cable telephony;
- Video servers at price points that allow viable business plans;
- Transactional billing software;
- Salvo switches: Switches that allow simultaneous turn-ons and turn-offs by large numbers of subscribers.[2]

This chapter will return to specific issues concerning the component equipment of networks. First, we should get an overview of the way the system works.

The Interactive Network—Downstream

The downstream portion of the network carries signals from a central point (headend in cable systems and the central office in telephone networks) down to the subscriber site. The process begins with both analog and digital signals coming into the headend via cable, satellites, and terrestrial broadcasting. In advanced TV systems, the incoming signals are processed and packaged for fiber optic cable. The process of sending television downstream over a wired system is shown in Table 7.2.

The information is processed for the wired system (demodulated) and divided into analog and digital signals. Handling many of the analog signals is relatively simple. For example, local TV stations and most cable channels are retransmitted directly to subscribers. Usually they are scrambled (optional), bundled, amplified, and split for transportation to neighborhood nodes.

However, if the programming is to be stored, it goes to the digital side of the operation, whether the original material is analog or digital. Analog signals are digitized—and the now-digital signals follow a

Table 7.2 The Interactive Fiber/Coax Network—Downstream

COMBINED PATH

1ALL) Receive ——>2ALL) demodulate and split analog (ANA) from digital (DIG) signals ——>

ANALOG PATH	**DIGITAL PATH**
——>3ANA) scramble (optional)	——>3DIG) digitize
——>4ANA) combine all analog channels	——>4DIG) compress
——>5ANA) amplify	——>5DIG) scramble (opt.)
——>6ANA) split to neighborhoods	——>6DIG) store some signals on server;
	——>7DIG) assign ATM addresses to all digital signals
	——>8DIG) time division multiplex
	——>9DIG) modulate
	——>10DIG) combine all digital

COMBINED PATH: ——>11ALL) combine analog and digital signals

——>12ALL) modulate electrical-to-optical on neighborhood laser

——>13ALL) send signals down fiber backbone and feeder pathways

——>14ALL) at neighborhood node, demodulate optical-to-electrical for coax cable

——>15ALL) add 90 volts of power

——>16ALL) amplify signals (up to 4 times)

——>17ALL) tap coax feeder and send signals down coax drop to home gateway

——>18ALL) segregate telephone from television traffic

——>19ALL) TV to settop box: picks out signals addressed to it, de-scrambles and decompresses them, and displays navigation system for subscribers to access system.

much more complicated pathway than the signals that were immediately retransmitted.

At the headend (or central office) a series of complex signal re-processing steps ensure delivery of a colorful, rich digitized TV picture to consumers. Each digital channel is compressed and, in some systems, scrambled. If it's an on-demand product (for example, a movie) it will be sent to the media server (a kind of video jukebox) and stored, waiting for a subscriber to call it up. All digital signals pass through the server where they are given a unique identifier, so subscribers get the right product.

That identifier is read by a specialized processor called a switch. Most designs call for a product called an "ATM switch" (Asynchronous Transfer Mode), an exceptionally fast and efficient switch. Later in this chapter, we will discuss ATM switches at length. For now, it is sufficient to know that this new innovation makes it possible for large amounts of information to be sent rapidly to individual households with virtually no errors.

When the information stream passes out of the ATM switch, it is time-division-multiplexed for further transport toward the consumer. (Multiplexing means mixing several signals together into one information stream for transport over a single channel. See Figures 7.3 and 7.4 for depictions of multiplexing.)

Signals going to the same neighborhood are then all modulated and combined together and re-combined with all the analog signals. At this point, the analog/digital information stream is converted from electrical to optical signals for transport over the fiber portion of the system.

It is only now that the signals are actually leaving the headend! They travel down the fiber optic cable backbone, split off into feeder cables, and continue to the neighborhood node. There, the signals are re-converted from optical to electrical, 90 volts of power are added to them to power the amplifiers (and any telephone service that may be provided). Amplifiers are "cascaded," or placed every few hundred yards, up to 4 times.

As the signals pass subscribers' homes, they are "tapped"; that is, split off to the gateway outside the home. The telephone signals are segregated from the TV signals, which go into the settop box. The settop box for an advanced television system picks out the signals specifically addressed to it, descrambles and decompresses them, and displays any necessary graphics (such as navigation software) for subscribers to access the system.

The Interactive Network—Upstream

The process for sending information upstream from the subscriber's home or business is quite similar to the downstream process, but it is not an exact mirror image, as Table 7.3 reveals. Some steps of the up-

Table 7.3 The Interactive Fiber/Coax Network—Upstream

DIGITAL UPSTREAM PATH

1) Send signal via remote control device, computer (modem), or telephone to settop box or gateway ——>2) digitize, scramble, compress, encode signal into ATM cells, and modulate in code division multiplex format ——>3) transmit through site gateway over coax drop cable to neighborhood node ——>4) amplifiers every ½ mile or so ——>5) at node, remodulate signal ——>6) convert electrical RF into optical signals ——>7) send signal over fiber to headend ——>8) convert optical to electrical RF signal, demodulate and descramble for ATM switch——>9) ATM switch sorts signals for transport to server or to router into public switched telephone network (PSTN)——>10) when appropriate, split off billing and account management information ——>11) convert telephone traffic from electrical to optical signals and multiplex for transport to telephone network.

stream process take place in different locations than occurred on the trip down. Also, the user's signal is never analog, so it doesn't have to be digitized. If subscribers wanted to send video upstream, they would need to pre-digitize it locally before transmitting it.

A large question facing cable systems is how to migrate from one-way to two-way capability. Over the past two decades, the average bandwidth of cable systems has been increasing, from about 200 MHz to 300 MHz and 450 MHz. Today, many systems offer 550 MHz capacity and the current generation of upgrade is to provide 750 MHz or 1 GHz.

Typically, cable systems leave the 5 MHz-40 MHz portion of that bandwidth empty; it is used for shuttling video from local sites upstream to the headend. For example, when the local Little League team plays it may be carried on the public access channel and the cable system may have a hookup so the video can be sent upstream to the headend from the remote location at the park.

Some cable operators are thinking of using that return bandwidth from 5 MHz-40 MHz to offer upstream services to subscribers. In May, 1995, Tele-Communications, Inc. announced it would create a service for its systems called "@Home" to provide 10 megabit per second access to computer networks by early 1996. Operators think that that now mostly-unused portion of the cable system could earn them valuable revenue. Like TCI now plans to do, they could use it to provide access to the Internet or other online services via cable modem, offer telephone service, and deliver interactive advertising.

There isn't much available bandwidth at this low end of the cable spectrum, however. Frequencies below 10 MHz are difficult to use and those above 20 MHz are susceptible to interference. To get around these problems, the Time Warner testbed in Orlando, Florida allocates the 900-1,000 MHz band for the return loop. It's really a gamble: If operators think interactive services will be the growth area, then putting them at the high end of the spectrum will allow them to expand. If they believe the number of one-way networks will increase, then they should put interactive services at the low end and leave room at the top for the one-way services to grow.[3]

Still another solution is to put in two separate cables, at least from the home to the fiber feeder. One cable would carry downstream services and the other transport upstream communication. This low-maintenance technique would eliminate the need for expensive electronic spectrum management and filters to separate the two streams, although it would cost more for the cable and associated amplifiers.[4]

The newest approach to solving this problem is called the "spectrum manager," or operating support systems (OSS). Currently, cable systems allocate specific spectrum assignments to specific channels.[5] In this new view operators should adopt dynamic spectrum allocation and management. This concept means that bandwidth would be assigned to a service as needed and only when needed. There must be a processor, detecting sensors, agile modulation, and automatic adjustment of frequencies and power.

The Components of Interactive TV Systems: Another Look

So far, we have considered the components of wired systems and how they work together to bring material to and from subscribers. The next sections examine in detail the individual components of the system, what they do, how they do it, and how they are incorporated into advanced television systems.

Archives: media servers and mass storage

For programming distributors to provide a movie or TV show when a customer orders it, they must have a place to store it and a system for sending out the ordered program to the right place. The equipment where programming is stored is called "mass storage." The equipment that takes the order and routes the show to the view is called a "media server," a kind of "video jukebox."

We need to begin by comprehending the scale of this problem. Steve Rose writes: "The entire Internet ran successfully for years on a backbone of 45 Mb/second for the entire U.S. (and much of the rest of the world). Video-on-demand for a small town of 10,000 customers, assuming 20% on line (2,000 simultaneous digital, compressed video streams), requires a throughput of 8 Gb/second, or about 200 times the Internet backbone!"[6]

Media servers are part of a revolution in computing, in which the relationship has changed between the central site (where stored material resides) and the terminal that requests the material has changed. In the previous era of mainframe computing, the term that characterized the relationship was "master-slave" because no intelligence resided in the terminal. Today the term "client-server" describes how users on a network with intelligent terminals can call for information stored at the central site and use it locally.[7]

When a server stores video material for access by users within a company or organization, it is called a "video server." Video servers are already penetrating the professional broadcasting market where they are used for news editing and for ad commercial playback and insertion.[8] When a server is part of a commercial operation and billing records are part of its function, it is called a "media server."

The equipment is actually a special-purpose computer that makes possible some of the most potentially profitable services that providers anticipate offering. A customer order comes into the media server; the server locates the desired information in storage, retrieves it, and sends it downstream to the viewer. It then sends a message to the billing software to charge the consumer. The ordered material might include video of movies, television programs, and direct-to-home programs. It may be entertainment-oriented or informational in nature. Other content could be music, games, catalogs, lists, and announcements.

Media servers use all-digital processing and storage techniques. Current questions about media server design are: Is it possible to build a cost-effective server? What type of processing should be used? Where in the network should the server be located? What kind of storage should be used?

The cost-effectiveness of media servers is described in terms of "cost-per-bitstream" (CPB) or "cost-per-user," with CPB being the preferred term, because not all users access bitstreams. Cost-per-bitstream calculations are based on cost of storage, plus the cost of the special-purpose computer. In the spring of 1995, the cost per bitstream dropped from an average of $1,200 to $440 (including storage costs) when nCube came out with a new generation server. Continuing rapid advances in server technology are expected to result in a substantial drop in cost to less than $60 per bitstream in the near future.[9]

The question of the cost per bitstream is intimately tied to the underlying technology for media servers. Several types of computers can be made to operate as media servers, ranging from single PCs and workstations to mainframes. There are three favored architectures: "loosely coupled computers" (LCC), "symmetric multiprocessing" (SMP), and "massively parallel processing" (MPP).[10]

Servers based on the LCC model are big—too enormous for the average headend to accommodate because they are composed of so many machines operating together. The LCC model is best exemplified by the Microsoft approach. Tiger software, based on the Windows NT operating system, coordinates several hundred 486- or Pentium-level PCs on a bus, switch, or mesh. The linking method slows processing and limits the number of processors that can be interconnected.

An example of the SMP model is Silicon Graphic's Challenger design that is used in the Time Warner test trial in Orlando, Florida. This design calls for many processors, all linked to a single disk controller and shared memory. So far, SMP works better on paper than in practice as the Challenger has not been able to deliver on its promise of 100 bitstreams (to 100 customers). Another problem is that even if this media server can be made to operate for the smaller number of users in a test site, it will not scale up to meet the requirements of a 50,000-subscriber cable system—or even a 10,000-subscriber cable system.

The MPP design is similar to the LCC server, except that the processors are linked in such a way that they allow thousands of powerful chips to be assembled in a relatively small space. There are many advantages to the MPP design. It is the only server that is truly scalable up to the requirements of a 10,000 to 50,000-subscriber cable system. Its enormous processing capacity allows it to perform the switching function, eliminating specialized (and expensive) switching equipment. Finally, only MPP allows an unlimited number of simultaneous downstream flows of a single on-demand product from one storage cache.

This issue of simultaneous flows of one product has been a thorny problem for would-be providers of on-demand video. When a new movie comes out, everybody wants to see the new hit at once. At the video store, all the copies of the popular film are gone and customers have to be placed on a waiting list—by contrast, an interactive system just crashes. The MPP server is composed of dozens of processors. It breaks a movie (or any video) into tiny pieces of just a few seconds and shunts them into a fast-access buffer. Each piece is handled by a different processor, then replaced in cache memory. The information is reassembled into a complete bitstream as it leaves the server. Since only a few seconds at a time is being worked on, when another subscriber calls for that same movie ten seconds later, the material is still available. Combined with sophisticated mass storage equipment, the MPP can produce almost as many bitstreams as there are processors. Media servers by nCube, Cray, IBM, and Intel all use the MPP model.[11]

Once a media server is selected, the question of where to locate it arises. For a time CableLabs proposed locating giant servers in regional hubs, where system operators would access little-used material. Likewise, AT&T originally thought to archive its material in Manhattan, storing only a small portion of a program or film locally.[12]

Arguing against this model, Rose makes the following pragmatic observation: "The network operator owns the local network and must pay to transport material accessed outside the network. The most popular movies and programs will reside in memory locally. This means that the least popular products will be the furthest away, cost the most to transport—and the operator must charge less for this seldom-called upon material!"

At the same time, locating the media server in a neighborhood node means there must be a server for every 500 subscribers, a very expensive approach. Thus, although there is continuing experimentation to find the best location for the media server, chances are the cable system headend or the telephone company's nearest central office will be the appropriate site.

Finally, there are important issues of mass storage for those building interactive television systems to consider. In Table 7.4, mass storage is evaluated in terms of cost, speed, the amount of material that can be stored, and the amount of maintenance required.

Systems will use more than one type of storage. Often-used programming will reside on an array of magnetic disks, called a RAID array, standing for "Redundant Array of Inexpensive Disks." This ingenious storage solution allows operators to back up programming using only 25% more disk space. If there is a malfunction, the system

Table 7.4 Mass Storage Media

Medium	Cost	Speed	Amount of mat'l	Maintenance
Random access cache memory (RAM) (digital)	high	fast	small	low
Magnetic disk (digital)	moderate	moderate	large	low
CD-ROM (digital)	moderate	slow	moderate	moderate
Videodisc (analog)	moderate	fast	moderate	moderate
Tape (digital)	low	fast	large	high

has enough redundancy built into it that the flaw will not be apparent to the customer.

When a customer orders a program from the media server, the server looks first in the RAID array where the last-ordered material is still stored. As new material is called up, the oldest videos are pushed back into longer-term storage such as tape or optical disc. When a customer requests rarely-requested video, it will be fed into the disk array.

The most popular movies stay in the RAID, until their appeal declines. The advantage of the RAID over long-term storage media is that it allows access of multiple requests separated only by a few seconds. This use of RAID storage is called "cache-ing."

Mass storage is expensive. Each movie requires a massive 3 GB of memory, so that archiving 1,000 movies uses about 3 TB. When an operator decides she needs more mass storage, the cost increases linearly. In late 1995, one gigabyte of storage cost about $200. Add 100 gigabytes, and the cost is $20,000.

When a movie leaves the media server, it is transported to the address from where the request was made—and only that address. The ability to send material to a specific household is accomplished through switching. While storage costs are linear, switching costs increase exponentially, as we shall see.

They'd rather not switch

Suppose there were 2,000 ports to send out bitstreams of material to any one of 10,000 subscribers of an interactive system, a relatively small system. This means the switch would need to accommodate 2,000 × 10,000, or 20 million possible connections—a moderately complex switch.

Now suppose there were the same 2,000 ports sending material to 20,000 subscribers, twice the number as the first system. In this second case, the switch would need to accommodate 2,000 × 20,000, or 40 million connections. The number of people increased by 2,000—but the number of connections increased by 20,000,000. This is an exponential increase in capacity and it is reflected in the quickly mounting cost of added switching power.

A switch is a device that opens or closes a circuit or electrical path between a sender and a receiver. It was one of the first requirements of the infant telephone systems of the 19th century. The first generation communication system switch was human-mechanical: the operators who, like Lily Tomlin's character, Gertrude, sat at a switchboard, plugging incoming callers to the people they wanted to reach.

The first widely used mechanical switch was the Strowger automatic, step-by-step telephone switch. (Armond Strowger was an undertaker. It is rumored he invented the switch because the local operator's boyfriend was a competitor, so she routed all the calls for undertaking services to her sweetheart!)[13] Mechanical switches became ever more complex until their replacement by modern electronic switches in the late 1960s.[14]

Three kinds of switching are used to deliver messages from senders to receivers: circuit switching (CS), packet switching (PS), and asynchronous transfer mode (ATM), which combines elements of both circuit and packet switching. Table 7.5 compares the three types of switching.

CS handles messages that are time-sensitive, a format called "constant bit-rate" service. PS is effective for "bursty" traffic, like data transmission, where a computer may squirt out millions of bits per second for a few minutes, then stop for some period of time. ATM switching handles both types of traffic within the same network. William Stallings, a noted expert on network design says: "(ATM) is in a sense a culmination of all the developments in circuit switching and packet switching over the past twenty years."[15]

ATM is potentially useful for systems that carry video because it handles different kinds of traffic simultaneously and it is extremely fast and efficient. Also important is the fact that ATM takes advantage of the low-noise characteristics of fiber optic transmission and doesn't add a great deal of redundant data to ensure the information is re-

Table 7.5 Comparison of Circuit, Packet, and ATM Switching

Circuit Switched (CS)	Packet Switched (PS)	Asynchronous Transfer Mode (ATM)
Constant bitrate service (CBR)	Variable bitrate service (VBR)	Both CBR and VBR service
Time sensitive audio and video	Time independent, bursty traffic	All traffic
Telcos	Computer networks	Networks for video
Dedicated path	Multiple paths	Virtual path
Assumes noise, so redundant	Assumes noise, so redundant	Assumes no noise, so not redundant
If busy, caller must call back	If busy, network resends	If busy, network resends
No address	Address attached	Address attached
Fixed bandwidth	Dynamic allocation	Dynamic allocation

ceived without error. (CS and PS may add as much as 33% to the data just for addressing and error correction. In systems where the amount of traffic is already enormous, the reductions ATM allows are highly valued.)

Let's look more closely at the ATM datastream. The digital signals are repackaged into small, equal-sized "packets," called "cells." Each cell is made up of a byte. (Remember from Chapter 2, a byte is 8 bits!). In ATM lingo, a byte is called an octet—"oct" means eight. There are 53 octets (or bytes) in a cell, which is the basic transport unit of ATM. Of the 53 octets, five are "headers," which carry the unique identifier, the destination address, and error correction information. The other 48 cells make up the actual information, called the "payload."

It is the small, uniform cell size of ATM that allows the extraordinary speed of its bitrate—intended to operate in the range of up to several gigabits per second. Equipment now available permits data rates of 2.4 gigabits per second, with higher rates in the offing, carrying both constant bitrate (CBR) data and bursty (VBR) traffic. This figure compares to the speed of 140 megabits per second of the fastest circuit-switching technique and 2 megabits per second for the fastest packet-switching device.

The major problem with ATM switching is that standards are still in the process of being set by the ATMForum. While this body has been extremely successful, it still takes a certain amount of time to reach conclusions about such a complex, highly technical method. One sticky problem is how to map MPEG-2 compressed bitstreams onto the ATM cell protocol. In the absence of standards, CableLabs has taken the lead to coordinate vendors so that an ad hoc standard can be set for implementation.

The extraordinary speed of ATM and its flexibility in the types of data that it can carry make it ideal for the fiber optic networks under construction. The next section will consider these advanced networks.

Basics of fiber optic networks

All the systems presently under construction use fiber optic technology for only part of the network. Both cable and telephone networks use fiber in the "backbone" (the largest part of the system) because it is where all the upstream and downstream signals come together. Many cable systems have also replaced coaxial cable with fiber in the "feeder" lines, which carry traffic to some major portion of the network.

Cable systems continue to replace coax and many of them are now putting in fiber to the neighborhood node that serves approximately 500 households.[16] One attraction of fiber optic systems is the extraordinary reduction of their costs: between 1980 and 1985, their cost went down by a factor of 100.

When operators upgrade existing systems, they simply replace coax with fiber, leaving the existing "tree and branch" structure in place (see Figure 7.4, further in this chapter). When systems are constructed from scratch, some builders are choosing "SONET" ring designs. SONET stands for Synchronous Optical NETworks. This technology has already been quite successful and beginning in 1996 both long distance and local telephone companies will quadruple the speed of their SONET networks by replacing OC-48 (OC stands for Optical Cable) cable with the larger capacity OC-192.[17] A more detailed discussion of ring network structure appears later in the chapter.

We have already covered the processes that occur in a wired network, as well as the equipment that is needed. The only components unique to a fiber network are: the fiber optic cable itself, the transmitting and receiving devices—laser diodes and photodiodes—and regenerators and amplifiers, which strengthen the decaying light signals. (The decay of light signals is called "attenuation.")

The optical signals carried over a fiber optic network are light waves. Normal light waves spread out and dissipate as they travel. However, a special kind of light wave, solitons, do not attenuate. In 1993, AT&T Bell Laboratories researchers transmitted solitons over 13,000 kilometers at 20 Gbs per second, double the previous world record.[18] At some time in the future, optical networks may carry solitons.

Fiber optic cable   Fiber optic cable is basically a waveguide—a container composed of long strands of glass that carry light waves, surrounded by light-proof, reflective cladding. It offers many advantages. "Modulating light on thin strands of glass produces major benefits in high bandwidth, relatively low cost, low power consumption, small space needs, total insensitivity to electromagnetic interference and great insensitivity to being bugged."[19]

The first suggestion for silica fiber as a transmission medium was made in 1966 by researchers K. C. Kao and G. A. Hockam, working for Standard Telephone Laboratories in London.[20] The first AT&T field trial took place in 1977 in Chicago, and development of the technology has been rapid since that time.

The production of purer silica strands and reflective cladding material decreased the attenuation of the light signals, resulting in increased bandwidth over ever-longer distances without having to regenerate the signal. The bandwidth of fiber optic cable was originally 50 MHz. Today, it is in excess of 10 GHz. The spacing of regenerators was 50 kilometers in 1986, 70 km. in 1988, and 130 km. in 1991.[21] Functioning equipment is now available that will send and receive at 10 Gb/s, transmission at 20 Gb/s has been demonstrated, and equipment that will operate at 40 Gb/s is under development in laboratories.

To provide a sense of scale, Russell Dewitt of Contel is quoted in Harry Newton's "Telecom Dictionary": "For the future, the ultimate potential of a single mode fiber has been estimated. It is about 25,000 Gbps (25,000,000,000,000,000 bits per second). At that rate, you could transmit all the knowledge recorded since the beginning of time in 20 seconds."[22]

Tom Bowling argues that fiber optic technology should be regarded as an entirely new medium. Figure 7.1 shows how Bowling plotted the growth of bandwidth since communication by semaphore in the early 1800s. His point is that historically the world's demand for bandwidth is voracious, will likely continue, and will be made possible by fiber. Writes Bowling: "It is all new and different. The technologies, the content, the selection of programming, the need for

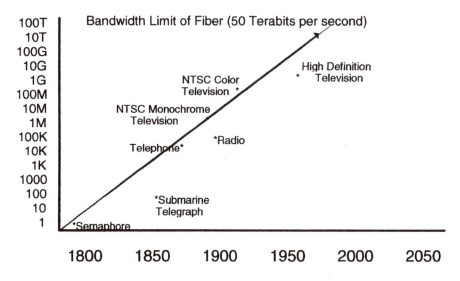

Figure 7.1 Information demands over time.

sophisticated search capabilities, and the methods of marketing, billing, and research will be entirely alien to the older television and cable industries."[23]

Improvements continue. In June, 1995, Boston Optical Fiber, Inc., announced that they had produced a clear cable capable of carrying nearly as much high-speed communication traffic as glass—but made of plastic. General Motors, Honeywell, and Boeing will produce GIPOF, or graded-index plastic optical fiber, at a fraction of the cost of similar cable made of silicon glass. The product should be available sometime in 1996.

Laser and photodiodes—tunable and otherwise   Fiber optic communication works by sending light signals from a source to a receiver. A laser diode translates electrical signals into optical signals and transmits them down the fiber optic cable. At the receiving end, a photodiode reverses the process and converts the optical signals into electrical signals.

One big advance has been the development of a piece of equipment called the erbium-doped fiberoptical amplifier, the EDFA. Regenerators, repeaters, and amplifiers all worked by intercepting the light signals, converting them to electrical signals, amplifying them, then converting them back to optical signals to continue on their path to the ultimate receiver. By adding the rare earth, erbium, to a stretch of optical fiber, the fiber itself becomes an amplifier that boosts the signal. Thus, the EDFA amplifies optical signals directly, without conversion, making the process faster and eliminating a possible source of noise and breakdown.[24]

Another advance is laser emitters which can be tuned electrically to produce light waves of different wavelengths. These emitters are called "wavelength tunable lasers" or "wavelength agile lasers." This development is important because using a coding technique called wavelength division multiple access (WDMA) allows a large number of parallel channels of information to be sent down the same fiber at the same time. At the receiving end, a photodiode is tuned to the desired reception frequencies and retrieves only the messages directed to it.

Three problems limit the acceptance of all-fiber systems: 1) it is difficult to connect two fibers together in the field, whether because of a break or a planned expansion; 2) the cost of photonic switches and amplifiers is very high because of the limited deployment of working fiber systems. Greater usage will result in lower costs; and 3) optical transmission equipment is complex and precise. For example, a laser optical termination unit requires the alignment of the laser diode and the optical fiber within microscopic tolerances.

More flexible systems would reduce the cost of equipment considerably. At the University of California at Berkeley, researchers are working on technology to build micromotors that will make automatic precision adjustments. Hewlett-Packard is supporting this research to speed the development of optical networks.[25]

Some systems have installed fiber optic cable alongside their coax lines, but have not hooked up the equipment needed to "light it up." Operators may not want to fund an all-fiber network when they are installing an upgrade or they may want to wait for lower equipment costs. Such buried fiber optic cable that is unused is called "dark fiber."

Although costs for fiber optic cable have been dropping rapidly over the past 15 years, they remain quite substantial. Thus, it is still less expensive to use coaxial cable for the last link into the home to the settop box.

A new generation of settop boxes

The current installed base of settops is analog and "addressable," allowing the cable company to send a pulse to them to block and unblock premium and pay-per-view channels. The new generation of settop boxes will be mostly analog, but have some new digital features and are called "hybrid analog/digital" units. Scientific-Atlanta's 8600X allows for graphic displays of electronic program guides, accepts different software downloads that change what the settop box can do, and displays information quickly. General Instrument's modular, flexible, and upgradable CFT 2000, has these same features, plus a port for security, expandable memory, and a radio device for upstream pay-per-view movie orders.

In addition to the infra-red signal input for remote control devices to send information upstream, the new units have many more communication options, as many as 6. The increased number of input/output ports include: a connector for high-speed telephone hookup; a jack for ordinary phone lines; a threaded receptacle for cable hookup; a connector for PC connection; a plug for an S-VHS VCR; and RCA plugs for composite video and audio.[26]

The all-digital settop box is simply another name for a powerful special-purpose computer. In fact, many digital units will contain Motorola's RISC (reduced instruction set chip), the same processor that runs the PowerPC. The unit itself is composed of two modules: 1) the network interface module (NIM), which can be swapped out according to the specific network configuration involved; 2) the processor itself. The new settop will be capable of accepting both downstream and upstream information, compressing and decompressing it, scrambling and descrambling it, and either multiplexing it directly or addressing it for multiplexing at the next upstream site, curb, or neighborhood node.

Other settop appliances are also in the works. The multiple input/output ports speak loudly of the number of potential hookups that settop makers believe will be connected to their equipment. The most urgent development at present is to construct a cost-effective cable modem: a device that would connect a PC to the broadband cable network in order to provide fast broadband access to the Internet and other on-line services. Both telephone companies and cable companies see PC connection as a new revenue stream. In May, 1995, CableLabs put out a Request for Proposals to provide a cable modem. They filled more than 800 requests for the forms, highlighting the large number of companies designing this modem.

Beyond the settop box

Nothing has changed television viewing in the past 50 years so much as the remote control device (RCD).[27] Interactive television systems call for a new generation of RCDs that are much more complicated to use than the current, familiar remote.

For one thing, interactive TV requires many more buttons than people are used to. Viewers must be able to browse menu pages, highlight the available choices one at a time, make a specific choice, and send their choice upstream. For video on demand, the RCD needs VCR-type buttons (rewind, fastforward, pause, etc.), as well. In addition, the RCD must be comfortable to hold and the buttons need to give a tactile response to the viewer's actions—a sense of a "click" or "snap." Prominent manufacturers of interactive RCDs include ICTV and Zing Systems, who have invested years of design and testing cycles to develop their products.

Information Processing in Advanced Wired TV Systems

The entire collection of hardware components in the emerging interactive television systems can be seen as parts of a giant information processing system—receiving input, processing it as throughput, and transmitting or transporting output. The next section will examine some of the changing aspects of information processing that distinguish advanced television systems from current television.

Innovations in modulation and multiplexing techniques

Recall that modulation refers to how a signal is varied to reflect the information that is conveyed. The digitization of the TV image has resulted in new ways to modulate TV signals. This discussion about modulators will give readers a good idea about how detailed, yet imaginative the thinking must be for companies to implement advanced TV networks.

In wired systems, there are three types of modulation: the established NTSC system, "QAM," and "VSB." The NTSC method modulates analog signals, using AM (amplitude modulation) for the picture and FM (frequency modulation) for the audio. QAM and VSB modulate digital signals. QAM stands for "quadrature amplitude modulation," while VSB is the acronym for "vestigial sideband."

QAM is a well-known public domain modulation method, currently used in most cable systems. VSB, developed by Zenith, is very new and not well-supported in technology. However, in head-to-head tests, VSB consistently performed somewhat better than QAM, so it will be the recommended standard to the FCC by its Advisory Committee for Advanced Television Service.

The multiplicity of standards was initially a problem because cable systems receive signals from many sources. The three modulation schemes required headends to have each type of demodulator to reprocess the incoming video for retransmission to their viewers—a very expensive proposition. Fortunately, Richard Prodan, Vice President of Engineering at CableLabs, invented a universal demodulator for installation in settop boxes that will handle an incoming signal modulated by any of these three techniques. This product should be on the market in 1996.

However, there is an additional consideration that will emerge as advanced systems are rolled out for commercial use. The NTSC standard of 6 MHz television assignment has led designers to plan for 6 MHz modulators in the new systems. Steve Rose, consultant for Pangrac & Associates, has hands-on experience putting together cable systems. He argues that the new networks need 30 MHz modulators, a product that is available for some applications in telephony, but has not been considered for advanced wired broadband systems.

His argument is based on the difficulty of accommodating the amount of equipment needed to modulate the sheer number of channels and signals in advanced television systems. A 6 MHz modulator for digital signals allows a throughput of about 27 megabits per second (Mbp/s). A single compressed channel requires about 4 Mbp/s, so current modulators handle 6 channels. The new systems dedicate a fiber for each neighborhood node serving about 500 households. Allowing for 20% utilization of downstream video-on-demand (at 4 Mbp/s each), then 100 streams are carried to viewers, requiring a 400 Mbp/s capacity:

$500 \times .20 = 100$
$100 \times 4 \ (Mbp/s) = 400 \ Mbp/s$
$400/6 \ (channels \ per \ modulator) = 17 \ modulators$

These numbers mean the cable system will need 17 modulators per neighborhood. A medium-sized system with 50,000 subscribers will have 100 neighborhoods would need 1700 large 6 MHz modulators—over 20 rack walls of equipment!

Rose figures that using 30 MHz modulators will reduce the number of devices (and the space they require), provide more potential bandwidth to any subscriber, and reduce costs by at least 25%. Now, each 30 MHz modulator accommodates the 155 Mbps of an OC-3 fiber optic line and the operator can send 30 to 34 channels of payload per modulator. Three 30 MHz modulators will handle slightly more than the 400 Mbps required for each neighborhood. This means system operators will need 300 modulators for the 100 neighborhoods, still a substantial amount of equipment, but taking up 4 rack walls as opposed to the 20 rack walls demanded by 6 MHz modulators.

Multiplexing signals for transport

The purpose of multiplexing is to load multiple message streams onto a single channel in the most efficient manner. Multiplexing means putting signals from the same source together over one conduit. Multiple access is quite similar; it means bundling signals from different sources to go over one conduit.

Both multiplexing and multiple access are important in wired systems. The TV and digital services coming to the subscriber are all originating from the headend of the cable system (or the central office of a telephone company)—so they are multiplexed for transport. The reverse is not true: when customers send signals from their homes, each signal originates from a different household. So multiple access is how they are bundled for transport upstream.

There are four ways to perform either of these functions:

1. Frequency division multiplexing   FDM   in use
   Frequency division multiple access   FDMA
2. Time division multiplexing   TDM   in use
   Time division multiple access   TDMA
3. Wavelength division multiplexing   WDM   in use
   Wavelength division multiple access   WDMA
4. Code division multiplexing   CDM   pending
   Code division multiple access   CDMA

Frequency division, time division, and wavelength division techniques are well-known and used in many systems. Most recently, another method has been developed, called code division multiple access (CDMA), a spread spectrum technology first invented by actress Hedy Lamarr! So far this system is used only for over-the-air cellular telephone applications, but there is discussion about using it for the return path by cable systems.[28]

The best way to understand the different techniques is through two metaphors: the footrace and the train.

Frequency division multiplexing is like a footrace with different runners running side-by-side, but staying within their assigned lanes, as in Figure 7.2. Frequency division multiple access is similar, but all the entrants came into their lane, already running, from different directions. FDM and FDMA are used for analog signals.

Wavelength division multiplexing and multiple access are similar to FDM and FDMA. Recall that wavelength is the inverse of frequency: as frequency increases, the length of the wave decreases, and vice versa. So the footrace metaphor works for WDM and WDMA as well. The advantage of WDM and WDMA is that both digital and analog signals can be transmitted on a single fiber by using separate lightwave emitters, and separating their frequencies sufficiently.

Time division multiplexing is like a train, with boxcars headed for different destinations behind the engine in some random order, as shown in Figure 7.3. So after the engine, Signal A is going to the Smith's household, followed by Signal B headed for the Jones' house, followed by Signal C going to the Brown's home and Signal D on its way to the Green's TV. After these, we might find signals B, A, B, D, C, B, C, A, etc. Time division multiple access is the same train, except that the signals all converge from different locations, as well as moving toward different destinations. TDM and TDMA are used for digital signals, especially with fiber optic cable.

Several cable operators, Time Warner, Rogers, and Cablevision, are considering dense WDM-technology on 1550 nanometer fiber: one wavelength could deliver telephony services; another would bring digital video; the third would provide video-on-demand channels; and the fourth wavelength would be reserved for the return path. Network vendors such as AT&T/Network Systems, Northern Telecom, and NEC America all plan to offer WDM for high-speed telephone systems in 1996.

In January, 1995, IBM introduced a new multiplexing product called MuxMaster, which employs a prism to break white light into multiple colors. Each color transmits data at a preassigned wavelength permitting up to 20 simultaneous streams on a single fiber—a 20-fold increase. Eventually the company hopes to open the full 20,000-plus gigahertz bandwidth of the optical fiber.[29]

Software: the brains behind wired interactive television systems

The greatest complexity of two-way wired broadband systems stems from the requirements of processing information. CableLabs, the re-

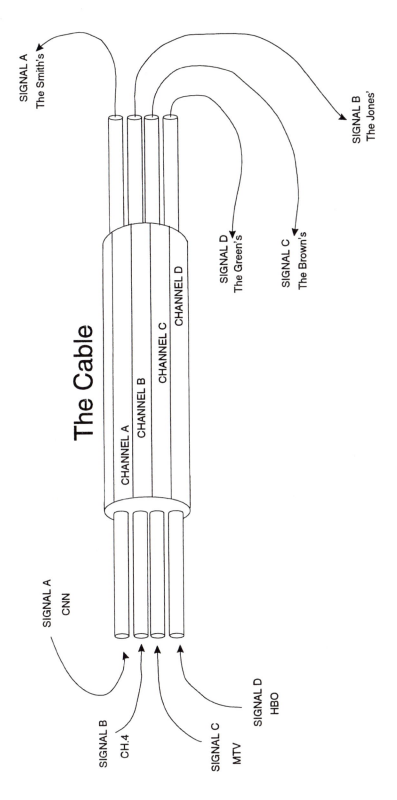

Figure 7.2 Frequency division multiplexing. Source: James Bromley.

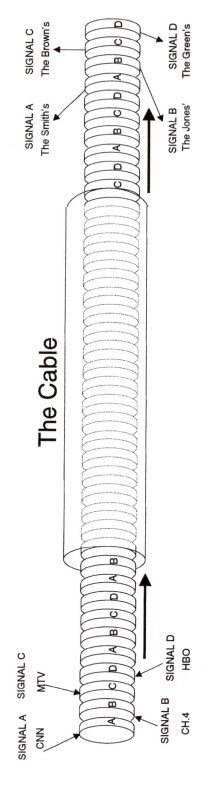

Figure 7.3 Time division multiplexing. Source: James Bromley.

search organization funded by the cable industry, has made software for the new systems a priority. In an effort to enlist the best efforts of computer network developers, CableLabs initiated discussions with the UniForum Association, an influential group of network software experts.

Any interactive television system needs five major software products: operating system, operation support system, transactional billing, server, and application software. This list does not include many software processes, such as signal processing—analog/digital/analog conversion, modulation, and multiplexing.

There are several as yet unresolved questions regarding software, including:

1. where network resource allocation should reside—in specialized management software or in the operating system;
2. how resources should be allocated—fixed and simple or dynamic and complex; and
3. who owns the resource control layer—network operators or software providers;

One approach is that all these functions should be located in the operating system. The other notion is that a "middleware" layer should translate between multiple operating systems and the various pieces of hardware from the server to the settop.[31]

Operating system software The most commercially competitive arena in software products is the development of an operating system. The operating system for the computer gave Microsoft its success, far greater than that of the hardware companies whose products it served. The primary reason is that control of the operating system ultimately means control of applications—and these are what give the product its value to the end user. In the same way, many industry observers believe that whoever controls the operating system of an interactive television facility will exercise the power over how applications are mounted on the system.

Microware, a company which designs software for industrial applications, has an early lead in capturing the market to provide an operating system, beating such giants as Sybase, Oracle, and Microsoft to the punch. Microware's head start is due to the flexibility of the company's DAVID multimedia operating system.[32] Microware's product was chosen by General Instrument, which has orders for more than 2 million settop boxes. Other customers include Bell Atlantic, IBM, Philips Consumer Electronics, Goldstar, Samsung, and NYNEX.

In 1995, Microware joined with Macromedia, whose "Director" program is a powerful, widely-used multimedia development software tool. This alliance will bring many popular applications to the Microware operating system.

The goliaths to Microware's David in the operating system software market are IDS, a software co-venture of AT&T and Silicon Graphics, Digital Equipment Corporation, Oracle, Sybase, and Microsoft. Microsoft's approach is slightly different from the rest, in that the company sees the software as "a distributed, resource-management-oriented operating system that can do constraint-based scheduling and resource guaranteeing on the fly," says Sanjay Parthasarathy, a group manager for the company's Advanced Consumer Technology Group.[33]

Other software producers see network management, applications management, and the client interface as specialized programs operating at different levels. Ben Linder, VP, Marketing, Oracle New Media, puts it this way: "I think it's way too early in this industry to strangle ourselves by limiting ourselves to a standard, well-defined hardware platform and a standard, well-defined operating system to dominate both ends of the industry."[34]

Client navigation systems   Another hotly contested area of software is the consumer interface, often referred to as the navigation program or program guide.[35] The leading systems are StarSight Telecast, developed by a consortium of cable companies; TV Guide on Screen, a spinoff of the print TV listings magazine; Prevue Networks, which got its start as an on-screen listing service for cable systems; and Stargazer, developed by Bell Atlantic and favored by many telcos.

There is a powerful incentive to create a popular navigation program: They will be the control point that makes or breaks a given program service offered on an advanced television system. In spite of the potential power, it has proved to be quite difficult to develop an attractive, easy-to-use, efficient, fast navigation system.[36] Many people believe that, just as it took 10 years for Windows to emerge as the standard user interface for PCs, it could take a number of years to develop effective navigation systems.

The key features offered by navigation programs are fast forward and rewind ("VCR functionality"), point and tune (to channels), point and record, program reminders, program theme sorting, pay-per-view information, search by title, lockout selected programs, changeable backgrounds, bill viewing, barker channels (promotional programming), and favorite channel selection.[37]

Billing and management software   The absence of billing systems that address the characteristics of selling information and communication services in a dynamic information environment is an impediment to the commercial rollout of advanced television systems. Telephone companies have far better systems to bill individual accounts for transactions than cable systems, which base revenues on monthly subscriptions.

However, both these industries are inexperienced in dealing with fast-changing competitive markets. As a result, their pricing and billing policies are based on collecting revenues, rather than building a customer base. In addition, they are more influenced by regulatory pressures than market pressures. These factors cast doubt on the ability of cablecos and telcos to adopt flexible, user-friendly, responsive billing systems.

There are several approaches to meeting the challenge of metering, pricing, and invoicing for interactive television network services. Cable giant, TCI, is completely re-designing its customer service operations to include billing for transactional services. It will have taken 4 years to develop the combination wired/wireless infrastructure that will handle information from the company-owned systems around the nation into 10 customer service centers. One goal is to reduce the cost of this type of billing: Cable bills cost 52 cents to process while transactional telephone bills cost more than $1.50.[38]

Another way to handle "back office" functions is to bring in a software package: for example, the Interactive Commerce Management System ICMS, supplied by Sun Microsystems and European giant, Thomson Multimedia. Company literature notes that ICMS will "manage and process consumer interactions across broad-based interactive networks . . . deliver a full range of marketing services for market research and promotions . . . dynamically manage prices, create orders, generate a bill, and pass the electronic order on to the existing legacy system for fulfillment. The ICMS platform will enable the development of a wide variety of transaction-based Open TV interactive services such as near video on demand, interactive ads and home shopping."[39]

Network Configurations

In the next section, we examine a number of ways the components can be assembled to carry out the functions of an advanced TV system. We begin by looking at two models that offer alternative designs of the fiber portion of the system.

Competing network models: hybrid fiber/coax (HFC) and switched digital video networks (SDV)

In late 1994, word began circulating that U.S. West planned to change the network design they had previously announced for their commercial rollout of interactive television systems.[40] In mid-1995, Bell Atlantic followed with a similar change. Both companies had initially planned to deploy the "hybrid fiber/coax" (HFC) design, widely adopted by cable systems across the U.S. The two telcos decided to drop the HFC configuration, and to build systems called "switched digital video networks," or SDV.

Both types of networks use fiber optic cable for the system's backbone and feeders to the neighborhood node. However, HFC uses coaxial cable from the node to the home terminal. By contrast, SDV continues to use fiber optic cable from the node to the curb, driving fiber even deeper into the network.

More alphabet soup is commonly used in conversation and appears in the literature to distinguish between these alternative configurations: "fiber-to-the-node" is FTTN; "fiber-to-the-curb" is FTTC; and "fiber-to-the-home" is FTTH. More general terms for this advanced architecture is FSA, or fiber-to-the-serving-area, or Fiber Deep. An all-fiber network is sometimes called a passive cable network ("PCN"), or passive optical network ("PON"). Finally, telephone companies use the term "FITL" (fiber-in-the-loop) to designate networks that extend fiber to a service area that includes no more than a few hundred customers.

There is another important difference between HFC and SDV. HFC is a single integrated network, while SDV is actually two networks with the fiber-to-the-curb network overlaying an all-coaxial cable network. HFC, pioneered by Time Warner, was invented by Dave Pangrac while traveling on an airplane. He conceptualized this elegant design and drew it on the back of a napkin. Most cable systems continue to reap the benefits of Pangrac's efficient design for handling analog and digital signals.

By contrast, the two-level SDV system the telcos plan to use will deliver analog channels over the coax system, which can be offered as a package to consumers. Customers who elect to buy the digital services will be hooked up to the SDV-FTTC network and get a hybrid digital/analog settop box. They will be able to receive downstream signals and send messages upstream, using an ATM switch for data, telephony, and digital video signals. As all video becomes digital, the coax portion of the system will be dropped.[41]

Both HFC and SDV can be designed into several different configurations. The most common networks are: tree and branch, bus, ring, and star, each with variations. The tree and branch is a type of bus network that is currently found in most cable systems. The star network is favored by telephone companies. Both cablecos and telcos use ring designs when building SONET networks. Finally, bus systems are often used for computer local area networks (LANs). As cable operators build advanced networks, extending fiber optic cable deeper into their systems, they are experimenting with mixed configurations: trunk/star, ring/star, and star/star/bus.

Tree-and-branch systems   Figure 7.4 illustrates how this type of network design acquired its name. Signals arrive at the headend—these are the roots. They are bundled into a single stream, the trunk, and transported to major regions of the service area where they are delivered to the feeders, the branches of the system. Feeders take the signals along streets and drop cables tap into them to take signals to the home, which are the leaves.

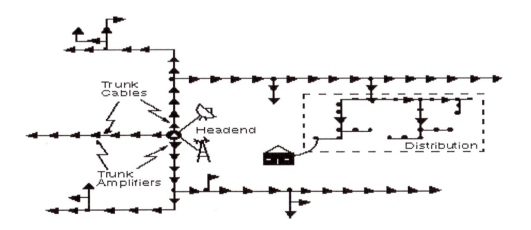

Figure 7.4 Network with tree-and-branch typology. Source: CableLabs.

Bus networks   Figure 7.5 shows a system designed according to this typology, with all its stations attached to a linear transmission network. All messages flow up and down the length of the bus and can be received by any station. Sometimes this configuration is called a multipoint system.

Ring networks: single, double, and triple   A ring network consists of stations joined together in a closed loop, as shown in Figure 7.6. Messages are addressed and sent in only one direction. When the message returns to its source station, it is removed from the ring.

SONET networks are built in the ring typology, in some cases with multiple rings. For example, the Cincinnati Bell Telephone MetroPLEX SONET network has a double ring with messages traveling in opposite directions. A variation is Cox Cable Communications' "ring-in-ring" design, developed to improve signal reliability and to reduce outages.

The SONET architecture is most cost-effective for operators who plan to connect their networks to the public switched telephone system, the "PSTN," in the near future. SONET works for systems that have many types of traffic from many sources, handing off the messages to many other networks. When considered strictly for delivery of interactive television, SONET may be considerably more costly than current hybrid fiber/coax architectures.[42]

Star and mini-star network typologies   In the star configuration, each station is connected to a switch at a central point—this is the ty-

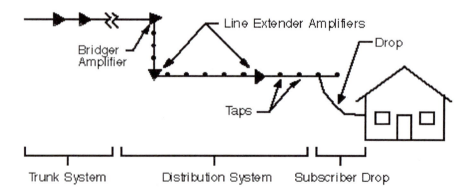

Figure 7.5 The bus network typology. Source: CableLabs.

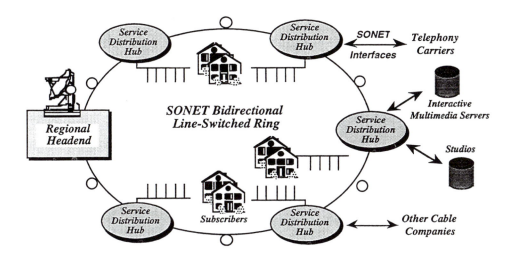

Figure 7.6 The network ring typology. Source: Nortel.

pology of the telephone companies. Even if a message goes to the station next to it, the message must pass through the switch. Now that cable companies are planning switched networks, some are adding mini-star networks of 150 to 500 homes at the neighborhood node of a tree-and-branch system.

The mini-star configuration is a common architecture because it takes advantage of the existing fiber-to-feeder network, which offers cost-savings when operators decide to upgrade their systems. In addition, in 1995, the mini-star fiber-to-the-node design achieved cost parity with fiber-to-feeder installations because it requires fewer expensive "active" RF amplifiers.

System Capacity

The following is typical of the way the people building advanced television systems discuss their projects: "For cable TV applications, consider this basic scenario: assume a master headend with a media server for the network. To handle video-on-demand (VOD) for just peak utilization of 25% of the network which serves 50,000 subscribers, 12,500 simultaneous video streams of data will be needed. Each video stream requires 4 megabits per second (Mbp/s) MPEG-2 format. Multiply 4 Mbp/s by 12,500 video streams and you have 50,000Gbp/s, or about 5 10 Gbp/s-capacity fiber optic cables."

The capacity of advanced television systems is measured in megahertz of bandwidth, i.e. 550 MHz, 750 MHz, or gigahertz for the largest systems, such as 1 GHz. Figure 7.7 shows the spectrum allocation in an advanced full-service 1 GHz network.

Downstream capacity

When a system is described as a "750 MHz" or "1 GHz" system, the numbers refer to the capacity that can be delivered from the headend to each home. The capacity of the backbone is the sum of the amount delivered to the neighborhood, times the number of neighborhoods. Every home served by that node receives 750 MHz (or 1 GHz in a system that size); the information delivered is the same to all homes.

Included in the mix of 750 MHz total capacity are both analog signals that every subscriber receives and digital signals that are addressed to the specific households which have paid for this premier tier of service. Typically, 200 MHz of the 750 MHz are reserved for digital services. This leaves the 54 MHz to 550 MHz for standard analog channels, which can continue to coexist with the new digital signals, resulting in no change for the subscriber on the day the system is installed.

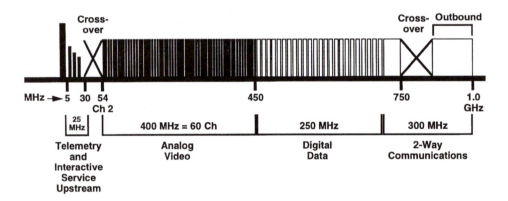

Figure 7.7 Spectrum allocation in a 1 GHz MHz system. Source: Carl Podlesney, Scientific-Atlanta.

Upstream capacity

In a 750 MHz system, all the upstream communication, including telephony, must fit in the 5 MHz-42 MHz at the low-end of the system's spectrum. However, not all that bandwidth is usable. There are problems at the lowest end, but ignoring those, the 40 to 54 MHz portion must be set aside in order to avoid interference with Channel 2, at 54 MHz. Allowing for spectrum that cannot be used, some 25 MHz is left for system monitoring and all upstream services.[43]

In a 1 GHz system the 25 MHz at the low-end of the spectrum can be used for upstream interactive data signals. At the upper end, setting aside about 150 MHz for guardbands (extra empty space) to avoid interference from the 700 MHz area where digital services stop, leaving the 850 MHz to 1.0 GHz area of the spectrum for telephony and high-speed two-way computer communication.[44] Since a compressed voice telephone call requires 4 kHz of spectrum, this 150 MHz would provide bandwidth for more than 37,000 simultaneous conversations.

Storage capacity

Another important type of capacity to consider in an advanced television system is how much material can be stored at the headend for subscribers to request on demand. Although there are subtly different storage requirements for different programs, the following figures represent an average.

People are used to fairly high quality video, so storing the average movie will require about 3 GB (gigabytes). Systems need to keep as many movies on hand as possible. Three hundred films would require about 1 TB (terabyte) of storage, 1,000 movies would take up about 3.3 TB, and 3,000 movies need 10 TB. An hour of television takes up about 1.5 GB and a half-hour program about 750 MB.

Probably no one system will need to store more than 1,000 movies at any given time. Systems can rotate both popular and historical material, working from a base of popular older films they have made a one-time payment to obtain.

The Wired World Meets the Wireless World

The last decade has seen significant advances in the technology of both wired and wireless systems. There is general agreement that rather than there being an entirely wired network, a union of the two will form a single interoperable, interconnected, seamless wired-plus-wireless infrastructure. The next chapter covers the fascinating, new wireless world.

Notes

1. Peter Brackett, Research Manager, Advanced Data Networks, Science and Technology Organization, BellSouth Communications, speaking on a panel at Digital World, '95, June 7, 1995, Los Angeles.
2. Ibid.
3. Roger Brown, "The return band: open for business?" *CED: Communications Engineering & Design* (December 1994):40–43.
4. Israel Switzer, "Methods of two-way service," *TV Technology* (September 1994):31.
5. Tom Kerver, "Turning TCI around," *Cablevision* (March 20, 1995):36–42.
6. I am indebted to Steve Rose for the many hours he spent teaching me about wired television systems. This comment was written in a proprietary report "Media server system overview." For a copy of the full report, contact Steve Rose at P.O. Box 100, Haiku, HI, 96708-0100.
7. Robert X. Cringely, "Thanks for sharing," *Forbes ASAP* (February 1995):49–52.
8. Richard Dean, "Media revolution revolves around video servers," *World Broadcast News* (April 1995):90.
9. "NVOD technology: commitment's the thing," *On Demand* (April 1995):35.
10. Ralph W. Brown, "Video Server Architecture," *1995 NCTA Technical Papers* (Washington, D.C.: National Cable Television Association, 1995):125–131.
11. Claire Tristam, "Stream on: video servers in the real world," *NewMedia* (April 1995):46–56.
12. Mary Barnham, AT&T Corporate Communications, telephone interview by author, April 1994.
13. Harry Newton, *Newton's Telecom Dictionary* (New York: Flatiron Publishing, 1994):981.
14. Pioneer New Media Technologies, Inc. Press release, *"Pioneer's PLUS: first year update,"* June 6, 1993, 600 E. Crescent Avenue, Upper Saddle River, NJ 07458-1827.
15. William Stallings, *Data and computer communications* 4th Ed., (New York: Macmillan Publishing Company, 1992):803.

16. B. Waldron and J. Harrison, "Re-engineering existing computer systems for fiber," *Communications Technology* (December 1994):74–79.

17. Carol Wilson, "Telco networks take the fast lane," *Inter@ctive Week* 2, no. 9 (May 8, 1995):35.

18. "Soliton waves double fiber-optic capacity," *New Telecom Quarterly* (2nd quarter, 1993):6.

19. Harry Newton, op. cit., 441.

20. K. C. Kao and G. A. Hockham, "Dielectric-fibre surface waveguide for optical frequencies," *Proceedings of the IEEE* 133, no. 7 (July 1966):1151–1158.

21. J. C. Carballes, "The impact of optical communications," *Alcatel Research* (1st Quarter, 1992):4–10.

22. Harry Newton, op. cit., 589.

23. Tom Bowling, "A new utility," *New Telecom Quarterly* (4th Quarter, 1994):14–17.

24. Harry Newton, op. cit., 849–850.

25. "Fiber-optics that work," *L.A. Times* (January 7, 1995):D-5.

26. Mark Magel, "The box that will open up interactive TV," *Multimedia Producer* (April 1995):30–36.

27. J. R. Walker and R. V. Bellamy, Jr. (Eds.) *The remote control in the new age of television* (Westport, CT: Praeger, 1993).

28. Roger Brown, "The return band: open for business?" op. cit.

29. Mark Mehler, "Network overload, unused capacity: fixing the optical fiber paradox," *Investor's Business Daily* (January 16, 1995):A4.

30. Matt Stump, "Multimedia Diplomacy," *On Demand* (March 1995):5.

31. Mark Magel, op. cit., 30–36.

32. Mark Berniker, "GI picks Microware's OS for digital set-tops," *Broadcasting & Cable* (May 1, 1995):35.

33. Peter Lambert, "Software Roundtable: six players show their hands," *On Demand* (March 1995):15–22.

34. Ibid.:16.

35. Richard Tedesco, "The main menu: who's on first?" *Broadcasting & Cable* (April 10, 1995):58–60.

36. Steven Schlossstein, "Intelligent user interface design for interactive television applications," *1995 NCTA Technical Papers* (1995):165–170.

37. Peter Krasilovsky, "Program guides face off," *Broadcasting & Cable* (February 20, 1995):51–54.

38. Tom Kerver, "Turning TCI around," *Cablevision* (March 20, 1995):36–42.

39. Press release. "Thomson, Sun select Broadvision to provide interactive electronic commerce management software." Contact: Thomson-Sun Interactive Alliance, Phyllis Scargle, 415/336-0514.

40. Richard Karpinski, "Up close: U.S. West's new video strategy," *Interactive Age* (November 14, 1994):53–54.

41. John McConnell and Jane Lehar, "HFC or SDV architecture? Economics drive the choice," *Communications Technology* (April 1995):34–40.

42. Roger Brown, "Does SONET play in cable's future?" *CED: Communications Engineering & Design* (September 1994):34–38.

43. John McConnell and Jane Lehar, op. cit., 34.

44. Carl Podlesny, "Hybrid fiber coax: a solution for broadband information services," *New Telecom Quarterly* (First Quarter, 1995):16–25.

8

The Wireless World

Introduction

The race between cable companies and telephone companies to build wired systems often distracts observers from the advances made in wireless technologies. New satellites, orbits, receiving dishes, modulation schemes, transmission techniques, and the digital revolution—all these developments have increased the possibility that advanced television may find its way into homes from over the airwaves, rather than over a wire.

The ability to use ever-higher parts of the spectrum to transmit information through the air is creating substantial increases in available bandwidth for wireless communication. While it may never equal the extraordinary bandwidth of fiber optic cable, there nevertheless will be more spectrum than anyone would have predicted as little as five or six years ago.

Wireless technologies must all be within line-of-sight of receiving antennas even when they come from a satellite in space in geosynchronous orbit, 22,247 miles away. Wireless systems fall into two categories: earth-based (terrestrial) systems, which this chapter will cover first, and space-based satellite systems.

The terrestrial systems to be discussed below include a variety of infrastructures: the television stations we have watched since the 1950s; multi-channel, multi-point distribution systems (MMDS); local multi-point distribution systems (LMDS); interactive video and digital services (IVDS); and low power television (LPTV). Before looking at the systems, the next section will present some basic information about wireless signal transmission.

Whether earth-based or space-based, wireless television is transmitted via waves that oscillate at some frequency. Chapter 2 covered the electromagnetic spectrum, which is essential to understanding wireless technology. Radio waves, traveling through the air at assigned frequencies carry the information that is translated into the images on viewers' television sets. This phenomenon is also called "wave propagation," as well as transmission.

Waves: The Invisible Infrastructure

Post-Newtonian modern physics has revealed a world that, at its most basic level, has both a material aspect and a nonmaterial aspect. As material, reality is made up of atoms—but atoms themselves are composed of sub-atomic particles whose properties must be described as "energy" as much as "matter." Early in the 20th century the famous scientist, Neils Bohr, proved that this energy results from the movement of electrons "jumping" from one atomic shell to another.[1] Similarly, the radio waves that travel through the atmosphere are caused by the purposive excitation and control of electrons, an expression of energy.

The energy that emanates from the waves is called electromagnetic radiation—hence the term "radio." The energy that causes the electrons to jump is provided by electricity and all wireless communication is transmitted by some amount of power. That power ranges in strength from the 50,000 watts used by commercial television stations to 100 watts for LPTV, low power television. Systems operating in the gigahertz part of the spectrum use even less power, from 10–20 watts for satellite signals, to fewer than 5 watts for LMDS, local multi-point distribution services.

Radio waves (which include television and microwaves) are transmitted on some particular frequency within the electromagnetic spectrum. Spectrum is allocated for commercial broadcast of radio waves in five frequencies: 1) the low and medium frequency range from 10 to 3,000 kilohertz for AM radio; 2) high frequencies of 3–30 MHz for FM radio; 3) very high (VHF) and ultra high (UHF) broadcast frequencies at 30–100 MHz for broadcast television; 4) ultra (UHF) and super high (SHF) frequencies of 1–12 GHz for microwaves; and 5) extremely high frequencies (EHF) at 30–34 GHz for cellular television.

All these transmission frequencies are much higher than the human ear can hear or the eye can see. In order for our perceptual system to process the information radio waves carry, they must be downconverted to our perceptual frequencies.

As we learned in Chapter 2, radio, or sine, waves have height (amplitude) and length (wavelength). However, they also have rotation (left or right), orientation (horizontal or vertical), velocity (186,000 miles per second), and frequency. Wavelength and frequency are inversely related: the longer the wavelength, the higher the frequency and vice versa. Each of these properties is important in one or more of the new, advanced wireless technologies.

In terms of transmission, radio and television are similar. Sound (and pictures) are converted to electrical energy and connected to a transmitter. The electrical signals are amplified and converted into waves, composed of a carrier signal that is modulated to carry information by its associated sideband frequencies that also travel with the carrier.

Electromagnetic signals can be modulated in one of three ways: amplitude modulation (AM), frequency modulation (FM), or phase shift keying (PSK). AM changes the height of the wave. FM changes how often the wave oscillates per second. PSK changes the timing of successive waves in relation to previous ones. In all cases, the carrier wave is modified to reflect the information it carries.

The modulated waves are sent out, or propagated, through an antenna that repels electrons—this force generates the electromagnetic waves traveling through the air. Think of how a pebble generates waves when tossed into a pond. The movement of the water results from the energy of the traveling pebble impacting on the stationary liquid. Similarly, the repelled electrons generate the modulated traveling sine wave.

Depending on the type of system and the environment in which it operates, the signals may need to be strengthened with repeaters. Or their direction may need to be deflected to serve a difficult-to-reach area with reflector equipment called a "beam bender."

The transmitted radio waves cause the free electrons in all nearby metal objects to vibrate in response—lamp posts, chain fences, and railroad tracks.[2] At the receiving end, the metal antenna is also affected. Its free electrons move down the antenna creating voltage. Inside the receiver, the voltage is amplified and filtered with a tuner that selects a particular frequency. The electrical signals are down-converted to frequencies that can be heard by the ear and amplified to the desired volume through loudspeakers.[3] Add a camera at the transmitting end, some processing complexity, and a television at the receiving end, and the same model will suffice for understanding how television is sent over the air.

Terrestrially-Based, Wireless Television Delivery Systems

One way to distinguish between the wireless modes of delivering television is the frequencies they occupy. The next section will begin with the familiar delivery system of over-the-air broadcast television, focusing on the enormous changes we will soon see occurring.

Free TV: over-the-air broadcast television

As with all complex innovations, television is the result of the work of many different contributors. The first person to suggest that pictures could be sent via telegraphy was a Scottish engineer, Alexander Bain, who also anticipated the scanning process of today's system. Sometimes credited as television's inventor, Philo Farnsworth patented a television system in 1927. The other individual often cited as creator of TV was Vladymir Zworykin who patented the iconoscope (a camera). Subsequent lawsuits over the patent kept TV held up in court until 1934. Finally, Zworykin's version was exhibited at the 1939 World's Fair, but went virtually unnoticed.[4] In 1941, the National Television Standards Committee, under the direction of the FCC, adopted the NTSC's standards for black and white television.

World War II intervened to put television on the back burner. In the meantime, radio had become extremely popular. Its commercial nature was well-established and a contractual network-affiliate distribution system was in place. It seemed natural that television would follow the same development.

Starting in 1946, early television was distributed over coaxial cable. New York and Philadelphia were linked in 1946, to Boston in 1947, cities in the Midwest hooked up in 1948, and the West Coast connected in 1950. By that time, there were 94 television stations and by 1951, there were 17 million television sets. Today, about 98% of the U.S. population has a television set and, on the average, sets are on about 7 hours a day. As of 1992, there were about 1,600 TV stations in the U.S.

Prior to 1995, the affiliate-network relationship was relatively stable. People had been watching CBS programming on their local station, perhaps Channel 6, since 1949. However, in mid-1995, a large station group switched its stations from CBS and ABC to FOX. Before the affiliate realignment, ABC had 223 affiliates; CBS had 210 affiliates; FOX had 137 affiliates; NBC had 209 affiliates; PBS had 349 affiliates; and there were 481 independent stations.[5] After, FOX had more than

150 and CBS had fewer than 200 but it was clear that a count at any one point in time would be misleading, since the old stable system has changed, perhaps irrevocably.

Changes in over-the-air television broadcasting   A decade ago, the television system seemed like it had been there forever and would go on into infinity. In ten short years that perception has proved quite unfounded. The U.S. has already seen many changes . . . and there are many more to come in the next decade.

Actually, change has been altering the system for some time: a declining network share of the audience; the increasing value of the remaining audience; a realignment of affiliate stations; and the creation of additional networks. There are more developments coming in the near future: multi-channel digital broadcast television, digital data and information services, and new program genres and venues.

Beginning about ten years ago, the audience for broadcast television began declining. In April of 1985, the television networks could expect to have 90% of the available audience watching their shows. In April of 1995, the network share dipped to a new low of 57%, down from 61% the previous year.[6]

It would be natural to think that as the audience declined, the value of the networks would decline, but that isn't what has happened. Rather, networks have increased in value. The reason for this surprising consequence is simply that for advertisers who need to address a national audience, the networks are still the only game in town. At any given time of day, basic cable networks account for about 25% of the viewing audience fragmented across nearly 100 networks. For advertisers, premium cable networks don't matter because they offer no advertising opportunities. Radio stations don't pull a national audience; neither do newspapers, except for USA today, but few read it compared to viewers of network television. Thus, the remaining audience has become more valuable than ever—so networks charge even more for it.[7]

Networks are also perceived as more valuable because of their ability to both produce and distribute national programming. In late 1994, there was considerable buzz as ABC, NBC, and CBS appeared to be possible candidates for takeover—all at a price of at least $5 billion. Newspapers carried reports that Time Warner, Disney, and Ted Turner were all suitors for NBC; Barry Diller and Ronald Perelman wanted to buy CBS; Capital Cities/ABC investor, Warren Buffett, indicated he would be open to a buyout. The resulting buying frenzy drove the stocks of all the networks to record highs. Finally, Disney did buy Capital Cities/ABC and Westinghouse bought CBS.

One cause of the rise in value of broadcast companies was the 1994 action of the FCC to change the "financial interest and syndication rule." Previously, networks were prohibited from any ownership position in programming, making independent studios very profitable. The rule change meant that networks could produce and own more of their programming, rather than that produced by studios. This action deprived independents and studios of some part of the network distribution outlet, leading to the interest of Disney, Turner, and other studios in buying networks to keep open the market for their programming.[8]

Individual local stations have gained in value, too. The increasing importance of a national audience is also a reason for the dramatic realignment of local stations in the summer of 1994, after many years of relative stability. Then, stations owned by Ronald Perelman (eight CBS, three ABC, and one NBC affiliate) defected to the FOX network. Stations realize they are essential to networks' profitable operation and they are using that knowledge to demand greater service from networks. Beyond the business arrangements, the technology in stations is changing, too, as shown in Figure 8.1.

Through 1992 and 1993, many observers opined that the television networks were dead, or at least dying. This attitude gave way, however, to the belief that the over-the-air franchise for free TV had value and would continue to have value for some time into the future. As a result, 1994 saw the birth of two new networks, the United Paramount Network (UPN) with 75 affiliates and Warner Brothers Television (WBT) which is still building an affiliate structure. UPN is a programming service that allows its affiliates to air the programs at their discretion, while WBT is an "on-pattern" network that requires its affiliates to play the programming as directed by the network.

Both stations and networks have gained in value because of the additional 6 MHz of bandwidth the FCC seems prepared to hand over to local broadcasters to implement HDTV. As mentioned in Chapter 1, there is some indication the FCC will permit flexibility in the use of this additional bandwidth, opening up the possibility that stations could broadcast 3 or 4 additional digital channels, a service called "multi-channel digital broadcasting."

The added bandwidth may also encourage offerings of auxiliary data services transmitted over the air by local stations, termed "multimedia datacasting" by vendor WavePhore. Such services could in-

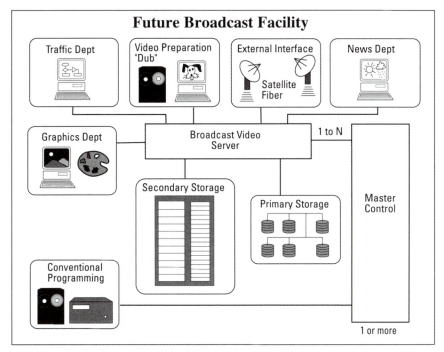

Activities within a broadcast facility will be connected to a central video server through high speed data networks.

Figure 8.1 The broadcast facility of the future. Source: Hewlett-Packard.

clude supplementary programming data, like statistics about sports events; supplementary advertising data about features, prices, and stores carrying an advertised product; distribution of computer software and video games; stock quotes; audio services; and other digital information services.[9]

Networks move toward the multi-channel, multimedia world Each television network has a strategy for dealing with the changes in the media world that it sees coming. Singly and collectively, they are actively engaged in defensive moves to shore up their existing businesses. They work effectively through trade associations, the National Association of Broadcasters and the Television Bureau of Advertising and they are quite protective of their resources. For example, they are currently engaged in a battle to protect the 1990–2110 MHz spectrum they use to microwave news footage back to their stations. New mobile and portable telephony services are trying to obtain that spectrum for their use.[10]

ABC—A strategy of engagement   More than any other network, ABC has positioned itself to participate in new technologies as a premier programming service. They formed Capital Cities/ABC Multimedia group to develop multimedia applications and programming for online services, interactive television, video-on-demand client/server programming, and standalone media such as CD-ROM and video games.[11]

The network has World Wide Web pages on the Internet and a presence on America Online. Program producers work with the network's publicity department to design and place pages where online users can access information about ABC's plans, promotions, and programs.

In mid-1995, ABC forged a partnership with software supplier Electronic Arts, forming a company called Creative Wonders. CW will publish titles for children's entertainment, education, and family reference works, drawing heavily from ABC-produced programming. The net also invested in DreamWorks SKG, the new studio started by Steven Spielberg, Jeffrey Katzenberg, and David Geffen. However, the purchase of Capital Cities/ABC by Disney makes this partnership uncertain.

Internally, ABC has invested heavily in tapeless digital video facilities in their headquarters in New York. The company will use the BTS Media Pool server and routers for several digital editing bays.

CBS—Last off the mark   CBS has monitored the development of the new media environment but has taken a slow, measured approach. They are concerned about the increasing amount of time young people spend with their computers but have doubts about how much interactivity the current audience wants.[12] They have devoted their efforts to promoting and strengthening the "brand" image of their network and its programming, making big investments in promoting the stars of their shows.

Like ABC, they see themselves as an expert developer and creator of programming. For CBS, multimedia is an ancillary business; however, working with the *New York Times* and Apple Computer, CBS' CD-ROM debuted in 1995. The network is online with Prodigy and has a site on the World Wide Web. The network has also put their affiliates on the Internet.

CBS excels in improving its internal functioning through advanced technology. LIDIA, Local Identification Inserted Automatically, is a digital system that lets affiliates display their call letters and logos during network promotion. CBS has also its own computer network, CBS Net-Q, which allows affiliates to log on and view the New York Master Control schedule, allowing them to schedule insertions or to turn them over to remote network control. In 1995, the media company bought a Hewlett-Packard digital video server and disk-based storage facilities to test for use in conjunction with Net-Q and the network newsroom.

FOX: Dares to be digital In addition to the FOX television network, Rupert Murdoch's media empire includes 122 newspapers, *TV Guide*, Twenty Century Fox Film Corp., FX cable network, 11 TV stations, 25% of British Sky Broadcasting satellite TV service, Star TV satellite system in southeast Asia, Indian ZEE satellite TV system, HarperCollins publishers, and Etak, Inc., an electronic map company. The company is also considering creating a small-dish direct broadcast satellite-service (DBS) beamed to Latin America.

FOX wants to leverage its film and television properties through extensions into interactive media. For example, viewers can access information about the highly-rated show, "The X-Files," through the Internet via the Fox-owned Delphi Online Service server. Fox Interactive will bring the property to market as a CD-ROM and an interactive game. Fox Interactive has responsibility for marketing all FOX video and multimedia products. The division also distributes product for Magnet Interactive Studios, an independent cutting edge title developer.

News Corp., the parent company of the FOX network, has made a dramatic commitment to new media. FOX bought the Delphi online service and Kesmai Corp., a developer of multi-player online games. The company is also engaged in publishing CD-ROM titles and electronic versions of its programming assets through its subsidiary, HarperCollins Interactive. In partnership with cable company Tele-Communications Inc., News Corp. formed TV Guide On Screen, navigation software for interactive television systems.

NBC: A multifaceted approach to multimedia Of all the networks, NBC has taken the most unusual path to new media. The firm began by investing over $5 million in Interactive Network, a service that provides playalong capability with over-the-air television. (IN will be covered later in this chapter under IVDS, interactive video and digital services.)

In late 1994, NBC brought news-on-demand to computer networks, a venture called NBC Desktop Video. For about $1500 a month, companies in the financial services industry can receive full-motion video from the news feed service. The content is business news and information taken from both NBC and its cable network, CNBC.

NBC is very involved in on-line ventures. Initially, the network's material appeared on America Online, Prodigy, and GEnie. In May, 1995, Microsoft offered NBC a deal they couldn't refuse: The software company would pay the broadcaster's expenses for developing computer versions of its TV programs for the Microsoft online network. In addition, the two companies agreed to collaborate on CD-ROMs and programming for interactive television systems.[13]

On its own, NBC Digital Publishing has ten CD-ROM titles in development, several linked to NBC TV programming. Finally, the network launched NBC On Site, a TV network designed for display in retail locations. The content is composed of 30-minute blocks of information-based programming, punctuated by 15-second commercials. Receivers are placed above product shelves with 12 to 24 monitors per store.[14]

PBS goes online   PBS was an innovator tying a program to a computer game when they released "Where in the World is Carmen Samaniego?" a show with geography education embedded in entertainment. The computer game allowed kids to extend their interest in the show by playing the game on their PCs. PBS has also worked on advanced satellite programming, developing materials for distant education projects to teachers.

In 1995, long distance telephone company MCI invested $15 million in the Public Broadcasting Service for new media projects. PBS will expand PBS ONLINE and develop interactive programming from existing PBS products. Potential services would be to let viewers download PBS shows and recipes from cooking shows, order related books, and make educational material available to teachers. A price structure has not been announced.[15]

Long-term outlook for over-the-air, free TV   Since the widespread deployment of cable, it has been recognized that broadcasters could only lose audience as the number of delivery mechanisms increases. Some broadcasters are hedging their bets with investments in cable networks, such as NBC's ownership of CNBC and ABC's interests in ESPN, ESPN2, A&E, and other cable networks. Capital Cities/ABC also owns cable systems and newspapers. However, no other network approaches the scope and reach of Rupert Murdoch's News Corp., parent company of the FOX network.

Broadcasters themselves concede that their future is complicated by the new media environment. "For the near term, broadcasting will continue to have the majority of viewing. But that viewing will be more and more supplemented by wired connections. The telcos will be permitted into the full-motion-video-to-the-home business, and the combination of their connections and cable's will mean an increasing percentage of broadcasting programs delivered by wire," observed Bob Wright, CEO of NBC.[16]

Interestingly, like most people, Wright is focused on the competition on the ground. Yet, as we shall presently see, there may be as much new competition from the air.

Scholar August Grant argues that the success of alternative TV distribution technologies depends on content, economic, and organizational factors. On the content side, new services need access to existing programs and services, since viewers prefer programs they know. This makes brand-name programs valuable, especially when they are available during an early viewing window. Economic factors favor services with lower up-front costs that are able to provide any needed specialized equipment to customers, since viewers have limited disposable income. Finally, organizational experience with the marketing of entertainment products, as well as business and personal links to the production community are important preconditions of success. All these factors can be strongly influenced by the regulatory climate, which can favor one delivery mechanism over another.

Wireless cable: multi-channel, multi-point distribution services (MMDS)

If television delivery via cable made over-the-air networks shake in their boots, 1994 was the year that alternative over-the-air technologies returned the favor to cable operators. Wireless cable began in 1973 as "MDS," a multi-point (many receivers) distribution service that sends a single-channel over the air. Viewers use a special antenna, a down-converter, and a settop box to receive the signals. Some readers may remember SELEC-TV, a now-defunct pay-TV service that utilized MDS technology. The more widely known premium service HBO used MDS in the early 1980s, eventually reaching 500,000 subscribers. However, multi-channel cable systems ultimately put most MDS companies out of business.[17]

The business languished until passage of the 1992 Cable Act, which for the first time guaranteed that cable programming must be made available at parity prices to any buyer. Before then, cable systems received favorable rates from suppliers such as Ted Turner who offers such popular networks as CNN, Headline News, TBS, and TNT. By 1994, wireless cable operators were cracking open the champagne bottles at their annual convention in Las Vegas. Even though they had only 700,000 subscribers nationwide, generating $117 million annually, they knew that wireless cable was on the verge of taking off.[18]

By the early 1990s, technology had developed so that MDS could become MMDS (multi-channel, multi-point distribution service). In other words, it became possible to transmit many channels over the air, using special frequencies. For the most part, the acronym has been dropped and this service is commonly called "wireless cable." Wireless cable transmits in the 2 GHz band of the spectrum, where it can send as many as 30 channels in its assigned bandwidth, covering about a 35-mile circle from the headend. They share that spectrum with MDS (the old single-channel multi-point distribution service assignment) and with four channels of ITFS, Instructional Television Fixed Service. The transmission area makes wireless cable especially effective in rural areas that are too underpopulated to support wired cable.

ITFS channels are licensed to nonprofit organizations for educational purposes. However, these nonprofits can use one channel and lease the remaining three to commercial operators, allowing these companies to offer customers 33 channels.[19] This practice is controversial and the FCC may eventually rewrite these rules.

Wireless operators quickly realized that digital technology would allow them to offer many more channels, between 150 and as many as 300, making them competitive with wired cable systems as they are presently configured. Of particular interest to MMDS businesses is the ability to implement pay-per-view, which is seen as a source of additional revenue. At the 1994 convention, six wireless companies formed an alliance to develop digital technologies that would let them send both television programming and digital services. In mid-1995, financing for digital conversion came from telcos, who sought to enter the video delivery market as quickly as possible.[20]

Wireless cable has the potential to offer interactive services. Zenith Electronics has developed a two-way settop device that will send digital messages back to the system headend. Technical specifics have not been announced, but the return path will be sufficient for ordering pay-per-view, opinion-polling, and play-along gaming with over-the-air television shows.[21]

Wireless cable isn't the only terrestrial wireless technology. There is an alternative called "cellular television," or local multi-point distribution service (LMDS).

New York's CellularVision: LMDS

" . . . the higher the frequency, the shorter the wavelength, the wider the bandwidth, the smaller the antenna, the slimmer the cell and, ultimately, the cheaper and better the communication," writes George Gilder about LMDS technology.[22] Local multi-point distribution service (LMDS) is a method for distributing many channels of television, using the super high frequency portion of the electromagnetic spectrum, at 27–29 GHz.

As Gilder observes, the use of such a high frequency means that LMDS signals are very weak, needing only 10 milliwatts of power. The signals are sent from small transmitters cells that are only 2 or 3 miles apart, broadcasting to a tiny "cell"; sometimes LMDS is called "cellular TV." The signals are received with antennas that are only about 4" × 4" square and produce exceptionally high quality TV pictures. The system is capable of rapid two-way transmission with sufficient speed and bandwidth for video conferencing. Since LMDS cells are only about 3 miles from any subscriber and the transmission frequency is so high, it doesn't take much power for subscriber to send signals back to the cell, using a very small transmitting antenna.[23] The small transmission area makes LMDS suitable for densely-populated urban zones.

Sending weak signals means they use substantial bandwidth— LMDS channels take up 20 MHz, in comparison to the 6 MHz of over-the-air or cable channels. However, since they are in the gigahertz frequencies, it really doesn't matter because there is so much bandwidth in that part of the spectrum. Remember that "giga" means billion. So the assignment between 1 GHz and 2 GHz is enough bandwidth to accommodate more than 160 standard broadcast channels and 50 20 MHz channels.

In addition, there is enough space between the channels to allow for a return path from subscribers, sufficient for videoconferencing. The FCC proposed assignment for LMDS includes 2 gigahertz of capacity between 27.5 and 29.5 GHz. This 2 gigahertz of spectrum will allow for more transmitted channels to be sent, as well as up to 400,000 telephone calls per cell, personal communication services (PCS), and data transmissions for transactional banking and shopping.

In late 1995, there were two cellular television systems actually operating: CellularVision in Brighton Beach, NY, and in Canada. More than 450 license applications for LMDS await FCC approval. CellularVision has important partners, including regional Bell operating company (RBOC) Bell Atlantic, Philips Electronics, and the David Sarnoff Research Center.

The delay in licensing is partly caused by a dispute over the 27.5–29.5 GHz spectrum allotment. Motorola, Teledesic, and Hughes want to lay claim to the same spectrum for their planned deployments of low-earth orbiting satellites (LEOS) and middle-earth orbiting satellites (MEOS). This chapter will cover these new satellite technologies in a later section.

Representatives from the International CellularVision Association for LMDS providers believe that spectrum can be shared; representatives from Teledesic and Hughes disagree. The FCC must ultimately make the assignment, and until the agency acts, LMDS cannot expand.

Interactive video and digital services (IVDS): the narrow FM return path

IVDS is a hybrid terrestrial/satellite-based approach to interactive television where a low-power FM radio signal is sent to a local cell site, then transported to a satellite and back down to the service provider. In the home, the subscriber has a transceiver settop box that also overlays text and graphics on the TV screen and a remote control device to transmit their responses, also via radio. Services include shopping, banking, and playalong gaming with broadcast programming.

The pioneer in IVDS is EON, a Virginia company, providing services for eight markets starting in 1995. They first acquired licenses in 1994 from the FCC through an auction process that let providers bid for the ability to use the 218–219 MHz part of the spectrum.[24]

A similar service is the Interactive Network, IN, which has 5,100 subscribers in California and Illinois. IN uses a low-power FM signal to transmit to a lap-held receiver—a computer in disguise. A small screen shows various choices the subscriber can make to play along with a football game or a "Jeopardy" show. However, the IN service does not use wireless for the return path. The player's responses are stored in the receiving unit. When the game is over, the user plugs in a phone line, the unit's internal modem automatically dials a central computer and downloads the answers. Winners of games are notified on their screens over the FM network.

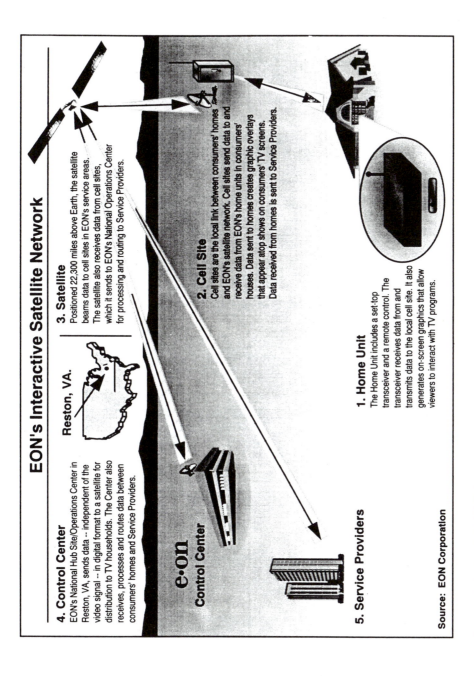

Figure 8.2 The EON IVDS system. Source: EON.

Low power television (LPTV): Home-grown television

Would you rather watch your child play Little League baseball in the local park or a mid-season game between the Braves and the Giants? A few minutes before you leave for work, are you more interested in the national weather picture or the chance of rain in your area?

Low power television, LPTV, is a wireless television delivery system that transmits at a maximum of 10 watts VHF or 1,000 UHF. The signals can travel as far as 35 miles, but the effective reach is more like 10 to 15 miles, and even fewer miles in areas with tall buildings.[25]

According to a 1990 survey of 102 LPTV operations, 46% of the stations cover a rural area, 22% broadcast in an urban neighborhood, and 17% reach a suburban population. The remaining 5% cover a combination of markets.[26] Today, there are about 1300 LPTV stations and the FCC has issued an additional 1,246 construction permits. Licenses are granted by auction, a procedure that has not been entirely successful. Rather than choosing people who have a desire to provide some kind of viable service, anyone who filed stood a chance of winning a license. As a result, only about one-third of the licensees have actually started an LPTV operation.[27]

Of those LPTV stations operating, religious organizations have about 200 facilities and commercial interests account for about 350 more. The rest either deliver over-the-air broadcast signals to difficult-to-reach areas, functioning as what are called "translators" or offer some combination of educational or community service telecasting.

Many LPTV stations offer local news, sports, weather, talk, children's programming, religious programming, and civic services such as council and local committee meetings. Most offer occasional live programming, but it is usually only a small part of an LPTV's overall programming.

The most recent trend in LPTV is multi-channel subscription service. Operators put together several FCC licenses in the same area and offer 10 to 15 channels as a low-cost alternative to cable service. Subscribers receive a settop box to descramble the signals but no special conversion is needed, as LPTV operates in the VHF or UHF frequencies.[28]

Although the technology itself is relatively simple, LPTV is an advanced system because it makes broadcasting personal by moving it to the neighborhood. There is no provision for interactive services or move towards digitization of the picture. However, LPTV is quite interesting as one more source of terrestrial television programming leading to the further fragmentation of the television audience.

In addition to earth-based sources of programming, there are also new sources of satellite programming. Advances in the processing capacity of small computers make it possible for new satellite design and placement. The next section will consider television delivered via satellite.

Space-Based Television Delivery

In 1994, space-based direct broadcast satellite service (DBS) became news. Full-page ads and well-produced commercials, extolling the benefits of 18" pizza-pan-sized satellite dishes appeared in newspapers and on television sets across the U.S. Sure, satellite dishes had been available for twenty years. But the older dishes were 10 to 15 feet across, required motors to aim them in the proper direction, received signals in the 3.7 to 4.7 GHz range, and cost thousands of dollars. By contrast, as shown in Figure 8.3, the new RCA dishes were small, stationary, high powered (11.2 to 12.7 GHz range), and cost only $700.

The smaller dish was made possible by a higher-powered satellite that provided a more powerful, focused beam to earth than that sent by previous "birds" (a term for a satellite). There are three different types of satellites, identified by where they orbit in space: geostationary earth orbiting satellites (GEOS), middle-earth orbiting satellites (MEOS), and low-earth orbiting satellites (LEOS), covered later in this chapter.

The terminology for the services provided by satellites is still being worked out, so it is somewhat confusing. TVRO stands for TV-receive-only on C-band earth stations; it comes only from GEOS, that transmit signals up from the earth to the satellite (uplink) at 6 GHz and down to earth (downlink) at 4 GHz. "DBS" stands for direct broadcast satellite, and this service comes from GEOS, as well. However, sometimes DBS means the same thing as TVRO; while at other times DBS refers to the new services from high-power GEOS that uplink at 14 GHz and downlink at 11 or 12 GHz, received on Ku-band earth stations. The terminology blitz gets worse: other terms for high-power DBS are "DTH" (direct-to-home) and "DSS" (digital satellite system).

The name "TVRO" for C-band, 4 GHz, large-dish service is gradually being phased out and replaced by the term "DBS." High-power satellite service is called DBS or DTH—the two terms are used

Figure 8.3 The RCA digital satellite dish, manufactured for DIRECTV.
Source: DIRECTV.

interchangeably and about equally often. Hopefully, this alphabetic
confusion will be resolved as these services become more familiar.

How Satellite Communication Systems Work

Satellites themselves range considerably in size and cost. For the government's space program, they can weigh as much as 1,000 to a ton,
carry payloads up to 10,000 pounds, and cost upwards of $50 or $60
million. Since 1985, however, there has been a trend towards smaller,
less expensive space vehicles called "smallsats."[29] They weigh as little
as 100 pounds and the launch price runs between $500,000 to $15 million, with costs moving down towards $100,000.

Think of a satellite as a cable headend in the sky. Signals are collected at a broadcast center from satellite, fiber optic land lines and videotape. The programming is digitized, encrypted, and transmitted up (uplinked) to the satellite then transferred to a transponder and downlinked. An uplinked signal comes from one site; the downlink goes to a potentially infinite number of receiving dishes; thus, this type of system is called point-to-multi-point. If the signals are encrypted and can be decrypted only at one site, then satellites can also function as point-to-point systems.

The area where the signals can be received is called the satellite's "footprint."[30] The footprint from a satellite 22,247 miles away covers about one-third of the entire world, (except for the poles), a hemisphere, a region, or even a fairly small area depending on how strong the signal is. The stronger the signal is, the more tightly focused it can be. One problem with signals that come from satellites is that they are not local, so viewers can get TV shows from around the world—but not from their local station.

Signals transmitted from satellites are simply microwaves, located in the super high frequency (SHF) portion of the electromagnetic spectrum, between 3 and 30 gigahertz. The signals received by earth stations are measured in EIRP, "effective isotropic radiated power" and expressed in dBw, the number of decibels per watt.

No matter how strong it is initially, the traveling signal spreads out and weakens as it moves, so that by the time it reaches earth it is usually very weak indeed. "Detecting a transmission from a 100 watt transponder is like clearly seeing a typical light bulb from a distance of 22,247 miles," writes one author about the challenge of receiving a signal from a geostationary satellite.[31] Signals from MEOS or LEOS do not attenuate as much because they are closer to earth.

Satellite transmissions are powered by solar panels, which constitute much of the weight of the vehicle. The information processing units aboard are called "transponders." Satellites now have up to 48 transponders with several 36 MHz-wide video channels transmitted on each transponder.

The receive site

The earth station, or receive system on the ground, consists of a dish, an amplifier, a feed, and a downconverter, as shown in Figure 8.1. The dish is an antenna that collects and concentrates the signals at the "focal point." Some dish owners are equipped with an "actuator" which is a motor that moves the dish so that it can point at and receive signals from different satellites.

The feed collects the signals reflected off the dish it faces, a process called "illuminating" the dish. The feed automatically detects how the signals are polarized: horizontally, vertically, or circularly. Attached to the feed, a small processor called the low noise block downconverter (LNB) converts the signals to an electrical current, amplifies it, and downconverts it to a lower frequency for transport to the indoor satellite receiver. The new digital services require special LNBs that are "phase lock looped" (PLL) to receive a precise frequency without drifting.

From outside the home, the downconverted electrical signals go to the inside receiver via coaxial cable. The receiver includes a tuner to separate the channels and modulation/demodulation equipment to process the signal so it will play on the owner's television set.

Limited interactivity is possible if all that is needed are a few pulses, such as those needed to order a pay-per-view (PPV) movie. The return path is usually a low-power FM signal that goes to a local receiver. The local receiver sends back a message to the settop converter to open it up to receive the channel carrying the requested material. Some systems use the telephone for a return path. The incentive to provide on-demand or near-demand service is to allow impulse buying which greatly increases the take rate for PPV.[32]

However, for a satellite service provider to offer telephone or two-way broadband service in addition to downlinked programming, considerably more power and costly equipment is needed. One way to add interactivity is to link the satellite system by terrestrial microwave, local cells, or even telephones. However, if the signal must ultimately reach a GEOS, one difficulty is the length of time it takes for a signal to travel 22,247 miles. For time-sensitive video and voice data, this delay can be a problem. However, as we will see later in this chapter, closer-to-earth MEOS and LEOS do not have this disadvantage and they offer the opportunity for interactivity.

GEOS: The "Deathstar"

"Deathstar" is what cable operators call GEOS, because the new DBS/DTH programming services confront them with their first truly substantial competition. Interestingly, the ability of GEOS to reach a national audience also makes them competitive with broadcast networks because they are the only other potential packagers of a national audience.

GEOS stands for geo-stationary earth-orbiting satellite, which means that the satellite appears to stay still when it is positioned in an orbit 22,247 above the equator. This precise distance, the geostationary arc, is named the "Clarke orbit" because it was first proposed by science fiction writer Arthur C. Clarke in 1945.

The Clarke orbit allows line-of-sight transmission to anyplace on earth except the poles using only three satellites. Further, the earth stations' antennas never move, so these large dishes don't need to be motorized. Often the transmission is focused to send a stronger signal to an even smaller area than one-third of the globe. In fact, downlink antennas can focus transmission to global, hemispherical, zone, or spot beams.

Getting GEOS into orbit isn't easy. Nevertheless, commercial launches have become more common from French Guiana, China, and Russia. A measure of the technical difficulty, however, is that of the last 5 Chinese launches, 3 have exploded, destroying the satellite it was carrying as well as the rocket. A limited number of liftoffs also occur from Vandenberg and Cape Canaveral in Florida. To compete in this growing market, ITT Corporation plans to build a commercial spaceport near Vandenberg Air Force base in California.[33]

The space available for geostationary satellites is very limited. Ostensibly, 2 degrees (916 miles) separates each satellite in the Clarke orbit, limiting the number of "parking spaces" to 180. However, there are actually 695 birds already in the geosynchronous orbit although there is little chance of the vehicles colliding. Nevertheless, competition to claim a spot is fierce as demand for international communication grows. The allocation of satellite slots has become increasingly difficult and rancorous and the picture is even more complicated by the presence of orbiting "space junk" and "dead" satellites whose useful life of 10 to 25 years has ended but are still in place.[34]

The cost of renting a satellite transponder full-time to send television signals runs from $1 to $5 million per month, and time can be rented in units of $\frac{1}{2}$ hour at a time. Most satellite traffic in the developed countries is video, while telephone calls are routed over fiber optic cable. However, there is little installed fiber in less developed countries, so they still use satellites for their telephony. The increase in the number of international TV channels is creating a boom for satellite companies. For example, contracts for video services have tripled in the past 5 years for Comsat World Services.[35]

Administering the satellite system

The regulation of slot assignment in the Clarke orbit is carried out by the International Telecommunications Union, through its control of the assignment of spectrum for transmission. The ITU sponsors administrative conferences, called World Administrative Radio Conferences (WARCs), to examine spectrum allocation. In the U.S., the FCC establishes the rules for applications, evaluates requests, and forwards their recommendations to the ITU.[36]

The first global satellite organization was INTELSAT, a not-for-profit cooperative headquartered in Washington that administers and managed the first satellite network. INTELSAT puts up satellites, then makes service available to 170 countries, both members and nonmembers. It is still one of the largest networks and many of the members of INTERSPUTNIK, organized by the former Soviet Union during the Cold War, have now joined.

There are a number of regionally-based systems: ARABSAT, EU-TELSAT, CONDOR (Latin America), and RASCOM (Africa). In addition, a number of countries have launched their own satellites, including India, Indonesia, Brazil, Mexico, France, and China.

A relatively new development is the launch of privately-owned satellites, with many more on the way. In 1988, the Pan American Satellite Corporation launched its first of two satellites, which provide domestic and international video, voice, and data service to Central and South America. The Orion Network System put up its first bird in 1994, serving the United States, Europe, and Africa. Both PanAmSat and Orionsat have the express purpose of bringing service to underdeveloped countries. "My objective is to bust open that monopoly [IN-TELSAT] and bring service to people with no phones [and] no TV," stated Reynold V. Rene Anselmo, the founder of PanAmSat.[37]

C-band TVRO: The first service from GEOS

Within the United States, satellites have been widely used for the past decade to distribute television programs from production centers to television stations and cable headends. Credit for the first use of satellites for non-live program distribution to cable headends goes to Home Box Office, which adopted the technology to reduce the costs of getting their shows to the 1,800 cable systems which carried the HBO service. Previously, the channel made expensive copies of the shows, called "dubs," and sent them by courier or mail. Twenty-four hour distribution, while expensive, proved to be much cheaper than "bicycling" the tapes, as the earlier process was called.

By the mid-1980s, many programming services followed HBO's lead and sent channels and programs to cable headends and television stations over satellite. Program distribution is still an extremely important and growing aspect of the GEOS business. In early 1995, two telephone companies won a court victory, allowing them to receive video programming from satellite feeds.[38]

Before long, technically clever entrepreneurs figured out how to downlink the satellite signals for private use. As the number of people with TVRO dishes grew to 1 and 2 million, programmers realized they were losing substantial subscription fees. Within a few years they began to scramble their signals, initiating an ongoing infowar between program suppliers and pirates. "The airwaves should belong to the people. If a TV signal comes trespassing onto my property, I should be free to do any damn thing I want with it, and it's none of the government's business," says one pirate.[39]

Today, there are about 3.5 million TVRO owners in the U.S. alone and many more in foreign markets. These large, motorized dishes can receive 350 channels, many of them still unscrambled, from dozens of satellites. However, when TVRO owners want to receive such popular programming services as CNN, HBO, and Showtime they must pay for service from a channel-bundler and a decoding settop box from General Instrument. (See Chapter 2 for a description of compression and scrambling.)

VSATs: Very Small Aperture Terminal Systems

VSATs are uplink dishes at a "hub" that send up signals to a satellite, which then retransmits them so they can be picked up by many small earth stations. Many VSAT networks are designed for point-to-multi-point communication, such as the Fordstar network. The company's headquarters can send training videos, graphics, voice, and data communications simultaneously to all their dealerships, worldwide.

Retail outlets such as Wal-Mart, Toys 'R' Us, and K-Mart also use VSAT networks for credit checks and inventory control. In these applications, the store contacts the headquarters by landline, where the request for information is processed by a computer. The computer sends the answer (also by landline) to the VSAT, which uplinks the answer to the satellite and it is instantly beamed back down to the retail outlet—all in a few seconds.[40]

VSAT systems transmit data in the Ku-band of the spectrum, sometimes called "12/14 GHz systems" because they send uplink messages out at 14 GHz and receive downlink messages at 12 GHz. Some VSAT systems are entirely wireless. In these, the VSAT "squirts" data over a microwave transmitter to a hub, rather than traveling over landlines. The hub bundles messages and uplinks them to a satellite, which then beams them down to the VSATs. This means that within a few seconds, geographically dispersed sites all have the same information. VSAT networks also allow for videoconferencing and other interactive uses because the local site transmits only to the hub, not all the way to the satellite. One application is in Alaska, where the state supports distance education with interactive video.[41]

Ku-band DBS/DTH: direct-to-home/digital satellite system

It's taken 12 years for the high-power (12/14 GHz) satellite transmission services to get off the ground. The FCC authorized eight orbital slots back in 1982, but substantial technical development was required before services could actually launch. An important advance was the availability of digital technology.

In June of 1994, when the 18-inch, pizza-pan-sized RCA-brand DSS dish was shipped to retail outlets in 150 markets, every unit was sold out within the first 72 hours. Within 6 months there were 500,000 subscribers.

The new high-power satellite technology, with its simple installation and smaller size and price, attracted many customers. The dish could be smaller because of the higher signal power, 100 to 400 watts, compared to the less than 60 watt C-band signal. (C-band TVROs downlink at 4 GHz and uplink at 6 GHz, so they are sometimes called 4/6 GHz dishes.) Another effective sales point was the digital transmission, which vendors argued provided vastly superior picture quality.

DIRECTV, the first digital DBS service used MPEG-1 digital compression, because MPEG-2 was not yet ready. The service will be converted to MPEG-2, allowing even more dramatic improvements. (Chapter 2 contains a full discussion of compression techniques.)

About half of DSS customers live in non-cabled areas, so the satellite service offers them multi-channel television with small, convenient, lower-priced equipment. However, about 39% of the DSS customers live in areas that are passed by cable. Early research showed that in addition to the improved digital picture these people chose satellite service over cable service because they wanted to "dump their cable operator" due to their perception of poor picture quality and service.

The cost per subscriber to deploy high-power DSS systems is $275 per household if there are 1 million subscribers; if there are 10 million customers, the price tag drops to $25. Much of this lower cost is due to the transfer of equipment from the provider to the subscriber, who spends about $750 to own their receiving equipment: dish, LNB, and home satellite receiver.

As of June, 1995, there were three program services up and running, DIRECTV, USSB, and Primestar, with two more services in the offing: EchoStar and AlphaStar. Each service was aimed at a specific target audience.

DIRECTV: the pioneer of high-power satellite services DIRECTV, a unit of Hughes Electronics Corporation, launched their service in June, 1994, and now offers 175 channels from 3 satellites, DBS-1, DBS-2 and DBS-3. The Total Choice TM package costs $29.95 and features 65 cable channels, 7 ENCORE channels, and 29 audio channels. Figure 8.4 shows how the DIRECTV system operates.

In addition, the service features up to 60 pay-per-view channels. Selected hit movies start every half hour and cost $2.99 each. Sports and special event programming is also shown. DIRECTV's research tells the company that their subscribers consider the price fair and comparable to videocassette rentals providing higher image quality and stereo, CD-quality audio.

Within just a few months, DIRECTV was an acknowledged hit with consumers.[42] By November, 1994, the service had signed up 400,000 subscribers, with more than 2,000 coming on board every day. This rate dwarfed the sales of other recent consumer innovations, such as the VCR, big-screen TV, and audio compact disc. By February, 1995, the marketing campaign had been so successful, a survey showed that nearly 45% of the U.S. public was aware of DSS availability.[43]

Since Hughes Electronics' breakeven point on its $750 million investment is 3 million subscribers, the company has hopes of meeting that goal as early as 1996. The price for the consumer-purchased equipment is expected to decrease by about $100 a year, expanding the system's appeal. DIRECTV believes they will have about 10 million

Figure 8.4 Overview of the direct broadcast satellite (DBS) service of DIRECTV.

subscribers by the turn of the century. The annual revenue of Hughes Electronics will exceed $3 billion if these predictions are correct.

USSB: The lesser-known pioneer USSB, a subsidiary of Hubbard Broadcasting, is the longest-standing FCC permit-holder for high-power satellite television delivery. The shortage of satellite transponders has limited the service's program offerings, forcing USSB to lease 5 transponders on GM Hughes Electronics' DBS-1 satellite.

USSB advances a concept called the Mini-Mass Concept, which addresses their ability to draw an audience from across the nation. The company observes that locally-based stations and cable operators cannot support programming that addresses only a small percentage of viewers in the market. However, satellite programmers, through their ability to reach viewers across the country, can pool "mini" audiences into a single "mass" audience. USSB argues that this feature of satellite-casting will revolutionize television programming and make it more like the magazine industry, which is able to profitably serve niche audiences with narrowly-defined interests.

Initially USSB's programming offerings were very limited. Their package included only 6 cable network offerings, without such popular services as CNN, CNBC, USA, ESPN, and most of the other familiar program services. By the end of 1995, that number grew to 30 channels. They do have multiple showing times of premium pay channels, such as HBO and Showtime, but have only 2 channels of pay-per-view.

Primestar: Cable MSOs hedge their bets   In 1990, six of the U.S.'s largest cable MSOs (multi-system operators) partnered with G.E. American Communications to head off the high-power DSS systems they knew were coming. The cable companies (Comcast, Continental, Cox, Newhouse, TCI, and Time Warner) launched a medium-power analog delivery service that required slightly larger 36" dishes than their 1994 DSS competitors. Between 1990 and 1994, Primestar acquired only 70,000 subscribers but the debut of high-power service acted to spur sales of all satellite-delivered programming including this medium-power service and low-power C-band TVROs. By May, 1995, the company delivered television to nearly 400,000 subscribers, about 20% of them from cabled areas.[44]

Primestar initiated a $55 million advertising blitz when DIRECTV and USSB debuted.[45] The service bundled about 50 channels including both basic foundation networks, like CNN, TBS, USA, Discover, and multiple offerings of premium networks, like HBO and Cinemax. Primestar marketing efforts were directed to subscribers outside cabled areas.

The DSS competition, with 150 channels, stimulated Primestar to upgrade its package, so in 1994, the service bundled 73 channels. In 1996, Primestar plans to imitate its competitors. It will expand its programming to more than 200 channels through shifting to higher-power, incorporating digital technology, and added transponder capacity from a new satellite. A snag developed when the FCC reclaimed the expansion frequencies.[46] Primestar disputed the ruling and it may well be reversed. Even if it stands, other options exist.

Primestar will be around for the foreseeable future—and so will even more competition.

EchoStar & AlphaStar: The price competition   EchoStar scheduled its launch of high-power DBS service in late 1995, transmitting over a Martin-Marietta satellite, using 11 FCC-assigned frequencies. This company will employ MPEG-2 compression from the beginning for its initial offering of 75 channels of programming. The company's marketing strategy is to provide niche programming at a reduced package price of about $19.95.

EchoStar plans to specialize in international channels and superstations from New York, Chicago, and Los Angeles, most of which have been dropped from cable systems. They are also looking at data services, such as stock quotes, software delivery, yellow pages service, and daily newspaper delivery.

EchoStar believes several advantages accrue to the later-to-market DBS providers. They will benefit from the consumer education the earlier marketers have paid for; they will start with MPEG-2; and they will be HDTV-ready. Finally, EchoStar doesn't have 20 years of investment sunk into the technology. DIRECTV will have $5 million invested per delivered channel. By comparison, EchoStar will invest about half that amount.

AlphaStar, owned by Tee-Com, organized itself in March, 1995. The company announced that it planned to launch a satellite in December, 1995, to deliver television to subscribers in both the U.S. and Canada. As of mid-1995, the AlphaStar marketing and programming plans had not yet been disclosed.[47]

DIRECPC: Internet in the sky   In late 1994, Hughes Electronics Corporation introduced DIRECPC to deliver datastreams to multiple sites via satellite. Customers install a $1,495 add-in card into their PCs that lets them take in information from coaxial cable. The cable is attached to a 24-inch satellite receiver dish that receives the over-the-air signals.

The most important advantage the system offers is the speed of data transfer, 12 megabits per second. Thus, Hughes Electronics charges between 25 cents and $1.00 for each million bits downloaded, rather than by the minute. IBM and Reuters Newmedia were two of the first users of DIRECPC. IBM plans to distribute most of its software update and maintenance services via satellite to their own offices as well as clients.

As the equipment prices indicate, the initial target for DIRECPC was the business market. However, the popularity of online services has led the company to believe there may be a residential market as well. In mid-1995, Hughes launched its third satellite, allowing them to expand their programming, and to add data and interactive services to all their subscribers.[48]

Hughes Spaceway   In addition to its DBS service and data delivery service, Hughes is also planning a multiple satellite system, Spaceway, that will eventually include as many as 17 vehicles, at a cost of $3.2 to $6 billion for the fleet. The service will offer fixed broadband and data services. However, Spaceway is very expensive, requires enormous fixed bandwidth, and adopts a mature technology in a conventional way. Thus, when Hughes goes to the ITU for licensing, the company may not get the frequency allocation it will be seeking.

MEOS and LEOS: New Satellites for a New Century

As mentioned earlier, the Clarke orbit, 22,247 miles above the earth, is the distance from which a satellite is seemingly stationary over the same place at the equator. This positioning allows global coverage with only three satellites using inexpensive fixed antennas on receiving earth stations.

Today, there's another alternative. Advances in computer processing power make it possible to locate satellites closer to earth and to send and receive signals from a much smaller area. When the message is addressed to a location outside the range of one satellite, it is switched to another satellite in the fleet, one that can reach the designated address.

Low-earth orbits extend from 644 km (402 miles) to 1,600 km (1,000 miles); medium-earth orbits start after the 1,600 km limit. MEOS and LEOS have advantages and disadvantages and in recent years serious disagreement has arisen over which approach produces better results.[49] For example, the higher the satellite, the larger the area covered by the satellite footprint, so the fewer the number of satellites are needed in the fleet—this favors MEOS. However, a larger distance also means longer latency for signal traveling time, so for very fast data exchanges LEOS are better.

The European Space Agency commissioned the Mitre Corporation to study LEOS systems. They found one technical problem, in that to achieve global coverage some satellites would have to be placed at very low elevation levels, making it difficult for users to link with them during certain times of the day. The Mitre report found that these systems are unlikely to be operating before the turn of the century, despite their announced start dates of 1997 and 1998.

MEOS and LEOS technology are well-understood; the former USSR launched a number of both types of satellites. Two MEOS systems are in the works and several organizations are planning to use LEOS to create communication networks, deploying many small satellites (smallsats) for global coverage—the so-called Big LEOS systems. All these systems have in common that they will offer global service and plan to tap the potentially huge market of people unable to get wired telephone service or who prefer mobile service.

The Federal Communications Commission began accepting applications for multi-satellite systems in January, 1995, and issued their first 4 licenses for service links (downlinks, satellite-to-earth) in March, 1995. Licenses for feeder links (uplinks, earth-to-satellite) and inter-satellite links will be issued after ITU's the World Administrative Radio Conference in October, 1995. As of that date, there were dozens of applications in the pipeline for both MEOS and Big and Little LEOS projects.[50]

The cost of a LEOS or MEOS system ranges from $200 million to $1 billion. This wide disparity in cost exists because each system is quite different. The total startup expense depends on how many satellites are used, the orbit (lower orbits require less power), the number of transponders, and the number of links to other satellites.[51]

MEOS systems

Orbital Sciences received the first FCC multi-satellite service license for service links for its OrbComm 36-satellite system. OrbComm will initially provide mobile telephone and position-determination services.

Another MEOS system was licensed in March, 1995 to TRW and Teleglobe, Inc., Canada's premier international communications carrier. The partnership's license allows them to operate a 12-satellite MEOS system named Odyssey. It will provide phone, fax, and paging services to and from anywhere in the world using a small pocket telephone.

LEOS systems

An impressive corporate consortium led by Motorola and McDonnell-Douglas including Sony, Mitsubishi, Lockheed, and Sprint was licensed by the FCC in March, 1995. The $3.4 billion Iridium system will orbit 66 satellites 420 miles up in 1997. Iridium already has 8,000 people working on it, and Scientific-Atlanta has delivered 10 $1.5 million LEOS for deployment in 1998.

Iridium will use time division multiple access (TDMA) technology to provide interactive video, voice, and data services. Critics argue that TDMA is an inefficient use of the spectrum, lobbying for the new code division multiple access (CDMA) frequency-agile technique. They claim CDMA offers a 20 to 1 advantage over TDMA and would substantially reduce the cost of service to the consumer. In spite of its promise, the technology has not yet been tried in a large commercial system. Rollouts began in a few cities in the U.S. and in China in late 1995 and the results will be known in 1996.[52]

Another LEOS system is called GlobalStar, whose main proponents are Loral/Qualcomm, joined by Hyundai of South Korea. GlobalStar will cost $1.8 billion, and place 48 LEOS 750 miles in space. The service will start out providing only narrowband telephone services, using the innovative CDMA modulation. Ultimately the system will be able to carry any sort of data transfer, including video. From the outset, this system will be interactive.

By far the sexiest of the LEOS plans are those of the "boomer billionaires," Bill Gates and Craig McCaw. Their $9 billion system, Teledesic, will offer global low-cost, interactive, high data rate, fixed terrestrial broadband service. The hardware will include 840 refrigerator-sized LEOS orbiting at 438 miles high, to be deployed by 2001 or 2002. Subscribers will receive signals through windowsill or rooftop antennas or with a receiver the size of a notebook computer.

Two other services were licensed by the FCC in March, 1995. Whether they ever actually launch is still in question. Constellation Communications, Inc. is a partnership between Bell Atlantic Regional Bell Operating Company (RBOC) and Telebras, the Brazilian telecommunications giant, which does not appear to be very robust. They propose a $2.2 billion, 46-satellite system to be launched by 2007. The other group, Mobile Communications Holdings Inc. (MCHI) was licensed for a 16-satellite system, Ellipso. This organization released cost and service information privately to the FCC, so their plans are not known.

HALE UAV platforms: GEOS at 14 miles up

This chapter ends with more speculative technologies: HALE, UAV, and RPV. These acronyms stand for "high-altitude long-endurance," "unattended autonomous vehicle," and "remotely piloted vehicle," respectively.[53] They all refer to vehicles that fly above commercial air space, in "proto-space" that starts at 13 miles and ends when outer space begins at 70 miles.

Strong, light, solar-powered HALE UAVs (or RPVs) would fly for several days at a time using their power to stay above a fixed location on earth. The unpiloted craft can be controlled from earth and destroyed in the event of malfunction. Transponders on these craft would support between 10 and 100 beams, with the capacity for 100,000 cellular telephone lines that would support up to two million subscribers. The allowable operational coverage would be a diameter of 500 kilometers, or about 300 miles.

This technology is new and experimental. However, all these new platforms demonstrate the widening world of wireless communication, as ever-higher portions of the electromagnetic spectrum become available for communication uses.

So far, we have examined the features of the technology, and seen how they fit together in complex wired and wireless systems. In the next chapter, we will look at the information that flows through these advanced systems, the entertainment programming, and information services that creative people bring before the audience.

Notes

1. Gary Zukav, *The Dancing Wu Li Masters* (New York: William Morrow & Co., 1979):12–13.

2. John G. Truxal, *The Age of Electronic Messages* (New York: McGraw-Hill Publishing Co., 1990):309.

3. Jarice Hanson, *Connections: technologies of communication* (New York: HarperCollins College Publishers, 1994):100.

4. Ibid.:127.

5. E. T. Vane and L. S. Gross, *Programming for TV, Radio and Cable* (Newton, MA: Butterworth-Heinemann, 1994).

6. Stuart Miller, "Soft soil sifting in webs' rose bed," *Weekly Variety* (April 4, 1995):30.

7. James Flanigan, "TV networks evolve from dinosaurs to darlings," *Los Angeles Times* (October 5, 1994):D1, 2.

8. John Lippman, "Prime-time targets," *Los Angeles Times* (October 12, 1994):D1–D4.

9. Chris McConnell, "Turning data streams into revenue streams," *Broadcasting & Cable* (April 10, 1995):32.

10. Chris McConnell, "Broadcasters fending off spectrum grabs," *Broadcasting & Cable* (July 25, 1994):82.

11. "Networks migrate to multimedia," *Broadcasting & Cable,* Telemedia Week (April 1995):14–19.

12. Gerald M. Walker, "CBS television: implementing the future," *World Broadcast News* (April 1995):32–37.

13. Julie Pitta, "Microsoft lures NBC away from rival on-line services," *Los Angeles Times* (May 12, 1995):D5.

14. Shane Ginsberg, "The digital revolution will be televised," *Digital Media* 5, no. 1 (June 5, 1995):3–12.

15. Elizabeth Rathbun, "MCI funds PBS new media ventures," *Broadcasting & Cable* (March 27, 1995):14.

16. Don West and Steve McClellan, "Bob Wright and the NBC nobody knows," *Broadcasting & Cable* (March 6, 1995):37–47.

17. Christopher Stern, "FCC moves to strengthen wireless cable," *Broadcasting & Cable* (June 13, 1994):11.

18. Dana Blankenhorn, "Wireless cable operators form alliance," *Newsbytes* (June 22, 1994).

19. Christopher Stern, op. cit., 11.

20. Christopher Stern, "Telcos hedge bets with wireless wagers," *Broadcasting & Cable* (May 1, 1995):14. And Mark Berniker, "PacTel joins wireless migration," *Broadcasting & Cable* (April 10, 1995):35.

21. David Tobenkin, "The wireless system that could," *Broadcasting & Cable* (May 1, 1995):20.

22. George Gilder, "The new rule of the wireless," *Forbes ASAP* (April 11, 1994):99–110.

23. Richard Karpinski, "It's in the air: broadband goes wireless," *Telephony* 225, n. 9 (August 30, 1993):16.

24. Aviva Rosenstein, "Interactive television," in August E. Grant (Ed.), *Communication Technology Update,* 3rd Ed. (Boston, MA: Butterworth-Heinemann, 1994):55–66.

25. Mark J. Banks, "Low power television," in August E. Grant (ed.), *Communication Technology Update,* 3rd ed. (Boston, MA: Butterworth-Heinemann, 1994):107–115.

26. Mark J. Banks and M. Havice, *"Low power television 1990 industry survey."* Unpublished report of the Community Broadcasters Association (December 14, 1990).

27. Jarice Hanson, op. cit., 200–201.

28. Stephen E. Coran, "Low power subscription television," *Wireless Broadcasting Magazine* (April 1995):14–16.

29. Rick Fleeter, "The smallsat invasion," *Satellite Communications* 18, no. 11 (November 1994):27.

30. Frank Baylin, *Miniature satellite dishes: the new digital television* (Boulder, CO: Baylin Publications, 1994).

31. Ibid.:16.

32. "Buy rates skyrocket for DBS pay-per-view," *Interactive Video News* (April 3, 1995):1–2.

33. James F. Peltz, "ITT plans commercial satellite 'spaceport'," *Los Angeles Times* (January 27, 1995):D1, 6.

34. Mark Stein, "Satellites: Companies, nations fight for spots in space," *Los Angeles Times* (September 20, 1993):A1, A16.

35. Sean Scully, "Satellite business looking up," *Broadcasting & Cable* 123, no. 26 (July 12, 1993):48.

36. Raymond Akwule, *Global Telecommunications* (Stoneham: Focal Press, 1992):57–87.

37. P. Apodaca and J. Shiver, "Southland firm spotlights FCC auction woes" *Los Angeles Times* (August 20, 1994):D-1, 2.

38. Jesus Sanchez, "Two phone giants open cable TV's door," *Los Angeles Times* (February 12, 1995):D1, 2.

39. Charles Platt, "Satellite pirates," *WiReD* (August 1994):8, 122.

40. George Lawton, "Deploying VSATs for specialized applications," *Telecommunications* 28, no. 6 (June 1994):27–30.

41. Brad Kayton, "Very Small Aperture Terminals," in August E. Grant (ed.), *Communication Technology Update,* 4th Ed. (Boston: Focal Press, 1995):307–317.

42. James F. Peltz, "Hughes DIRECTV already a rival to cable," *Los Angeles Times* (November 8, 1995):D-1, 11.

43. "TCI survey tracks DBS awareness," *Specs* (Louisville, CO: CableLabs, April 1995):3–4.

44. C. T. Veilleux, "Primestar in more Wal-Marts," *HFN* (May 8, 1995):59.

45. Rich Brown, "MSOs take direct approach," *Broadcasting & Cable* (June 27, 1994):26.

46. Chris McConnell, "Primestar's miles-high problem," *Broadcasting & Cable* (May 8, 1995):92–94.

47. "Two new DBS ventures set to challenge for piece of increasingly crowded sky," *Video Technology News* 8:7 (March 27, 1995):1, 3.

48. Roberta Bhasin, "Heaven sent," *Convergence* (March 1995):27–32.

49. Joseph N. Pelton, "Geosynchronous satellites at 14 miles altitude?" *New Telecom Quarterly* (2nd Quarter, 1995):11.

50. "Orbital free-for-all," *Information Week* (October 16, 1995):16.

51. Rivka Tadjer, "Low-orbit satellites to fill the skies by 1997," *Computer Shopper* 15, no. 3 (March 1995):49.

52. Steve G. Steinberg, "Spread-spectrum technology," *WiReD* (April 1995):72.

53. Joseph N. Pelton, op. cit., 12.

9

Content: The Creative Challenge

Introduction

In the past decade, the number of distribution channels for "content," that is, entertainment and information products, has steadily increased. The number of outlets grew from three networks and a handful of independents in larger markets to four networks, an array of independent stations, ever-larger cable systems, C-band DBS, Ku-band DTH, and now telco delivery, MMDS, LMDS, and LEOS. This does not even consider the rise in international demand.

The proliferation of channels has changed the marketplace for content in several important ways. Greater demand resulted in the production of more material. For example, at the 1995 National Cable Television Association convention, 98 new narrowcast channels awaited carriage by cable systems.[1] At the same time, it also brought about a dramatic increase in prices for content.[2] Expenditures by the broadcast network rose from about 64% of their gross revenues in the 1980s, to 80% in the early 1990s.[3]

The kinds of entertainment and information products also changed. From the beginning, television programs appealed to a mass audience. As cable systems added capacity, however, there was demand for material that would appeal to segments of the audience with special interests. Sports, religion, entertainment genres (comedy, science fiction, action, mystery, etc.), news, public affairs, culture, documentary, and business are all examples of entire cable channels that are devoted to specialized topics.

Interactive systems require their own special conceptualizations and formats. So far, developers recognize four types of content: pre-packaged transient, pre-packaged evergreen, user-created added data, and user-created content. Pre-packaged content is the text, graphics, audio, and video that consumers can access. Much of this material is transient, replaced within each week. If it is relatively permanent it is called "evergreen," meaning it will be of interest over some length of time. Pre-packaged content may be structured so that users can add to it, gradually building a user-created database. The content may also be entirely user-created, such as an online "chat room" or a telephone conversation.

A perennial romantic aura surrounds content. At the many conventions, demonstrations, expositions, and shows that feature new communication technologies, the phrase "content is king" has become a cliché. The lure of content origination has blinded many companies to its costs, both financial and organizational, at least where visual material is concerned. Although it may be cheaper to make original audio content, there are many advantages to licensing already-created visual intellectual property, rather than funding its development and production, as well as providing the venue for its distribution. Licensing can be a less expensive, less risky way to acquire visual content, but it lacks the mythic power conferred by the creative process.[4]

"Content" is a compromise term among the disparate industries involved in advanced television for the material that appears on a viewer's screen. Advanced television systems are converting writers, directors, producers, and programmers into "content providers."

Linguistic compromise is sorely needed. In the television industry, the images and audio presented to viewers are called "programming," referring to programs and shows. In the CD-ROM industry, the images, audio, and text that users get on their screens are called "software" and programming is a written document containing codes for the rules and variables that structure the operations of the computer. In the computer industry, software doesn't mean what's on the screen—it's the computer programming. Go figure!

In the linear entertainment business, the script is a blueprint composed of the scenes and dialogue that will make up the finished product. In multimedia, the script is software programming in an authoring language that defines what is going to happen as the user progresses through the material.

In television and film, the person who oversees a project is called a producer; in the multimedia and video game fields, that individual is a developer. In show business, the finished product is a program, a show, or a film; in multimedia, it's called a "title"; for online projects, it's an "application."

The linguistic problem is not simply that the creative groups speak different languages; the bigger problem is that they often use the same words to express different underlying objects and processes. Beyond language, the computer-based creative companies (video games, CD-ROMs) have a different culture than the traditional creative companies (TV, film). The disparate cultures lead to profound disagreements over what content should look like, where it should come from, and how it relates to its audience. (Even here, the defined reality is contentious: in the computer industry, the audience members are users; in the entertainment industry, they are viewers; in the gaming industry, they are players.)

Difficulties specific to advanced television systems also exist. Every creative effort has a history to draw upon; but for the content-providers of these new technologies, it's a very short one. Much is unknown: formats, appeals, flow, styles, uses, abuses, and limits.

The technology itself is a frustrating blend of opportunity and obstruction that changes on a yearly basis—sometimes even more rapidly. Two alternative technologies and creative paradigms must be taken into account when considering the content of advanced television: networked and standalone (or packaged) platforms.

Networked platforms include cable and telephone company delivered content over wired and wireless systems for which content providers develop "networked multimedia applications." Standalone (or stranded) platforms include the television set connected to a game player (IMPs, interactive multimedia players), personal computers with substantial memory (4 MB RAM, more than 100 MB hard drives) and, increasingly, a CD-ROM drive. For those platforms, developers create "packaged multimedia applications."

This chapter will cover the current technology platforms, products, and processes for developing content for advanced television— interactive TV, CD-ROMs, online services, and video games. In many cases, content is developed within organizations and we will describe the alliances formed to create products. However, there is also a strong tradition of independent creative work and the chapter will examine the differences between these two creative environments.

Finally, the chapter explores how creative work both changes and stays the same when it applies to advanced television systems. It looks at the creative process and how it is harnessed to provide products for the entertainment industry.

Interactive Television

Interactive television (IATV) is a networked environment where consumers access a server for material. On-demand television requires a level of interaction that is quite low, amounting to enabling the consumer to do nothing more than send pulses to request desired programming. These pulses may order the server to send a movie or a video, to display an information array, or to show a catalog. At a slightly more sophisticated level the customer may be able to tell a computer that he or she wants to buy something. In most implementations, the consumer then gets a telephone call from the merchant to verify the order and make delivery arrangements.

However, as we will see in the next section, there are many more possibilities like games, communication, and interactive storytelling. Over time, creative people will develop them.

The search for the "killer app"

Killer app is a frequently used term that is short for "killer application." It means a product that is so appealing, it entices consumers to buy the whole system in order to get the one service or program.

The overarching areas of applications in IATV include entertainment, shopping, information services, telecommuting, education, and communication services. Implementation of services in each of these categories can take many different forms. For example, every service can be subdivided to target niche groups based on demographic variables of age (young, adult, or senior customers), gender, ethnic or racial background, or interests (religion, hobbies, etc.).

Entertainment applications are gaming, playalong with broadcast TV programs, sports, music, adult, gambling, news, travel, and movies. Video-on-demand (VOD) which provides programming in a few seconds, near-video-on-demand (NVOD) which provides programming within a few minutes to a half-hour, and expanded pay-per-view (PPV) all fall in this category. Entertainment services might also include automatic commercial zapping, electronic program guides and digests, and automatic program recording.

Forecasters have long believed that VOD, or even NVOD, would be the killer app that would pay for the enormous cost of building wired interactive systems. However, tests have demonstrated that on-demand services may not bring in sufficient revenue. These results have led other observers to suggest that perhaps the killer app isn't so much a singular service; rather, a bundle of interactive services would raise the necessary revenues. Tom Grieb, vice president and general manager of GTE mainStreet expressed this view: "It's very much like cable TV. Everybody subscribes for a different set of features."[5]

Vincent Grosso of AT&T argues that there aren't so much killer apps as there are killer attributes. That is, what will make interactive systems successful is the way they are designed, rather than their subject matter, service area, or packaging.[6]

In 1994, a new version of the killer app emerged—online access over cable. In the next sections, we will consider the various programming and service packages and the technology platforms to which they are tied. Maybe one of them will be the killer app.

Technology platforms and current products

Since VOD and NVOD figure so prominently in the discussion of interactive TV systems, they must be considered in some detail. These are services where subscribers choose a movie or television from a library. The viewer, let's call him Bill, uses some handheld device that brings up a list of movies or television programs on the TV screen. Using an arrow-key, Bill arrows down to a selection he wants. He pushes another button to order his movie or program and presto! it appears.

Presto? Exactly when the movie appears has preoccupied testers since the 1970s. The amount of time between when Bill orders his program and it comes to his screen is called "latency." There is always some latency, even if it is a nanosecond. If the latency is five minutes or less, the service is called VOD. If the latency is longer than five minutes, usually 10, 15, 20, or 30 minutes, the service is called NVOD. The shorter the latency, the more costly the implementation.

To determine what kind of service to offer and how many channels should be set aside, programmers need to know how long a viewer will wait for an ordered program and how much he or she will pay for it. The data is analyzed in terms of "buy rate" or "take up rate," sometimes cited as a measurement over a baseline of ordering pay-per-view selections. Suppose that 5 out of 100 subscribers will buy a current popular movie, priced at $5.00, with a 2-minute wait. What will the percentage be if it is priced lower but there is a 5-minute wait? A 10-, 15-, or 30-minute rate? At each time increment, how much does the price need to drop for the offered program to attract the same number of viewers?

These questions began with pay-per-view (PPV), deployed on many cable systems in the mid- to late-1980s. Usually PPV runs on 2 to 4 channels and sometimes there is a continuous loop "barker channel" that runs promotions, movie trailers, and information about price, ordering, and starting times. The viewer calls a telephone number up to 15 minutes after the movie starts. The computer sends a downstream pulse to the consumer's settop box, signaling it to descramble the PPV movie.

PPV movies run consecutively, with one or two movies per channel, so the viewer must wait at least 90 minutes for the next showing. By contrast, NVOD takes up many more channels because its start times are 10, 15, 20, or 30 minutes apart. (Sometimes NVOD is called by the apt term, "stagger-casting.") The running time for most movies is about 100 minutes, but some run as short as 90 minutes and a few run even longer. Thus, a movie that starts every 30 minutes takes up 3 or 4 channels (with 4 being average); one that starts every 20 minutes takes up 5 channels; one that starts every 10 minutes will require 9 or 10 cable channels.

The first commercially-deployable VOD technology launched in November, 1994, a joint venture of EDS and the Dallas-based Interactive Network. Trialing over a Sammons Communication cable system, the EDS file server stores the material and their software manages billing, transactional, and database services. The IT Network has installations in more than 50 markets including a shopping, information services, educational services, an electronic program guide, and Yellow Pages service.

In the absence of widely available two-way broadband systems, many content-providers have developed applications for systems where cable or over the air programming delivers the downstream television and the telephone or FM radio carries the upstream interactivity. Audio Services, Inc. (ASI) and Cableshare, Inc. designed a telephone/cable interactive service. Viewers can shop, bank, play games, and receive news-on-demand through their cable box. A dedicated telephone line carries the user information back to the headend.[7]

Another technology is the unique Interaxx Television Network. Interaxx adds a video CD-ROM player to the settop, which provides the graphical interface and information. The consumer gets a new CD-ROM with updated information about every 3 months. Cable carries downstream programming and upstream instructions.

Interaxx is one of several multi-service providers. It offers shopping from more than 40 retailers, video games, classified ads, and an electronic program guide. They may also carry interactive gambling and lottery services.[8]

Another multi-service distributor is Videoway, in Quebec, Canada. It is the longest-running service in the world and although most of its services are graphics-only, the company, Videotron, has more experience actually providing interactive services to customers than any other organization in the industry. Videoway subscribers can send e-mail, shop at home, play games, print out coupons, access information, and use their settop as a PC modem.[9]

Videotron carries ACTV (the actual name, not an acronym), an IATV programming service that specializes in multiplexed presentations. For example, a sporting event is taped by several cameras. ACTV picks up the feed from four cameras and shows them on four different cable channels or multiplexes all four onto one channel. The viewer uses the remote to move from one camera angle to another, unaware there are actually multiple channels.[10]

In addition to sports, ACTV produces educational programming based on the same principles. The company films as many as four different versions of the same children's program. Suppose, for example, that an emcee introduces the show and then says, "If you are a girl, press 1; if you are a boy, press 2." Then: "If you are under 6, press 1; if you are over 6, press 2." This strategy results in the following alternative programming:

	Boy	Girl
4–6	Channel 1	Channel 2
6–8	Channel 3	Channel 4

This technique allows a seemingly personalized message to youthful viewers.

Telephone company GTE is a multi-service content-provider offering a suite of services marketed under the mainStreet brand. mainStreet includes 80 interactive services spanning shopping, education, games, sports, financial, news, and community information. Implemented in Carlsbad, CA, and Boston, MA, the service costs $9.95 per month. In late 1994, the telco expanded beyond "repurposing" existing programming to develop original material such as talk shows, game shows, and local services like restaurant and entertainment guides. In 1995, GTE experimented with allowing subs to vote for their choices during the Academy Awards.

Another currently deployable technology is offered by Zing Systems. The creative team writes an interactive script around cable, satellite, broadcast, or videotaped programs, and commercials. The interactive opportunities give subscribers the ability to playalong, guessalong, or learnalong with the resulting TV program. A special remote prompts the settop to decode the interactive material for the consumer. USA Network, the Sci-Fi Channel, and the Encore Channels all allow Zing Systems to develop interactive material from their programming.

Of the potential interactive television services, the most attention has been paid to shopping. MicroMall, under development at Digital Equipment Corp.'s Digital Media Studio, is a "turnkey solution" that allows retailers to offer their products and services over DEC server-to-settop platforms. Product providers can update their catalogs easily and frequently to add and delete new items and prices. Lands' End was the first merchandiser to come on board MicroMall.[11]

A virtual supermarket, ShopperVision, will operate on Time Warner's Orlando system. It will stock 27,500 grocery and drug store products, requiring the digital storage of 192,500 photographs. The consumer can virtually "cruise" down aisles, click on products, and turn the products to read the labels.[12] CUC International with AT&T, Nordstrom's, Bloomingdale's, Spiegel, and Macy's are just a few of the retailers planning interactive shopping services. Cable giant Viacom is developing shopping extensions for MTV, VH-1, and Nick at Night that could be made interactive. The Box, an interactive music channel, has a shopping component on the drawing boards.[13]

Another important content area is interactive games for adults. Several companies are planning interactive gambling. ODS Technologies' platform went into trial in 1994, allowing home betting on horse racing and possibly on other types of sporting events.[14] TV Answer will offer viewers a datastream of FM radio signals from Sports Ticker, receiving sports scores, breaking sports news, statistics, and specialized sports stories that appear on the television screen.[15] Other sports content includes ABC's alliance with NTN Communication to develop interactive playalong games from ABC Network's broadcast lineup. The effort will start with ABC college and pro games, such as Monday Night Football.

ABC is active in using its programming for IATV, a strategy they call "leveraging the ABC brand." Interactive programs will use material from ABC News and daytime soap operas. However, the network considers it very important to maintain the look and feel of the original program to get the most from the branding strategy.[16]

Geared to a broader audience, the Game Show Network, a Sony-owned channel, launched at the 1995 NCTA convention. They have two interactive games, "Decades," and "Race for the Numbers," which viewers play from home, dialing a toll-free number and using their telephone keypad to record their plays. The shows let viewers win thousands of dollars in prizes.[17]

Kids are the biggest game players and the Sega Channel delivers video games over cable channels to them. SC rotates 50 games per month offering four of them as "play-per-day" specials. The technology by General Instrument uses three cable channels, delivered from a Sega Game Server. The entire game is downloaded into the Sega Game Adapter in the subscriber's home, so all play is local rather than networked.[18] SC has proved quite popular in trials.

More interactive programming aimed at children is in development. daVinci Time & Space has partnered with National Geographic Television, the Discovery Channel, the Cartoon Network, and the Children's Museum of Boston to create new content. They plan to start with linear TV programming, integrating advertising into the product. The company plans to take part in an IATV test trial in 1996. Other entrants in the kid IATV content market will be Nickelodeon and FOX Interactive.

There is some programming development in the area of news and public affairs for advanced TV systems as well. Time Warner's "News Exchange" is a news-on-demand service that will digitize information from ABC, NBC, CNBC, CNN, and *Time, Fortune, Entertainment Weekly, People, Life,* and *Sports Illustrated* magazines.[19] For its operation on the Time Warner Full Service Network in Orlando, subscribers will be able to access local news sources as well. In addition, TV Answer will implement polling and real-time town meetings over its FM-return loop system.

Table 9.1 summarizes content development for IATV. In the near future, IATV content will increase dramatically through a technology that allows CD-ROMs to play over advanced TV systems. An alliance between software makers, Microware and Macromedia, opens the way for developers to adapt their existing and future titles from CD-ROM to IATV systems. Macromedia makes popular multimedia authoring software and Microware provides server-to-settop software. As we will see in the next section, this is an alliance of some importance.

In the trenches

In 1993, talk about IATV systems centered around the technology. In 1994, people started discussing content. By 1995, content providers

Table 9.1 Programming for IATV

Company	Type of service	Content
Suites of services:		
Audio Services, Inc. (ASI) and Cableshare	Telephone/cable (tel. return path)	Shopping, banking, games news on demand
Interaxx	CD-ROM on cable system	Shopping, games, elec. program. guide classified ads, IA gambling and lottery
Videotron	Text-based services and ACTV multi-channel programs	Sports, children's educational
mainStreet (GTE)	Cable systems	80 services: shopping, education, games, news, sports, financial, local information
Zing Systems	Cable systems	Playalong, guessalong, and learnalong with TV
Shopping:		
DEC & SGI	Look only	MicroMall: catalogs
ShopperVision	Look and order	Many retailers
Viacom	Tailored to system	MTV, VH-1, and Nick at Night shopping extensions
The Box	In development	Music shopping
Games:		
ODS Technologies	No info released	Betting on horse races
ABC+NTN sports college	Tailor to system	Playalong with ABC programming: pro, college games
Game Show Network	Tel. return path	Two games: "Decades" and "Race for the Numbers"
Sega Channel	Download games to game machine	Play at home on dedicated game machine

Table 9.1 *Continued*

Company	Type of service	Content
Programs:		
ABC	Tailor to system	Will adapt news and soap opera material
daVinci Time & Space partners	Tailor to system	Kids' programming
FOX Interactive	Tailor to system	Kids' programming
Time Warner	Tailor to system	Digitize footage from ABC, NBC, CNBC, CNN, *Time*, and other publications.

were complaining bitterly about the multitude of "APIs," the application programming interfaces.

An API defines how an application hooks into the system software and calls services that are transported across networks. It is called "middleware" and is the translation/gatekeeper software that stands between the application on the network and the settop and the operating system. It makes sure all the software components can talk to one another when an application is run.

In test trials, every site has its own unique hardware and software configuration. The companies trying to have their APIs accepted as standard are Apple, Interactive Digital Solutions, Microsoft, Microware, Oracle, PowerTV, and Sybase.[20] These multiple APIs force content providers to write interactive sequences for each implementation. Software rewrites are time-consuming and expensive, prompting creative people to demand standardization across networks.

The alliance between Macromedia and Microware is a welcome one to the content developers because Macromedia's Director is one of the most standard authoring programs and Microware has a headstart on capturing the server-to-settop software market. This offers the possibility to content providers that there may be standardized APIs in the offing.

The creation of programming for IATV is not without humor and the flavor of the activity is reflected in the lexicon that appeared in a recent column:[21]

bicam /v. to think with both sides of the brain. "She can't spell to save her life, but she bicams like a champ."

enchilada /n. a TV show full of the codes needed for interactivity. "'Jeopardy is an enchilada in the test markets."

shelfware /n. a useless computer program

steak 'n' sizzle /n. the complete script and software package of an encoded IATV program.

testies /n. test markets, usually small ones so no one will notice if the product bombs. "That enchilada was a big hit in the testies."

Interfaces and navigation software

The interface is the visual and functional field that lets the subscriber access the available programming and services. Navigation is a set of procedures embedded within the interface that the subscriber uses to move around the system. Ideally, the graphics and text that make up the interface and navigation software are consistent across the applications. Usually, they are based on an operating metaphor, such as VideoMall, Neighborhood, or StarGazer, that allow differentiation within a single theme. Figure 9.1 is an example of a sophisticated navigator.

The interface/navigation program is an extremely sensitive product. The organization that controls the interface is in a powerful position to influence the choices the viewer makes—and to collect information about the viewer and the viewer's actions. The use and abuse of this information is of concern. And burying a given application under an avalanche of menus can kill it.

In IATV, the interface and navigation are often referred to as an electronic program guide (EPG). The major developers of EPGs are StarSight Telecast, TV Guide on Screen, Prevue Networks, ICTV, and Bell Atlantic's StarGazer. Worried about losing their position as the channels of choice, the broadcast networks have pressured EPG creators to assure them of prominent spots within the EPG. At least one of them, StarSight, is considering allowing program services to improve their visibility—for a fee.[22] An example of how StarSight guides viewers is shown in Figure 9.2.

There have been problems creating EPGs. Original designs confused consumers and required too much memory from settop chips. In 1994, the Prevue Guide and TV Guide On Screen both reduced the

Figure 9.1 An on-screen menu from Bell Atlantic Video Services' video-on-demand market trial provides consumers with both graphic and text choices. A click of the remote control button takes them where they want to go on the system. Source: BAVS.

complexity of their EPGs to accommodate both the hardware—and the wetware subscribers.

Interface designer Joy Mountford puts it this way: "When you're designing an interface, you need also to ask, 'How can I marry the user's desire with a set of good constraints?' These constraints should allow them to explore. They should be consistent. They should be forgiving. They should be interesting. They should be clear. And they should allow the user to always stay in control."[23] She adds that the interface must give users a system, a scheme, for coping. It must allow them to exit at any time.

Steve Schlossstein has developed a step-by-step approach to the problem: 1) develop the interface concept; 2) identify the functionality of each interface model; 3) identify the user population; 4) identify high-level constraints, such as hardware and software; 5) create a storyboard sketch for each interface metaphor; 6) identify the usability goals; 7) plan the screen layout; 8) analyze the input device and design

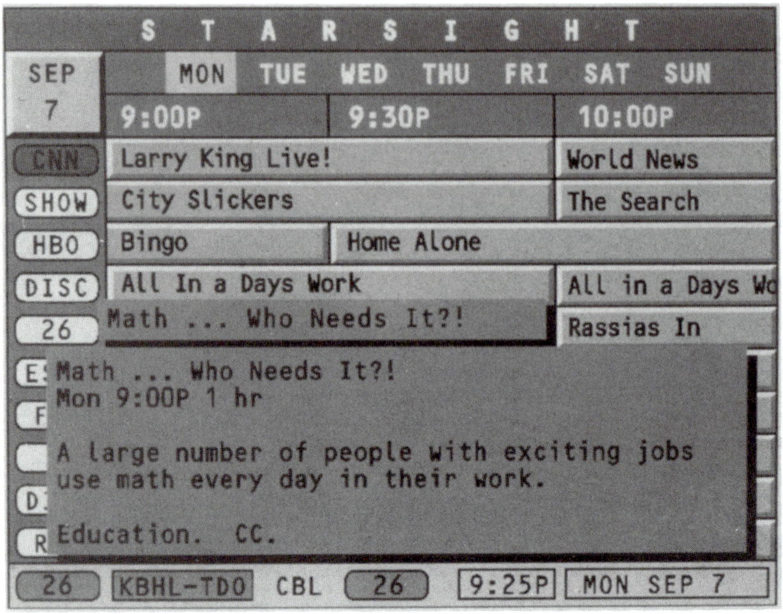

Figure 9.2 A StarSight navigator on-screen menu. Source: StarSight Telecast Inc.

appropriate action sequences; 9) develop prototypes of key screens that include the concept, the tasks, and the look and feel; 10) test; 11) re-design, based on test results; 12) conduct final review and usability testing.[24]

A distinctly different point of view is advanced by Kai Krause, who creates graphics tools for computer users. He maintains that the one rule of design is to "keep the user happy."[25] Krause refuses to patronize his customers by dumbing down the interface so that it loses all appeal and surprise. Rather, he builds in an evolution of learning so that as the user continues to interact, new messages and functionalities appear.

Media journalist Peter Lambert believes that current interfaces are still klunky and uninteresting. Virtually all observers agree that there is still much work to do before interface and navigation systems and EPGs offer subscribers the flexible, interesting, entertaining, and informative experience that would satisfy them.

CD-ROM

The technology that carries IATV is a network; the technology that delivers CD-ROMs is standalone—so far. However, many observers believe that CD-ROMs are a transition technology that will soon be replaced by some version of a digital video disc (DVD). Also, it may soon be possible to place the CD-ROM's content onto a server for multi-user interaction over networks.

The market for CD-ROM titles is small relative to revenues taken in from other sectors of the communications industry.[26] However, according to a report in the Financial Times on March 14, 1995, worldwide sales of multimedia PCs in 1994 were 10.3 million units, quadruple the sales of 1993 at 2.5 million units. Home sales predominated. In addition, it is also important to look at CD-ROM titles because this is where the current generation of creatives are learning and perfecting interactive storytelling and design.

Telephone company GTE is pursuing this strategy. The company put 20 titles into development, targeting both cartridge and CD-ROM platforms by Mac, PCs, and 3DO players. "We're attempting to bridge the gap between the entertainment and technical communities, and instead of talking about what we do, we're shaping the interactive entertainment market now," said Tom Casey, GTE Interactive Media vice president of marketing.[27]

Despite the low sales volume for CD-ROM-based entertainment, a number of prominent Hollywood stars have appeared in them. Clint Eastwood is involved in making a retrospective of his film career. Dennis Hopper, Grace Jones, Ned Beatty, and Margot Kidder have all taken roles in products for this medium.[28]

Technology platform and current products

Since 1994, many computers sold for home, business, and school use come equipped with CD-ROM drives. Increasingly, they are also placed in public kiosks that offer information or provide the equipment to conduct transactions. On the average, kiosks serve between 50 and 100 people a day, sometimes more.[29] The CD-ROM can display attractive video sequences that help people find their way or show products in a flattering environment.

Two other developments in CD-ROM technology are important. The emergence of the digital video disc (DVD) is about to revolutionize the amount of information they hold—at least 2 hours and 15 minutes of high-quality video on a TV or computer screen, long enough for a movie. (This compares to about 1.5 seconds of broadcast-quality images on today's CD-ROM.) Agreements on a single format between Sony, Philips, Toshiba, and Time Warner was reached in late 1995. DVD will employ MPEG-2 compression to achieve high storage capacity and will come to market in 1996.[30]

Another important technological development is the availability of recordable CD-ROMs. The first CD recorders (CD-Rs) cost over $10,000, and declined to $5,000 within 5 years. In 1995, several companies priced their CD-Rs at less than $2,000 and, simultaneously, low-cost software authoring and mastering came on the market as well. Not only were the drives less expensive, recording speed is faster and the quality is better.[31]

An implication of recordability on CD-ROM is that the new DVD products must also offer this feature. The manufacturers of the DVDs have indicated they will offer DVD-R soon after product launch.

CD-ROM titles are divided into three areas: entertainment stories and games, education and edutainment, and productivity software. Of the top ten selling CD-ROM titles in mid-1994, 6 were entertainment, 1 was edutainment, 2 were education, and 1 was personal productivity.[32]

Encarta—Education	7th Guest—Entertainment
Myst—Entertainment	5 FT, 10 Pak—Entertainment
Rebel Assault—Game	Mixed Up Mother Goose—Ent.
Quicken—Productivity	Eco-Quest—Edutainment
Mega Race—Game	Ultima VII: Speech Pack—Education

Entertainment titles are the most popular. In the first quarter of 1994, CDs and TV game cartridge sales grew 26% to $81.3 million for that period. The best-selling entertainment titles are developed originally for the CD-ROM medium. In 1995, the popularity of "Myst" led to the release of pictorially-rich CD-ROM exploratory and role-playing products. "The 11th Hour," "Total Distortion," "The Dig," and "Quake" are all noted for their technical virtuosity, character development, and interactive sophistication.[33]

It has proved difficult for films to crossover to the CD-ROM market, as exemplified by Turner Entertainment's conversion of the FOX movie, "PageMaster." Nevertheless, Hollywood studios persist in attempting to repurpose their material, as part of an overall branding strategy. The reverse direction has been somewhat more successful, so the films "Mortal Kombat" and "Super Mario Brothers" did passably well at the box office although they were not the gigantic hits on the big screen that they were on the game player screen.

The latest trend is to use television shows as the creative basis for the silver discs. In 1995, Paramount Pictures licenses the shows "Frasier," "Beverly Hills, 90210," and "Melrose Place."[34]

An important category of content for CD-ROMs is "edutainment," a newly coined term to describe titles that present information in an entertaining way, mixing enjoyment with learning. It is the most rapidly growing segment of CD-ROM titles. A study by Denver-based QED Inc. showed that the number of K-12 schools with at least one CD-ROM-equipped computer doubled in 12 months, increasing to 29% of all schools. Sales in this category grew 128% in 1993 to more than $89.7 million in sales for the first quarter of 1994.[35] Educational titles enjoy another advantage—they may be "evergreen," selling for years, as opposed to the hit-driven entertainment market. However, potential educational title developers should beware because by 1994, there were 900 titles, with 150 new ones appearing every 6 months. Evidently these numbers aren't bothering the Reader's Digest folk— they've allied with Microsoft to distribute CD-ROMs with text derived from their publication.[36]

In the trenches

Developing content for CD-ROMs is a fascinating topic because of the range of subjects they address, the styles needed to address them, the potential for interactivity they offer, and the creative challenges they present. Each area—entertainment, information and education, and productivity—requires a different approach.

Entertainment-oriented CD-ROMs CD-ROMs for entertainment fall into 2 categories: storytelling and games, with considerable overlap between the two genres. There are several styles of games, but they have in common that they give feedback to the player about how he or she is doing at any given time and they allow a sense of mastery.

"Twitch games" are high speed exercises in hand-eye coordination that are too fast for CD-ROMs, so they are released on diskettes for computers or cartridges for game players. "Simulation games" provide a complete environment in which the player is immersed and that changes in response to the player's decisions. "Adventure games" are usually based on a fantasy and the player takes on a role. Simula-

Figure 9.3 Ocean of America's Flight Simulator CD-ROM that rivals the speed of game players and has very high quality graphics. Source: Ocean of America.

tion and adventure games can exploit the involving video and complex graphics of CD-ROMs and may not require the speed of twitch games, so more and more often they are released on CD-ROM.[37]

CD-ROM is also a natural format for titles that involve interactive storytelling because they can deliver an aesthetically-pleasing pictorial presentation. According to one Hollywood pro, Jeff Berg, "the missing link in some of the games right now is the whole area of character development and narrative plot. Software developers may be able to provide you with the most fantastic graphics, but finally you need a story. You need characters, you need a conceptual framework, and you need a theme."[38] Stylistically, interactive stories involve role-playing, exploration, puzzles, or some combination of these.

It is difficult to separate games from interactive stories. Generally, the more expensive the production, the closer it resembles a movie. An interactive story costs about $1.5 million to produce; a game usually costs less than $750,000. For example, "Voyeur," "Myst," and "7th Guest" are stories that allow a role-playing user to explore a fantasy world or a simulation, leads them into a puzzle and maybe a little adventure, and lets them win (or lose). Are they games or are they stories? Robert Weaver, president of InterWeave Entertainment, Inc., who directed Voyeur and Voyeur II, believes that all interactive entertainment is a game because the key to interactivity is mastery and gaining mastery requires knowing how well you are doing. For Weaver, that feedback process turns a story into a game.[39]

Education-oriented CD-ROMs   The dominant publisher of edutainment titles is Broderbund, with such products as "Living Books" and the Carmen Sandiego series. In mid-1995, Broderbund merged with Learning Co., which produced more strictly educational titles. Learning Co.'s works include "Reader Rabbit," "Student Writing Center," and "Learn to Speak" foreign language software.[40] Many education-oriented titles are designed for children, such as the VTech product shown in Figure 9.4.

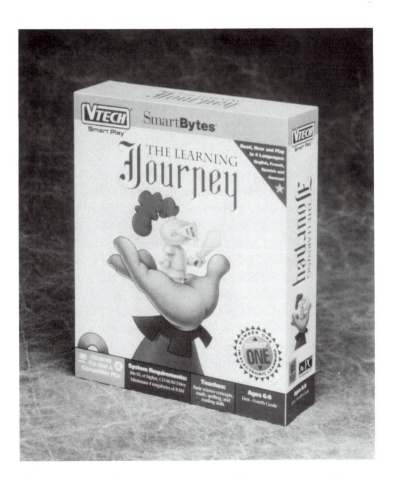

Figure 9.4 The Learning Journey by VTech is designed for preschoolers to help them learn to read. Source: VTech.

Creating educational titles requires experts in the area of the presented subject matter as part of the development team. Frequently, material must be licensed. For example, compilations of art, photographs, speeches, and videos of known works may involve considerable effort and significant fees to acquire the necessary material.

Online: IATV could start here

The "Net," that large, amorphous non-place called "cyberspace," the domain of connected computers, is the fastest-growing arena for new communication technologies. Daily newspaper reports trumpet the exponential growth of the number of people connected to the Internet, the network of networks that allows everyone to talk to everyone else. Ten years ago, who had heard of Prodigy? CompuServe? America Online? Today, each of these services claims more than 1 million members. More than 24 million Americans and Canadians are connected to the Internet: 74% are 44 or younger, 25% have incomes exceeding $80,000, 50% are professional or managerial, and women account for 44% of the users. In October 1995, Hambrecht & Quist, an investing research service, estimated the value of publicly traded Internet-related companies at $6 billion.

In 1990, audio first appeared online in the form of compressed files. Musicians can place their new tunes on the Net and fans can hear the music in real-time, bypassing the traditional industry structure. This development came about in 1992 with the World Wide Web. It began as an experiment by a laboratory in Switzerland, the European Laboratory Particle Physics (CERN) to access and transfer graphics files more rapidly using a server-type mechanism. By 1995, 10% of the enormous Internet traffic was WWW activity: More than 18 million Americans and Canadians used the Web in a 3-month period. The number of websites is expanding rapidly, numbering in the tens of thousands; in February, 1995, 200 new sites every week appeared on the net.[41]

The exuberant growth and the fast-changing hardware and software available to netnauts have resulted in many new kinds of online content. For example, the Library of Congress is planning to create a virtual library of its books, manuscripts, and photos for distribution over computer networks. The goal is to convert the most important holdings by the year 2000.[42]

Computer communication is moving rapidly towards ever-more broadband services. Fifteen years of text, 5 years of audio, now 3 years of still photos and graphics—it seems obvious that the next step is moving video. This progress has led many observers to believe interactive television could come into the home via an expansion of online capacity rather than through cable systems. Oracle and Intel are partnering to develop technology to distribute video over ISDN to avoid waiting for broadband networks.

Currently, online content is both "live" and pre-produced, or packaged. By early 1995, many papers carried schedules for online events. In July, 1995, the Los Angeles Times printed the following schedule:

America Online: Wednesday, 7 p.m.—Director Irwin Winkler talks about his new film, "The Net"; Thursday, 7 p.m.—Musician Bruce Hornsby in an on-line chat; Sunday, 7 p.m.: Urb LeJeune, author of Netscape and the HTML Explorer, in an online conference.

CompuServe: Thursday, 6 p.m.—Author Cleveland Amory leads a discussion on cats, dogs, and animals of all kinds; July 31, 5 p.m.—Musical group, Hootie & the Blowfish in a live conference.

Prodigy/TimesLink: Thursday, 7:30 p.m.—Actor Dennis Miller in TimesLink online chat; Friday, 3 p.m.—Actor Tom Selleck in Prodigy online chat.

Internet: *If you want to read the complete works of Emily Dickinson, William Shakespeare, Mark Twain, and other famous authors online, try http://www.cloud9.netscharf/author.html; *The International Museum of the Horse is now online with a vast repository of equestrian lore. Connect to http://www.horseworld.

Questions remain about the viability of packaged online services. A study by Inteco Corporation found that there is a significant amount of satisfaction with America Online, Prodigy, CompuServe, GEnie, and Delphi. People take advantage of free trial offers and then cancel the service. The 6.7 million subscribers to services represent only about one-fifth of all users worldwide and less than half of users in the U.S. Until now, the online environment has prospered as a place where the users themselves provide the content—rather than prepared and packaged content as with other communication products.

Still, many content providers are doing their best to prove that packaged content will be as popular online as it is on TV, radio, and CD-ROM, putting their talents to work to attract users to their products. The next section will explore those efforts.

Technology platforms and current products

There are two platforms supporting computer internetworking. As we saw in the chapter on wired systems, the narrowband, circuit-switched telephone infrastructure connects most point-to-point communication. Many organizations are connected by packet-switched local area networks (LANs). When messages go outside the organizational campus, they travel over wide area networks (WANs) or the telephone network.

The bandwidth limitation has resulted in the development of ever-more ingenious software. The software environment of the Internet is TCP/IP. TCP stands for "transmission control protocol" and IP means "Internet protocol." Together, they allow different types of computer networks, makes, and types of computers to communicate over a common pathway and to share readable messages.

Techniques to present both sound and visual material have grown quickly. RealAudio is an important development that allows online users to produce, distribute, and listen to on-demand audio clips. Before RealAudio, netters could get music, but they had to download it first and then play it locally off their computer's hard drive. The download took hours and the huge file demanded enormous disk space. RealAudio allows bitstreamed audio to play in real time on the user's computer using a 14.4-kbps or faster modem.[43]

New software technologies offer entirely new functionalities that influence what content is even possible. Cornell University's CU-SeeMe software already makes live videoconferencing over the Internet possible by compressing the files at very high rates and providing only thumbnail (small) video images on the computer screen.

Oracle Corporation is planning to market a $500 device that will let people who don't have home PCs download video material from the Internet. The technology would enable them to access videoconferencing services that company CEO Larry Ellison believes will exercise strong appeal.

The World Wide Web (WWW) not only allows the addition of photos and graphics to messages. It also provides a structure for individual "sites" that can be accessed by users from around the world.

The sites are called "URLs" (uniform resource locators) and identified by an odd-looking, rather intimidating, unique address. For example, the address for Mr. Showbiz, an online service that covers Hollywood and the entertainment business, is: "http://web3.starwave.com/ showbiz/."

A World Wide Web site exists as part of a recognized Internet domain. The Internet isn't located in a place—rather, it is dynamically spread across the memories of many computers. To become an Internet domain, the person or organization either registers with the InterNIC or gets service from an Internet provider.

The content exists in the memory of one or more computers configured as a server; the people who "dial up" are "clients" and they enter through a "port." The exchange of files with embedded audio and graphics is made possible with the software advance called MIME, Multimedia Internet Message Enhancements.

There are two types of online content: services and programming. The word service is ambiguous—it can mean a product offered to customers like shopping, banking, or information such as Mead Data Central. It also refers to browsers (programs that help users navigate the Net) and bundlers of online programming and services that charge a monthly subscription fee. Examples of online services are America Online, CompuServe, Prodigy, GEnie, AT&T's ImagiNation (for children), and many others.

To create the environment where users access content (whether programming or services), developers use specialized authoring languages. Since 1986, one universal language is SGML, Standard Generalized Markup Language, that specifies the structure of the content. The program used to create most WWW sites is HTML, Hypertext Markup Language. In addition to structure, HTML allows designers to control how the content will appear to the user.

HTML is the most widespread of authoring languages. It is relatively simple to use and widely available as shareware. It allows designers to use text, graphics, and scanned-in photos and to control colors, fonts, arrangement of elements, and so forth. The hypertext links are marked by an underlined phrase. When the user clicks on the phrase, they are connected to other areas of the website. In some cases, they link to another website altogether, automatically "telnetting" and connecting the user to another server.

Three-dimensional worlds are created with VRML and VRML+, Virtual Reality Markup Language. A host of other new authoring programs like Java and RealAudio have cropped up, giving creators exciting new tools to create new types of online content.[44]

The rapid proliferation of websites has made some kind of accessing scheme necessary. The software programs used to navigate the World Wide Web are called "browsers" or "spiders." There are dozens of spiders such as Netscape, TIA, NetCruiser, and Metacrawler.

The games begin   Text-based, networked, multi-user games, called MUDs (multi-user domains or multi-user Dungeons, depending on the source), are one of the most popular Internet offerings. "Hawaii" is the working name of a multiplayer online service that displays a graphic environment. It allows 32 players to race cars against each other online. Users use their modems (9600 baud and up) to dialup an 800 number and enter a race, competing against other players. The software will also support a single car that represents a team of users.[45]

Japanese game titans, Sega and Nintendo, avoided the online market until 1996. Sega has placed its bet on IATV, rather than online, and some observers agree with Michael Schrage that "if Japan looks back from the next century to review multimedia opportunities missed, the unwillingness of its top video game companies to get on the Internet will probably prove the biggest. . . . That's terrific news for American video games."[46] In 1996, both companies announced they would create websites and consider the development of online services.

CD-ROM/online hybrid services are yet another new development. Users receive a CD-ROM that is filled with complex graphics. They use their modems to connect to a server, where other users are also logged on. They all communicate through the server. The server also sends commands to the users' computers, telling the CD-ROM where in the environment they are "located."

This means that the information traveling down the telephone line are low-bandwidth data: 1) text messages between users, 2) information about the location of other users in the environment, and 3) control messages to the computer's CD-ROM drive. The hybrid CD-ROM/online platform gives users an exciting and visually complex environment that can change quite rapidly, while communicating with others in the same environment. One popular such service is the Interactive World's Fair by Worlds, Inc. which lets the CD-ROM supply the rich 3-D world users move around in.[47] We will describe this in more detail in the "In the Trenches" section of online content creation.

Entertainment online   Entertainment content is some of the most popular material on the Web. Users access the work of both little-known artists and highly sophisticated entertainment industry giants and professional product marketers. The Web is an inexpensive venue where "garage bands," musicians who have no professional representation, can reach an audience. Users find them on the Internet Underground Music Archive (IUMA), which stores videos, 30-second audio clips, and sometimes entire cuts, bios, photos, and club dates. Increasingly, online is a forum for the largest companies, too. Meatloaf released a comeback hit on the Net. Aerosmith released a song, "Head First," from a new album on CompuServe. Cutting edge groups like Towhead, Deth Specula, Eden Retread, and Mr. Paul put whole sets online.[48]

As mentioned briefly in Chapter 8, many television networks are also creating a cyberpresence on the web. ABC has an area on America Online where subscibers can get information about ABC news, sports, soaps, and entertainment fare. ABC's publicity department works aggressively with producers to create temporary WWW home pages to publicize their upcoming programs. CBS maintains a hub on Prodigy and has an Internet presence as well. Several CBS programs have Internet newsgroups devoted to their programming, including David Letterman and Tom Snyder. In August of 1995, Fox was the last major network to come online, but they believe theirs is the biggest and best entertainment website.

NBC has the most extensive online effort. In February, 1995, the network launched NBC Desktop Video, which allows PC users to receive live business and financial information from their PCs. Content from NBC News, CNBC, and the Private Financial Network is repurposed for the new service, which also airs live feeds of corporate announcements and significant annual meetings.[49]

By mid-1995, every major movie studio, music publisher, and broadcast network had created a homepage on the WWW, most with hypertext links to their new releases. Many cable networks are also online. CNN produces CNN at Work, allowing subscribers to view CNN over their PC's for $12.50 per month. The Discovery Channel and MTV have sites on America Online, ESPNet is on Prodigy, and The Weather Channel is on CompuServe and is considering expanding to Prodigy, AOL, The Microsoft Network, and the Internet.[50]

Even aspiring cable networks are online. The Jackpot Channel, RecoveryNet, Network 1, and the Hobby Crafts Network created websites because there was no space on cable networks for new programming. Through online sites they can test the attractiveness of their programming and use the data to promote their service to cable operators.[51]

Hundreds of newspapers and magazines have come online. Some analysts predicted that within the next year, nearly 3,000 newspapers will offer some electronic or interactive services. *The New York Times* offers an online paper and 3 sports magazines. Time Inc. launched electronic versions of *Sports Illustrated, Fortune,* and *People* on CompuServe, in addition to the presence of *Time* on America Online.[52]

Online marketing Almost from the beginning, the technically-sophisticated computer users who populated the network used it for informal equipment exchange. That early marketing activity is still present and computers, accessories, and repair parts are among the best-selling products on the Net. Flowers, books, candy, and consumer electronics are also popular buys with netnauts.

About 25% of their subscribers shop in the cybermalls of CompuServe, Prodigy, and America Online. CompuServe's Electronic Mall offers products from more than 130 merchants. Sales grew 80% in 1993 and individual sales averaged $71. Prodigy's PC Flowers is the second largest FTD florist in the U.S. and retailers on the service such as JC Penney and Land's End are planning to digitize their catalogs to allow users to access them online.[53]

Where marketers tread, advertisers cannot be far behind. It is no great surprise that advertising has come to cyberspace. Online usage changes the traditional communication relationship between advertisers and consumers. On television, the advertiser beams their message towards a passive, broad audience. Online, the advertiser targets a message towards an active audience of one-user-at-a-time.

The Net extends many benefits to marketing promotion and advertising activities. It is cost effective. The material is current and easily updated. Segmenting potential customers based on their interests is simple. Messages can be personalized and the customer can quickly provide the marketer with detailed information, at no cost.[54]

Of course, nothing is perfect. At present, it is very difficult to let customers know how to find the marketer's website. The culture of the Net can be very anti-marketing, as well, so that advertisers find they must be entertaining and informative, in addition to pushing their own marketing objectives.[55]

Prodigy currently has more than 100 advertisers who post information consumers can access, such advertisements and detailed product information. The advertising agency, Ogilvy & Mather, has one of the most ambitious interactive plans, with about 30 people working on a variety of projects including on-line ad campaigns, ads-on-disk, and ads-on-CD-ROMs.

It's instructive to note one service with a great deal of intuitive appeal that hasn't occurred because of security concerns: online banking and bill-paying. Currently, about 25% of U.S. banks allow banking from home over the telephone, while 10% offer banking over PC.[56] In May, 1995, Wells Fargo of San Francisco established the first Internet site to let customers check their balances and history of transactions. However, only 2% of banks anticipate providing service over interactive television systems. First Union Corp. and Open Market plan to team up to offer purely virtual banking (they have no actual, physical banks).

The difference between banking with IATV and banking online over the telephone is striking. Once security issues are addressed, online home banking will be widely implemented; 90% of banks expect to offer it by the end of the century. People will get spending money by online authorization onto a smartcard, a piece of plastic that works like money in machines and retail outlets.

MasterCard, working with IBM and others, has been able to reach a set of technical standards for secure online transactions. Visa and Microsoft Corporation are also pursuing a system to allow over-the-net shopping.

In the trenches

The creation of an complex online environment is a huge undertaking. When Ken Locker, whose career had included the vice presidency of Hollywood production giant Carolco Entertainment, decided to join Worlds, Inc. and build the Interactive World's Fair, he was very well prepared. His education included literature, technology studies at MIT, and working for new tech guru Jonathan Seybold, in addition to his solid Hollywood background.[57]

Worlds, Inc. spun off from Knowledge Adventures, an edutainment producer that distributes products on diskette and CD-ROM. Worlds, Inc. was set up to create online environments. Locker describes his work as: "People who think they are creating programming that people will view or receive don't understand the nature of interactivity or cyberspace. We are creating an environment, a universe, that people come to and interact in. The more opportunities for activities, personal contact and communication, transaction, exploration, and expression, the more people will like it and want to be there."

Worlds, Inc. developed an environment called the "Interactive World's Fair" (IWF), shown in Figure 9.5. Composed of 7 themed pavilions, subscribers can enter Science and Technology, Asia, Europe,

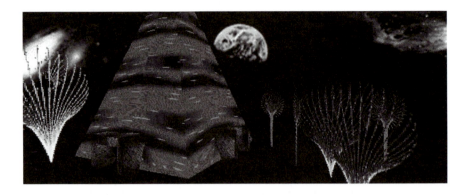

Figure 9.5 The Interactive World's Fair, a hybrid CD-ROM/online entertainment venue. Source: Worlds, Inc.

and more. Each pavilion has entertainment spaces, where online conferences and performances are going on, chat rooms, malls and shops, and galleries. Some objects have hypertext links to other WWW sites. For example, suppose in the Earth gallery guests come across the shot of the earth from the NASA weather satellite. They can click on the image and be transported to the NASA website where they can get up-to-date weather information.

Building this huge environment posed unique problems. For one thing, 7 pavilions is a huge cyberplace. To build the IWF, Worlds, Inc. teamed with Landmark Entertainment, a company that develops amusement centers and location-based entertainment sites. For example, Landmark built Caesar's World in Las Vegas, Nevada. Creative groups from the two companies used Landmark's proprietary construction software to keep track of the millions of software creations it took to build the IWF.

The elements that make up the pavilions were designed on the computer, then printed out so that artists could retouch them by hand. The artwork was then scanned back into the computer for integration into the environment.

The design of these kinds of worlds and even less ambitious works such as hypertext documents is not well understood. One critic, Howard Strauss of Princeton University, argues that WWW content designers need to relearn some old lessons of scholarship: "In the past we learned how to use footnotes, tables of contents, and indexes effectively, but in our electronic formats we seem to have forgotten all that. We use too many hypertext links, use them where they make no sense, ignore the difference between footnotes and tables of contents, build links to bizarre and unexpected places, ignore standard ways of linking, and confuse, rather than enlighten, with hypertext structures that make bowls of spaghetti seem like models of good organization."[58]

Video Games

Technological advancements are rapidly changing this $5 billion industry. Full integration of 3D, 2D, and full motion video is now possible, so video games are increasingly distributed on CD-ROM. Today, some gamers play on their PCs, although the majority still use dedicated, standalone video game machines that accept only products manufactured for that platform.

The original popularizer of video games was the Atari company, which launched "Pong" in 1972, and has seen four platform revolutions since then. In 1975, Atari's home game system brought video games into households.

The race for ever more powerful processing began when Nintendo marketed its 8-bit home player 9 years later in 1984. (The n-bit system describes the amount of data a processor can address at one time. The larger the number, the greater the amount of data it handles. This capacity is important for graphics; the higher the processing rate, the more complex and rich the system can produce in the same amount of time.)

Sega introduced the 16-bit "Genesis" player in 1986 and in 1993 many more companies made plans to enter the business with 32-bit and 64-bit systems. Atari created a standard for cartridge-based software but the company that built a truly mass market was Nintendo. The company's success was based on cheap platforms, engaging software, and ease of use. About 40 million U.S. households own a Nintendo game system.[59]

These 40 million game players, in addition to the owners of the other platforms, play video games on the TV. By contrast, only about 5 million play video games on their computers and about 40% of computer households also have dedicated game players. They're convenient and allow multi-viewers and players to participate because the TV screen is larger than that of the computer and is placed in a central location. Figure 9.6 shows a virtual reality game with its own headset.

The most popular games are "twitch" games that require hand/eye coordination and fast reflexes and center around speed and com-

Figure 9.6 Atari's virtual reality video game. Source: Atari.

bat, exercising a strong appeal to young boys. One reason many analysts believe it will take a few years for PCs to compete with dedicated game players is that access time for CD-ROMs is too slow to satisfy these speed demons. The cartridge-based games are able to respond more quickly. However, Pentium-based PCs are about to give game players a run for their money. Windows '95 has a friendly interface, 32 bit-capacity, and a standardized format.

In the video game market, content and platform are intimately related. Historically, a popular game entices consumers to buy the needed platform. Nintendo had Super Mario Brothers and then Donkey Kong Country. Sega Genesis sold into homes with Sonic the Hedgehog.[60]

The marketing of video games has developed rapidly. Most are sold in mainstream discount stores: Wal-Mart, Toys-R-Us, K-Mart, and Target (in that order) are the most popular sources. Racks of games like those found in Figure 9.7 are found on the floors of the discounters. By contrast, video games on CD-ROM have had difficulty finding distribution. Discount stores stock very few CD-ROMs, so computer stores remain the major outlets.

One problem encountered in video games is violence, particularly when that violence is directed against women. In "Harvester" by

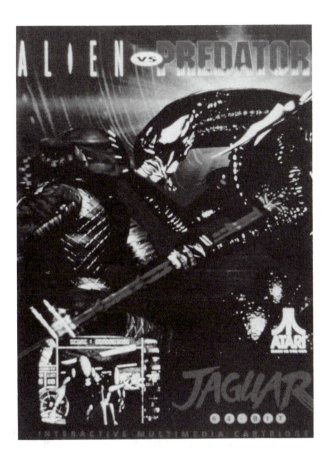

Figure 9.7 Atari's "Alien," a very popular video game. Source: Atari.

Merit Studio, players slice and dice their enemies with a chainsaw. In "Death Mask" by Electric Dreams, the climax of the game is shooting the head off the evil femme adversary. In "The Great Game," the player is a CIA agent who must torture an Asian female spy to get information about a plot to assassinate the President.

Research on the effects of such violence suggest the nature of video gaming makes gory depictions especially powerful. Brian Stonehill, director of media studies at Pomona College in Claremont notes: "People have long been put in the position of spectator at violent events. This takes you out of the role of spectator and into the role of murderer. And that's not just a little change, it's a big change."[61]

Threatened with parental outrage and the likelihood of congressional action, the Recreational Software Advisory Council Association

(RSAC) and the Computer Games Work Group joined to develop a rating system for video games. They established a process that required game developers to submit storyboards or video of the most extreme scenes before distribution. The RSAC defines blood and gore as: " . . . depiction of a great quantity of a sentient being's blood or what a reasonable person would consider to be vital body fluids, or depiction of innards, organs, and or dismembered body parts. The depiction of blood or vital body fluids must be shown as flowing, spurting, flying, or collecting in large amount of pools. To be classified as Blood and Gore, there must be more than just simple dismemberment; the dismemberment must be accompanied by tendons, veins, bones, muscles, etc." For example, "A sentient being is thrown in a tree chopper and is spewed out as hamburger." Whew![62]

So far, the Entertainment Software Ratings Board has rated 500 games. The classification scheme puts products into one of five age groups, graded with labels such as "mild animated violence," "realistic violence," "animated blood and gore," and "realistic blood and gore." The group also distinguishes between passive sequences, where the player watches someone else commit violence, and active aggression, where the player engages in violent behavior.

Technology platforms and current products

As mentioned earlier, video game platforms are described in terms of their processing power. Nintendo dominated the market with their 8-bit player until Sega came out with its 16-bit system two years later. Sega's move forced Nintendo to upgrade its platform to match the competition and the two companies have accounted for 99% of the 16-bit market until just a few years ago.

By 1994 the new generation of players were 32- and 64-bit and there were 5 competing platforms: Nintendo, Atari, Sega, Sony, 3DO, as well as CD-I and multimedia PCs and Macs. (This fragmentation contrasts starkly with the up-and-coming competitor Windows '95-equipped PCs.) Table 9.2 shows the 5 dedicated game players. This proliferation of platforms is expected to re-align the market shares once again, with Sony, Sega, and 3DO expected to be the big winners.[63] In the fourth quarter of 1995, Sony Play Station grabbed a 13% market share; this was double the share of its nearest competitor, the Sega Saturn.

The most popular video game titles as of May, 1995, were "Gex" (3DO), "Coach K Basketball" (Sega), "Road Rash 3" (Sega), "True Lies" (Nintendo), "The Need for Speed" (3DO), "World Series Baseball '95"

Table 9.2 Next Generation Video Game Platforms

Manufacturer/Machine	*# Games*	*Delivery Date*	*Price*
Nintendo Ultra 64	6	April, 1996	$250
Sega Saturn	20	Fall, 1995	$350–450
Sony PlayStations	20	Fall, 1995	$300–450
3DO M2	186	Early 1996	$400
Atari Jaguar	24	Late 1994	$149–159

(Sega), "Donkey Kong Country" (Nintendo), "NBA Jam" (Nintendo, Sega), and "Knuckles Chaotix" (Sega).

Coming up fast are a raft of sports-based games from Sega like "Virtual Fighter," "Daytona USA," and "NHL All-Star Hockey," and "Ridge Racer" for the Sony and 3DO platforms. One interesting development is the emergence of games based on films. The "Alien Trilogy" lets game players take the point-of-view of Sigourney Weaver's character. "Johnny Mnemonic" began as a William Gibson short story that became both a film and a video game. Both the "Alien Trilogy" and "Johnny Mnemonic" feature high-quality video, so are released on CD-ROM. However, "Judge Dredd," released at the same time as the Sylvester Stallone movie, was distributed on Sega and Nintendo platforms as well as on multimedia PC.

The above list makes it clear there is little video game product for females. Several companies, including American Laser Games, have titles in development to appeal to what will be a growing market segment.

In the trenches

Developing games requires designers who can enter the mind of 8–15 year old boys—and, increasingly, their fathers. The only popular game that appeals to both genders is "Tetris," whose players are equally divided between males and females. The emphasis on men means that games center around sports and physical prowess, a characteristic that extends the life of dedicated game players that can deliver the necessary speedy responses.

Developers use three platforms in game design: Mac, PC, and Silicon Graphics (SGI) machines. Their software includes Autodesk "3D Studio," Adobe "Photoshop," Macromedia "Director," Hash,

Inc. "PlayMation," Alias "PowerPlay," Wavefront "GameWare," and many others. The language of game creation centers around the number of "polygons," the building blocks of forms, and "texture-mapping," the surface appearance of the forms.

Production, especially when it includes video sequences, is different from traditional shooting and involves some adjustment on the part of actors and crew. Much shooting takes place on "blue screen" stages where the actors do their scenes in front of a special blue background. Computers throw out the blue pixels and replace it with any other pixels the operator calls up, so designers use it to place actors against computer-generated digital backgrounds that fit the theme of the game.

The cost of digital production is far less than building sets but the lack of props and scenery makes it quite difficult for the actors. They inadvertently walk through walls, off balconies, or grab a gun from mid-air, forcing re-shoots. Many lines are spoken directly to the camera, so actors may have to spend days delivering one-liners into a void. According to Robert Weaver, president of InterWeave, "producing games with video is the hardest thing I've done. It's so detail-oriented and the planning—it incorporates so many different disciplines."

Institutional Versus Independent Creative Work

The settings for creating content vary considerably. Mega-alliances between industry giants establish huge creative departments, while "garage band programmers" work in basements and garages. Indeed, media giants may not provide the most nourishing environment for creative talent.

Pulitzer-winning journalist Jonathan Freedman recounts that his last book was published by Atheneum, a small quality publishing concern. While he was writing the book in 1990, Atheneum merged with Scribners. The joined firm was then purchased by Macmillan. In 1991, Macmillan was sold to Simon & Schuster. Simon & Schuster was a previous acquisition of Paramount. So in 1994, when Viacom bought Paramount, it got Simon & Schuster along with the deal—and Atheneum was put out of business by its corporate master.

Freedman wrote in a column entitled "Fried green writers at the Viacom Cafe": "Megamergers impoverish American culture and thwart the search for truth. The billions of dollars invested in acquisition, debt service and technology reduce the money available to introduce new writers and subsidize provocative, if not wildly popular, nonfiction books. The demand for entertainment puts dollars above sense; diversity is sacrificed to the vast bland middle market."[64]

Marianne Paskowsky of "Multichannel News" finds that many creative people working in the new media departments of large companies are "downright miserable."[65] She wonders if creatives can flourish in the bottom-line oriented, risk-averse, innovation-avoiding corporate atmosphere of media conglomerates. Headhunter Michele James, of The Accord Group, who places new media creatives, believes that the people still working in their garages and basements are the happiest, while their corporate-cuffed colleagues languish.

In spite of the difficulties, some companies have invested millions to create new media. Bell Atlantic Video Services Co. (BVS) built an 8,000 square foot facility in Reston, VA that will be used to digitize material and to produce new digital content. In keeping with the culture of new media (and unlike that of the parent phone company), the Digital Production Center encourages the loose dress and behavior codes that characterize young techs and designers.[66]

Digital Equipment Corporation is another company investing in a facility for content creation. DEC's Digital Media Studio in Tarrytown, New York, just north of New York City, provides the resources required to design, develop, and release interactive applications for distribution over broadband networks. Development and design services will range from storyboarding through digital encoding and conversion.[67]

Three companies have taken another approach to content development, providing creatives with tools, rather than controlling the location and product of creation. Pacific Bell, TRW, and the Silicon Graphics Studio provide 2-way broadband networks to link creative people working in different sites. PacBell's Media Park network allows creatives to access stock images, video, music, talent, and model databases. It also allows transmission of film and video for remote screenings and editing.

PacBell also set up a link between Paramount Pictures and Kaleidoscope, a production company that specializes in creating movie trailers. This first service led to the connection of other studios so that today Imagine Entertainment, Twentieth Century Fox Studios, Warner Brothers Studios, Amblin Entertainment, Walt Disney Studios, ad agencies, and subcontractors are all on PacBell's fiber freeway.[68]

Silicon Graphics Studio links New York and Los Angeles creative studios, companies, and "boutiques," small specialty design and production groups. The link allows high-speed transfer of graphics, shared whiteboard and animation, including CMe-SeeU video.

The TRW HollyNet lets producers call up photos of locations, talent, wardrobe, and props. It also allows high-speed transfer of 30-frame per second video and near-film resolution.

The proliferation of fiber networks opens up the potential for distributed work groups, including partnerships between giant media conglomerates and highly creative, innovative boutiques. The creative explosion is documented by Dataquest, a consulting firm, reporting that there are more than 20,000 multimedia development companies, most of them in the U.S., and only half of them earning more than $100,000.[69]

Independents are inspired by the success of creatives working on their own, such as Robyn and Rand Miller, the two brothers in Washington State who developed the smash hit, "Myst," the 4-guys-in-a-loft behind Cyberflix, Steve Nelson of Brilliant Media, Greg Roach of HyperBole Studios, and others. There are thousands of would-be CD-ROM producers sweating in their garages, making what they hope will be the next hit title.

Throughout 1994 and 1995, many of these creative people and the small, specialized boutiques shops that produce interactive content were snapped up by large media companies. The smell of success has lured Hollywood talent agencies to search for new creatives, as well as towards their current clients who work in traditional media. Creative Artists' Agency, International Creative Management, the William Morris Agency, United Talent Agency, and Allied Performing Arts have all set up new media departments to package creative teams for publishing companies and conduit-providers.[70]

Interactive Storytelling

When asked what they believe interactive storytelling is, many people respond by saying that it lets the audience pick the end of the story. Today's developers have a far more sophisticated view based on the idea that endings are generated from actions. So it is by virtue of taking action that people affect the outcomes of the story. The audience doesn't so much view, read, play, or choose the story—they participate in it.

David Riordan, senior producer/creative director for POV Digital Entertainment, refers to it as the creation of a "variable state environment." Says Riordan: "The model sets up a variety of events which can occur in an environment and then the environment molds and changes to accommodate what the user is doing."[71] He tells storytellers to begin by setting out their themes, the logic of the created universe, and the traits and attributes of the characters.

The second stage, counsels Riordan, is logistical. The writer must lay out what the user can do, where they can do it, and how abilities and resources they acquire in one place can be used in another environment. In addition, he or she must specify what the player needs to do to get from place A to place B and what will happen if they leave without doing what they need to do.

The storyteller creates a design document which defines the graphic design and any sets that might be involved. The script is written last; Riordan says scripts run about 150 pages of dialog. He recommends economy of dialog, with attention given to how it is accessed and how it changes as the story progresses. He warns writers to allow players to bypass dialog if they choose since discs are experienced many times.

A more activist approach is suggested by Mikki Halpin: "Checklist of actions to take now: get an Internet address and use it. On the Net, read up on all the latest MPEG news and learn how soon you can use networking capabilities to post your film to various sites, broadcast it when you desire and create multimedia Mosaic or World Wide Web sites for projects less linear. Trade equipment and software to lower costs. Call Kodak and ask when they will offer one-hour CD-ROM pressing services. Use e-mail capabilities to lobby for equipment to be made available for mass use, not just in educational institutions."[72]

The Creative Process

"Where were you when the paper was blank?" is the familiar taunt of writers, pushed off the production of their own scripts in the film and television industries. No matter what arena they work in, the creative team faces the intimidating vista of—nothing. It is their job to bring into being something that has never before existed in this precise incarnation. The creative process itself is a sometimes agonizing alternation of expansion and boundary-setting, flourishing and limitation, growth and pruning. These stages are shown in Table 9.3.

Table 9.3 Steps in the Creative Process

Step	Task
Idea generation	Brainstorm, either individually or in teams, to create as many ideas as possible to select from.
Development	Take one or more ideas and flesh them out to make preliminary decisions about: platform, format, intended audience, subject matter, characters, plot, interactive structure, purpose, overall look and feel, budget, schedules, type of assets (text, graphics, photos, animation, FMV, film, narration, sound effects, music, dialog).
Design	Select one idea, per assignment. Put together necessary tools, depending on final platform. Assemble information about target consumers. Assign writer. Design navigation and interactive sequences. Storyboard key frames for all scenes, graphics, photos, and visual fields. Create schedules. Hire graphics, programming, music, production, writers, and content experts and, where necessary, get them started working.
Testing 1	Test a prototype version with a focus group of individuals similar in characteristics to intended audience. Execute other tests: User testing—how people use it; unit testing—how parts of program work; integration testing—how whole program works together.
Pre-production	Approve script. Make preparations for film, video, and photographic shoots. Cast all performers. Find field and studio locations. Order all equipment. Hire make-up, wardrobe, set design and construction, and prop people. Animation: Approve key frames. Graphics: Select style and text fonts and colors. Music: Approve composition. Sign contracts and firm up schedules.
Production	Approve graphics. Record music, narration, and sound effects. Videotape and/or film scenes with actors. Shoot photographs. Build animated scenes.

Table 9.3 *Continued*

Step	Task
Post-production	Edit film and video sequences. Mix and equalize final music tracks, sound effects, and narration (separately). Add special effects to film, video, and graphics. Takes about twice as long as conventional post-production.
Integration	Use authoring program to put all assets together into a single program.
Testing 2	Pre-master a "one-off" version of the program. Conduct alpha testing—in-house; beta testing—send out to potential customers, but give tech support; media testing—how program works on planned medium; stress testing—how platform works; configuration testing—how program works with the platform; content testing—content experts confirm accuracy of assets.
Mastering and duplication (for CD-ROM)	Create a master tape or disc. If a CD-ROM, first copy is nickel-coated, called the "father," used to create several "mothers." Mothers are used to make negative masters, "sons," that are the stampers for positive, mass-duplicated CD-ROMs. Mold replicas for distribution.
Assembly	Add liner notes to final medium. Package disc or diskette, add inserts, manuals, stickers, boxes, or registration cards. Shrink wrap or seal.
Marketing, release, and distribution	The product is ready. The execution of the final steps to bring it to market falls under the overall business plan. Chapter 11 will cover how companies conceptualize their business opportunities and plans.

Summing Up: Differences between Creating Linear and Nonlinear Products

As processing power and bandwidth increases, the creative process for interactive content is ever more complex, more expensive, and more demanding of people and equipment than it is for traditional programming. Even more important, nonlinear works transform an audience from viewers into participants and users and they accomplish this by allowing their actions to matter, to affect the outcome.

Creatives working in interactive media design both environments and characters to respond to participants and make the audience part of the story. This attitude of listening and making room for others is not the stance of the creators in the television and film industries and many doubt they will be able to make a successful transition from linear content to nonlinear content. The future is likely to: (pick one)

a. include you and me as garage band programmers;
b. keep TV and film creatives in caviar and cashmere for the rest of their lives;
c. demand the same quality, time, and spending that all fine art calls for;
d. replace existing linear media and cause books to fall into disuse;
e. add new media to the existing ones and cause a renaissance of reading.

Notes

1. "New cable channels," *Broadcasting & Cable* (May 8, 1995):28–30.
2. "1990: pivotal year for DBS hopefuls," *Multichannel News* (January 7, 1991):4.
3. John Lippman, "Networks push for cheaper shows," *Los Angeles Times* (February 19, 1991):D-1.
4. Michael Schrage, "Humble pie: Japanese food for thought," *Los Angeles Times* (November 4, 1994):D-1, 4.
5. Vince Vittore, "They'll like it! They'll really like it!" *Supercomm '95 Show Daily* (March 22, 1995):19.
6. Vincent Grosso, telephone interview with author, April 1994.
7. IT Network press release. "ASI to offer interactive television service in United States through new Cableshare licensing agreement," June 15, 1992. IT Network, 8140 Walnut Hill Lane, Suite 1000, Dallas, TX 75231.
8. Mark Berniker, "Interaxx plans trials in Florida, Washington," (May 7, 1994):29.
9. James Careless, "Interactivity in the here and now," *TV Technology* (March 1995):11.
10. Mark Berniker, "'Star cam,' player stats among ACTV sports features," *Broadcasting & Cable* (October 31, 1994):21.

11. "DEC, SGI court content developers," *On Demand* (December 1994/January 1995):36.

12. Christine Blank, "ShopperVision starts grocery shoot," *On Demand* (December 1994/January 1995):42.

13. G. Robinson, "Yo! MTV . . . shops?" *Los Angeles Times* (October 27, 1994):E1, E6.

14. Game Show Network advertisement, "Game Show Network goes interactive at NCTA," *NCTA Show Daily* (May 19, 1995).

15. TV Answer press release. "Sports Ticker and TV Answer team up to bring sports information to interactive television," June 9, 1993. TV Answer, 1941 Roland Clarke Place, Reston VA, 22091.

16. Mark Berniker, "ABC signs NTN to create interactive services," *Broadcasting & Cable* (October 24, 1994):30.

17. *NCTA Show Daily*, op. cit.

18. David Dolnick, "Digital game delivery systems," *Broadband Systems & Design* (June 1995):23–24.

19. Peter Lambert, "All the news that's fit to digitize," *On Demand* (December 1994/January 1995):44.

20. Peter Lambert, "Interactive TV content moves to on-deck circle," *On Demand* (June 1995):4–8, 14

21. John Vorhaus, "Two way tube talk," *Los Angeles Times Magazine, L.A. Times* (February 16, 1995):14.

22. Harry Jessell, "Level playing field on program guides," *Broadcasting & Cable* (April 10, 1995):58.

23. Joy Mountford, "Essential interface design," *Interactivity* (May/June 1995):60–64.

24. Steven Schlossstein, "Intelligent user interface design for interactive television applications," *1995 NCTA Technical Papers:*165–170.

25. Erik Holsinger, "Kai's Power Tools," *Digital Video (DV)* (April 1995):37–40.

26. Andy Marx, "Where have all the numbers gone?" *Inter@ctive Week* (January 16, 1995):20.

27. Mark Berniker, "GTE Interactive playing games," *Broadcasting & Cable* (June 27, 1994):33–34.

28. David Colker, "Cyber stars of the next frontier," *Los Angeles Times, Calendar Section* (December 8, 1993):4–5, 82–86.

29. Mark Magel, "Friendly, inviting, informative: building a successful kiosk," *Desktop Video World* (March 1995):68–71.

30. Richard Doherty, "Ultimate CD? The digital video disc," *Digital Video (DV)* (May 1995):74–77.

31. Frederic E. Davis, "A CD-R in every studio, a CD-ROM on every desktop," *Multimedia Producer* (April 1995):39–45.

32. James A. McConville, "Taking the CD-ROM plunge," *Computer Merchandising* (June 1994):34.

33. David Colker, "Evolution Revolution," *Los Angeles Times* (May 13, 1995):F1, F14.

34. Mark Berniker, "Preiss bringing TV shows to CD-ROM," *Broadcasting & Cable* (February 13, 1995):28.

35. Eric Brown, "The Edutainers: thrills for skills," *NewMedia* (December 1994):50–56.

36. "Reader's Digest books in multimedia format," *Atlanta Journal-Constitution* (September 22, 1994):K-2.

37. Caitlin Buchman, "Back to the future: the art of interactive storytelling," *FilmMaker* (Summer 1994):34–39.

38. David Kline, "Interview: Jeff Berg," *WiReD* (March 1994):99

39. Robert Weaver made these remarks at a one-day conference, "Your future in multimedia: Interactive Hollywood, new media, new jobs, new markets," organized by Stuart Fox for the Academy of Television Arts & Sciences and the International Interactive Communications Society, North Hollywood, CA, Saturday, July 22, 1995.

40. Amy Harmon, "Software giants Broderbund, Learning Co. agree to merge," *Los Angeles Times* (August 1, 1995):D-2, D-11.

41. George A. Alexander, Seybold Report on Desktop Publishing 8:8 (April 4, 1995):3–5.

42. "A virtual Library of Congress," *New York Times* (September 12, 1994):B-1.

43. Mitch Ratcliffe, "Real Progress: the Internet as information utility," *Digital Media* vol. 4, no. 12 (May 10, 1995):19–22.

44. Julie Pitta, "Sun to re-emphasize its core business," *Los Angeles Times* (November 6, 1995):D6.

45. Papyrus Design Group press release. "Papyrus announces new multiplayer online service, code named 'Hawaii'." PDG, 35 Medford St., Somerville, MA, 617/868-5440

46. Michael Schrage, "Why Sonic the Hedgehog needs to jump onto the Info Highway," *Los Angeles Times* (November 3, 1994):D1–D3.

47. Joan Van Tassel, "The WWWorld's Fair," *WiReD* vol. 3, no. 8 (August 1995):43.

48. Marie D'Amico, "Home music network," *Digital Media* vol. 4, no. 2 (January 1995):18–19.

49. Mark Berniker, "NBC Desktop Video to deliver news to PCs," *Broadcasting & Cable* (July 8, 1994):26.

50. Matt Stump, "Program nets go interactive," *On Demand* (March 1995):24–29.

51. Tim Clark, "TV Channels go online," *Inter@ctive Week* (September 1994):10.

52. "Time Inc. says it will launch 3 magazines on CompuServe," *Information & Interactive Services Report* (October 7, 1994):6.

53. Chris McConnell, "On-line services blossoming," *Broadcasting & Cable* (June 6, 1994):60.

54. Tom Lehman, "Doing business on the Internet," *NTQ* (1st Quarter, 1995):46–54.

55. George White, "On-line mice aren't stirring," *Los Angeles Times* (November 25, 1994):D1, D2.

56. Karen Kaplan, "Banks seek to branch into homes," *Los Angeles Times* (June 14, 1995):D-4, D-7.

57. Ken Locker, personal interview with author, May 1995, Beverly Hills, CA.

58. "Hypertext bowls of spaghetti," *EduTech Report* (May 1995):1.

59. "Why Nintendo is not going away . . . (anytime soon)," *DFC Interactive* (April 1995):1–3.

60. Julie Pitta, "Let the video games begin," *Los Angeles Times* (May 9, 1995):D1, D5.

61. Amy Harmon, "Fun and games—and gore," *Los Angeles Times* (May 12, 1995):A1, A22, A29.

62. "Blood and Gore," *Harper's* (December 1994):18.

63. Louise Yarnall, "Cyber-cinema," *Los Angeles Times, Westside Magazine* (February 13, 1995):12–14.

64. Jonathan Freedman, "Fried green writers at the Viacom Cafe," *Los Angeles Times* (May 14, 1994):F2.

65. Marianne Paskowski, "New media blues," *Multichannel News* (June 12, 1995):50.

66. Peter Lambert, "BVS' video vision," *On Demand* (December 1994/January 1995):6–14, 50–51.

67. Digital Equipment Corp. press release, "Digital announced three-tiered strategy for Digital Media Studio interactive content and application development business," June 5, 1995, contact (508) 841-2609.

68. Andy Marx, "PacBell and Hollywood's fiber freeways," *Inter@ctive Week* (November 7, 1994):25.

69. Amy Harmon, "Joining the multimedia Gold Rush," *Los Angeles Times* (September 30, 1994):A1, A24.

70. Andy Marx, "Agencies bank on brave new business," *Inter@active Week* (November 7, 1994):78.

71. Buchman, op. cit., 34–39.

72. Mikki Halpin, "Detours to Utopia," *FilmMaker* (Summer 1994):40–41.

10

Trials . . . Triumphs and Tribulations

Introduction

When Alexander Graham Bell invented the telephone, a leading journal of the day noted that it had no practical application. The inventors of radio believed it would be used as a way of broadcasting private messages by average people, slipping into a booth to invite their friends over for dinner. The company that invented the videotape machine asked consumers, "How would you like to record television shows?" and decided not to manufacture it when consumers answered, "Why would I want to record that junk?" Japanese manufacturers asked the correct question, "Would you like to be able to watch current movies at home, anytime you want to?" and went on to make millions.

Similarly, American companies predicted no future for the fax machine; today, the Japanese are the major manufacturers of this successful technology. People believed that a handful of computers would satisfy the world's demand; observers didn't think there would be any demand at all for copying machines.[1] On the other hand, the failed PicturePhone, teletext services CAV/CLV laserdiscs, and 8-track audio cassette technologies were all launched in a rosy glow of optimistic forecasts.[2]

The above examples illustrate the difficulties in estimating the demand for and use of communication technologies. Virtually every major company involved in developing communication products has experienced the consequences of incorrect predictions.

To forecast as accurately as possible, companies conduct extensive testing. Avoiding the high cost of actually producing and launch-

281

ing a product, only to fail in the end, provides substantial incentive for such tests despite the fact that they are time-consuming and expensive.

Building a test site costs much more than constructing the final system. One reason it is so expensive is that there are no economies of scale. Economies of scale occur because prices for large amounts of materials are lower, labor can be used more effectively, and processes can be streamlined for efficiency. Another cause is that a test system is likely to have a unique configuration—or perhaps even several testable configurations. In spite of the high costs, many companies are spending huge sums to conduct such tests. The next section explores the reasons they have been willing to incur the expense.

The Goals of Testing

Building and testing interactive television is not for the faint of heart or the lean of budget. Given the level of uncertainty surrounding advanced television systems, the goal of every type of testing must be to learn. The education process begins with the technology—its very complexity makes imperative the construction of a small-scale version.

The four main aims are: 1) to understand the technical aspects of the proposed system; 2) to establish working relationships with a team of vendors and suppliers; 3) to gain experience with all aspects of the business, gaining economies of scale and scope; and 4) to develop a viable business model for the offered services. Companies test on their own, independent of competitors in their own industry, because they want proprietary data and also because it allows them to develop their own relationships and expertise.

So far, testing has concentrated on the technical aspects. Just building advanced television networks has proved extremely problematic. Technical difficulties accounted for delays of Time Warner's site in Orlando, FL, U.S. West and Cox testbeds in Omaha, NE, and Viacom's trial in Castro Valley, CA. In Rochester, NY, the VOD test of Rochester Telephone and USA Video crashed one week after launch due to too much information being pumped over the network—it took a week to get it back online.[3]

Aside from the purely technological knowledge, trials allow sponsors and vendors to learn the equipment, people, and processes involved. Of these, sometimes the human relationships can make the most difference. Correcting incompatibilities takes time and effort and can suck the profit out of a supplier's participation. There is often much mutual finger-pointing to avoid both blame and correction costs unless a real commitment to the project goals exists. A notable example of such loyalty is Silicon Graphics, which provided the servers for Time Warner's test in Orlando. SGI modified its hardware and software to meet the evolving technical requirements, acting as a constructive partner throughout the process.

Another important advantage of testing is building a learning curve. Participating in a trial gives the sponsoring company knowledge about the demands of the new technology and offered services. Organizations create repetitive operating cycles to cope with uncertainty.[4] Test trials allow people to anticipate the work and to design systems that will enable them to recognize and handle recurring situations when the actual system is in place. They also highlight unusual, one-of-a-kind problems so the organization develops procedures for responding to unique events.

Testing allows organizations from different industries to learn special skills they haven't previously performed. For example, cable companies have no experience with transactional billing or working with such detailed customer accounts. Telephone companies have extensive systems for customer accounts but lack entertainment expertise. Power companies provide services their customers need but know less about marketing to fulfill people's wants and desires.

Once the technical side of testing allows delivery, the focus shifts to acquiring the information required to develop a viable business model. At the heart of the issue is consumer demand: Who uses which products at what price and what marketing strategies will maximize sales?

It sounds easy enough, but this simple proposition opens up an almost endless number of questions and data streams to answer them. For example, there may be an enormous difference between the products and services that attract subscribers to sign up initially and the ones they will actually use and pay for.

Finding the right offering is quite challenging, too. Products and services must be packaged into various bundles and priced at different levels; when companies plan many offerings, designing the packages is extremely complex. While a company will fail through overpricing, underpricing will lengthen the time to profitability.

Part of this effort involves learning how consumers behave when they use the system. Do they use it several times a day or sign on and stay on? Do they use many or just a few services? What determines which services consumers choose? How can consumers be aggregated and how should appeals be structured to reach the various groups.

Testing will answer many questions, but not all of them. Even after the trial, there will be the uncertainty caused by the "law of large numbers," additional challenges simply due to expansion.

Types of Testing

Hardware and software—conduit and content—meet for the first time in the test setting. Broadly, there are four types of tests: Technical, use, market, and diffusion tests that take place in several settings.

Trials begin in the laboratory as demonstrations, alpha tests. Once the equipment runs reliably and has a comprehensible user interface, then researchers can bring in consumers to see how they react. The test then moves to the field with heavy technical and service support, called beta sites. Usually the programs and services are available at no charge or at some minimum cost to the consumer. Later, more realistic conditions are phased in—support is scaled back and more appropriate price tags are placed on products and services. Finally, conditions are systematically manipulated to learn how consumers will act under alternative real-world scenarios. At each stage, more and more subjects are required to gather reliable data.

Testing usually begins with the technical aspects, just to get the system up and running, capable of carrying content. The technical issues will be of interest through the length of the trial. Thus, the different aspects of the tests—technical, usage, market, and diffusion—run concurrently rather than progressively.

Phase 1: Technical tests

Ideally, technical tests begin as laboratory demonstrations and proceed from a small number of installations to more and more individual reception sites. The problems experienced by the Rochester Telephone Company/USA Video trial is an example of where prematurely introducing complex technology to customers can cause acute system failure.

It is important to distinguish between the "failure" of the software and hardware and the "learning success" of the test. Equipment failure is always a very real possibility in a technology trial and it can be extremely useful. The search for the causes and corrections is a vital learning experience for everyone involved. Indeed, many technical issues, such as system integration and maintenance, can only be tested under real-world conditions where engineers can observe scale effects.

Questions in this phase include: Will the technology work? How much will it cost? What are the potential economies of scale and scope? Are the components compatible? What maintenance will be required? Are there network level effects? Will heavy use crash the system? How reliable and durable is the hardware? Do the vendors stand behind their systems? Do they provide sufficient support? Will the equipment intended for consumers pass the "pablum test?" (That is, will the technology continue to operate when the baby spits up on it!).

Phase 2: Ergonomics and use tests

In tests of "ergonomics," sometimes called "human factors research," researchers seek to learn how people interact with the equipment. Like technical tests, they begin under controlled conditions in the lab and then move into the home or business site. Micro level details, such as the location of the control knobs, the size and weight of the remote control device, and the placement of the settop box are all examined. Special needs of children and seniors may come into play, as well. Increasingly, integrating different services into the same piece of equipment becomes an issue. For example, the universal remote has gained popularity to replace the annoying multiplicity of remotes for the TV, VCR, and CD player.

Use testing is a broad category of testing which includes how people incorporate both equipment and programs and services into their lifestyles. In the laboratory, one research method is to ask subscribers to try out the equipment and access one or more programs or services, individually. Often they are observed and videotaped through a one-way mirror.

Another research method is the focus group. Here, the people who tried out the system meet in a group of between 8 and 15 people. A discussion leader conducts a session to get the group members to talk about their impressions of the equipment and planned programs and services. The leader must keep one, or just a few members, from dominating the conversation. Usually they are observed and videotaped as well.

Once testing moves into the field and goes into actual subscriber homes, the equipment, programs, and services are given to the customer or provided at a very low cost. This strategy allows usage behavior to emerge naturally, with as few external constraints as possible.

Researchers study the use habits and patterns of behaviors that subscribers engage in, using that information to categorize customers by those behaviors. The first thing most companies find out is that the focus group is not a very accurate predictor of actual use. The "virtuous consumer effect" comes into play, so people over-report their intentions to use the system for education and information and under-report their projected entertainment usage.

In actual practice, consumers segment into populations with patterns of usage. For example, users who use at-home banking and shopping services might be called Utilitarian Time Savers, while those who spend hours on the system playing multi-user, networked, interactive games might be termed Game Players. People who order interactive Spice channel, the Fitness Channel, and gambling could be Hedonists, and people who watch mainly over-the-air and standard cable channels are Traditionalists. Just as the labels Couch Potato, Channel Surfer, Grazer, Zapper, and Zipper apply to viewers of linear TV, interactive TV users also differentiate themselves by their usage patterns.

Research must also examine the relationships between the programs and services subscribers choose. Perhaps the same people who shops and banks would like to regulate their energy usage—the Control Freaks. Maybe the Shop-Till-He-Drops customer would like to sell items as well as buy them. Categorizing these patterns of usage may be even more important than uncovering some single "killer app" in determining the ultimate viability of the business model and the profitability of the system.

Tests of navigation programs, programming, and services—that is, content—offer special challenges to ergonomic and use testing. In the ergonomics area, on-screen text must be attractive, readable, and comprehensible. If there are several pages, designers must build incentives to continue at the end of each page. There must be some rationale for progressing to different levels of the information or service; the controls must be easy to highlight and operate.

Just a few of the questions to be answered in this phase are: How do the subscribers interact with the hardware? the software? How easy is it to learn the system? How easy is it to use? Do people have the right controls? Are the controls comfortable to use? How much interactive control over video content will customers require? Under what conditions do people access the system and does the hardware support actual usage conditions? What are people's cognitive models of how the system works? Are the applications compatible with the proposed cognitive model (Video Mall, Amusement Park, Carousel, etc.)? How do people reinvent the system by using it in novel and unexpected ways? How many applications do subscribers use, how often? What patterns of usage emerge, in terms of time of usage and types of applications accessed? How do patterns of usage vary with age, gender, income, education, and size of family.

Market tests

These tests explore consumer demand under specified conditions. Earlier research has defined the market in terms of consumer attributes and behaviors. In these tests, the goal of the research is to define market segments by appeals, products, and prices. Products are packaged in alternative forms, offered at different price points, and marketed with different sales appeals and promotional strategies.

The usual procedure is to offer alternative product/price packages to different neighborhood nodes within a system or to test homes with different systems. Consumers choose whether to buy and how much to buy, just as they would do if the offerings were available universally. The more reliable the data needs to be, the fewer incentives should be extended. Marketers may actually change the conditions to further refine their findings, but there are limitations on how much and how often they can make changes. However, the participants are aware they are part of a test, so they may be asked to fill out questionnaires projecting what they might do under real-world conditions.

These modifications, dictated by the needs of market testing, will result in changed usage patterns so the use studies must continue. The earlier work provides a baseline from which to evaluate the degree of variation and continuing usage research tells the provider how sensitive consumer behavior is to the packaging, pricing, and promotional elements in the marketing mix.

Interactive TV providers see themselves as adding a tier of digital services to traditional linear television delivery. It is instructive to look at the cable television marketing model to understand why and how the market tests for interactive TV are proceeding.

Originally, cable operators simply provided over-the-air channels through cables to subscribers who were geographically unable to receive those signals. In the late 1970s, they began retransmitting "superstations," the distant signals from far off cities. For example, in 1978, subscribers to the community antenna system in Hailey, Idaho, were receiving WTBS from Atlanta. In the early 1980s, basic cable services, the so-called "foundation services," became available and shortly thereafter, premium cable networks were born.

As these different types of programming emerged, cable operators began "tiering" the offerings, bundling program services together into a single pricing category. Consumers must buy a basic package in order to access higher tiers. Marketing cable was accomplished through "lift," selling the movies on premium channels to get the entire subscription package into the household as well as the premium network itself. Only in 1991, during the Gulf War, was this model superseded when CNN became the killer app drawing in new subscribers.

The research questions in this phase of testing include: What programs and services do consumers want? How much will they pay for them? Should they be tiered, and, if so, how? What are the price points? How elastic are prices? What is the ideal use/price equation? What is the best product mix? Which services have broad appeal? Niche appeal? What non-electronic activities do the services replace or enhance? What are logical market segments? What appeals should be directed to which market segment? How much advertising will customers tolerate?

Phase 4: Diffusion tests

Companies conducting trials usually see this as the final phase of the market test. However, it is conceptually and procedurally different enough from first-phase market testing to warrant a separate consideration. Marketers design diffusion tests to learn how the offerings will fare under real-world conditions. This testing requires a fairly large number of subjects in order to make reliable assessments of responses and buy rates from customers who make up the different market segments. These segments may be defined by demographics; psychographics (attitudes about religion, family, social issues, and politics); lifestyle (interests, hobbies, scheduling); adopter category (innovators, early, middle, or late adopters, and laggards); or patterns of user behavior.[5]

An example of how much effort is expended conducting a diffusion test is offered by Your Choice TV. YCTV is a TV-program-on-demand service that allows viewers to order recent television shows. They tested their product in several markets, altering the product/mix between the different sites.

Here are the steps YCTV takes to implement the test: One week before, they run teaser spots (pre-debut material that hypes the imminent event) on the cable channels where the service will appear. Similar teaser mailers go to each subscriber of the system. The day before the service starts, YCTV holds a press conference to kick off the service and start a buzz (word of mouth communication). The next day, test homes receive a second YCTV direct mail promotion describing the product and the preview channel. To encourage consumer sampling, the package includes a coupon for a free viewing.

A few weeks after launch, YCTV invites both subscribers and nonsubscribers to participate in a focus group. They discuss their responses to the programming and the marketers offer additional promotions to the nonsubscribers. Finally, YCTV sends out a survey to everyone who bought the YCTV service, asking for demographic information, about their viewing patterns, what they liked, didn't like, and how they evaluated the service.[6]

Program testing

There is a symbiotic relationship between the marketing of specific programming and information services and the method of delivery. Companies providing the delivery can't sell their service without content; content providers need to get their products before customers. However, creators of programming and information services want to acquire experience with alternative delivery mechanisms and they want to learn about the demand for their specific offering in the context of competitive products.

Content providers need the same kind of information as conduit providers. They want data about how, how often, why, why not, and when subscribers choose their material. They need to have a basic understanding of market essentials so they can package, price, and present their service to maximize their take-up rates and profits.

The next section will cover trials in the U.S. by organizations in different industries, all considering entering the market as conduit providers for interactive television: cable, telephone, and power utility companies. We will look at what they've done, what they plan to do, results they've released, and conclusions they've reached. Results of content testing will follow under a separate heading.

Testing by Cable Companies

The first round of testing interactive television occurred in the 1970s. None of the testbeds were digital but, to varying degrees, they were interactive. Hi-OVIS in Japan was the most interactive, followed by a system in Biarritz, France. The least interactive was Warner's QUBE's system, in Columbus, Ohio.

Warner-Amex (later Time Warner) has been the historic U.S. leader testing advanced television and, even today, retains their lead with the Full Service Network testbed in Orlando, FL. In 1977, the company constructed the first interactive television site in Columbus, a city that has long been used as a site for consumer product testing by manufacturers of packaged goods. Downstream, it was a 30-channel interactive cable system; the return path was a narrowband response channel used for audience research and polling and ordering pay TV.

Ultimately, QUBE rolled out to become a commercial service, but it was only marginally successful. The real bonanza was that its testing of programming ideas led to the creation of MTV and the Nickelodeon networks.[7] In 1984, QUBE was quietly closed down as Warner Communication turned its attention to more pressing corporate concerns.[8]

In 1978, the Japanese government agency, the Ministry of Posts and Telecommunications, initiated the Hi-OVIS project. The technical test ran from 1978 to 1980, followed by market testing until 1986. Services in the trial included robot-provided video-on-demand, telephony, and local two-way video for such applications as remote attendance at civic meetings. In the mid-1980s, the Hi-OVIS project was terminated and never rolled out to become a commercial operation.[9]

A year later in 1979, the French government launched a wired service in the resort town of Biarritz, providing 15 cable TV channels, 12 stereo audio channels, and video telephony. Subscribers used a terminal that included a camera, monitor, keyboard, and telephone. "The terminal could be used as a video telephone to transmit still or full-motion images. Subscribers could consult on-line image and data banks and access any of the thousands of videotext services provided over the Teletel system . . . Subscribers were charged a monthly tariff for cable service and both a monthly fee and a usage-sensitive rate for switched services."[10]

The results of these trials were not encouraging. While people generally liked the services, none generated the excitement and enthusiasm that could drive their diffusion. Further, there were some barriers. Hype had surrounded the technology (just as it does today) and subscribers were disappointed at the actual implementation of the services. They were also concerned about privacy and security issues.

Current tests

Probably the most important test facility for the cable industry is CableLabs, in Louisville, Colorado. The nonprofit organization is the research arm of cable operators, headed by the much-respected Dr. Richard Green. CableLabs built a facility to test HDTV transmission over cable. A move to Louisville, CO, gives them the room to build a new, expanded test site intended for conducting technical tests. The Cable Test System (CTS) was completed in 1995.

Individual cable companies typically site their testbeds in an existing system and maintain it indefinitely. This strategy is a function of the geographic dispersion of the systems owned by multiple system operators (MSOs). Using that one system, or even part of a system, they rebuild it to meet the requirements of the test procedures. If research shows that a business case can be made for investing in a system-wide rebuild, only then will the cable company implement a given technology throughout the system.

In 1993 and 1994, half a dozen of the largest MSOs planned to conduct tests of interactive television systems. Several of them use Zing technology for its return loop, as shown in Figure 10.1. The two most ambitious tests were Time Warner's 1 GHz system in Orlando, FL, and Viacom's 1 GHz site in Castro Valley, CA. A slightly less ambitious project is the Cox 750 MHz testbed in Omaha, NE.

Table 10.1 provides a listing of many of the cable-company sponsored trials now underway in the U.S.

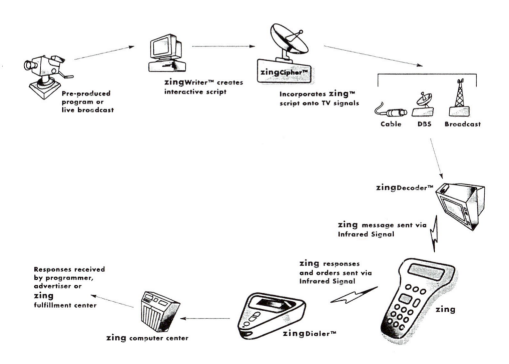

Figure 10.1 Zing Technology provides IATV's "active ingredient." Source: Zing Systems.

Table 10.1 Cable Company Interactive Television Test Sites

Co./Location/HH	Network Description	Type test/offerings
Cablevision/Long Island	HFC/Uses "Cornerstone" from Nortel; DEC equip.	Technical/telephony and high-speed data
CellularVision/ Brighton Beach, NY/ 100 HH	LMDS Wireless	Technical and market/VOD
Continental/Boston MA/12 homes	550 MHz	Tech, use and mkt./ GTE mainSt.; high-speed online access
Continental/Dayton OH	HFC	Market/NVOD and VOD
Continental/Exeter, NH	550 MHz	Market/NVOD
Cox/Hampton Roads VA/48 employee HHs	750 MHz HFC/NorTel "Cornerstone"; fiber ring-in-ring & FTF	Technical/integrated voice; video; data
Jones, Lightwave, MCI/Alexandria, VA	HFC; Sci-Atl "Co-Access" system	Technical, market/ Long distance telephony
Jones, Lightwave, Chicago, IL	HFC; Sci-Atl Co-Access system	Technical, market/ Local and long distance switched phone
InterMedia/Milpitas, CA	HFC/IBM servers; Zenith settops ICTV software	Technical and market/ Shopping; ads; VOD
NewChannels/ Syracuse, NY/apts & businesses	HFC; optical to SONET	Technical & market/ telephony; SONET compatibility
Sammons/Waterbury, CT	HFC/Vela Research principal vendor	Test 30-channel NVOD
TCI/Littleton, CO/ 300HH	Traditional cable system	Market/VOD vs. NVOD

Table 10.1 *Continued*

Co./Location/HH	Network Description	Type test/offerings
TCI/Mt. Prospect, IL/passes 9000HH	Dual 450 MHz systems for 900 MHz; Jerrold settops	PPV, NVOD, dig. svcs.; YCTV; home shopping; TV Guide OS navig.
TCI/Lansing MI	HFC	Technical and use/ Telecommuting and teleradiology (med)
TCI/Microsoft/Seattle WA/2000HH	HFC/Compaq server and settops; Msoft s/w; IA svcs by MedioNet;	Technical and market; Power mgmt; home security; digital services
Times-Mirror-Cox/ Phoenix AZ/12 companies	Metro area fiber network	Technical, use, mkt./ Desktop videoconf.; concurrent CAD; multi-user white-board; telecommuting
Time Warner/ Orlando/4000HH	1 GHz HFC/SGI servers; Sci-Atl settops; H-P printers; Andersen Cons. integrators	Tech, use, mkt. VOD, shopping, ads; dig. svcs., banking; networked games
Time Warner/Queens, NY/pass 10,000HH	HFC	Technical, market/ NVOD
Time Warner/ Rochester, NY	HFC/Uses "Cable-span" from Tellabs as principal vendor	Market/consumer acceptance of telecom svcs. over HFC system
Viacom/Castro Valley, CA	1 GHz/AT&T servers; Jerrold settops GI/Intel cable	Tech. and mkt. Atari 64-bit Jaguar games; online access StarSight navig.

By 1995, the picture had changed. The showcase Viacom trial was canceled and only two aspects of it went forward—a test of the Star-Sight navigation program and high-speed PC connection through a cable modem. Due to the difficulty and expense of building even test networks, the Cox/Omaha and Time Warner/Orlando testbeds were much delayed, but would ultimately go forward as planned.[11]

Focus on interactive television test: The FSN

The Full Service Network is the name given to the Time Warner test site the company committed to build in 1993, shown in Figure 10.2. Originally scheduled to debut in 1994, a series of technical difficulties delayed the launch of the FSN. The digital video servers didn't deliver the number of movies in the volume promised by their manufacturer, Silicon Graphics. Instead of 100 on-demand streams per Challenger server, the actual number proved to be less than ten. The software was much more complex than anyone had imagined. The delivery of the digital settop boxes was delayed into 1996. Media coverage of the de-

Figure 10.2 A screen from the Time Warner testsite in Orlando, FL. Source: Time Warner.

lays was generally spiteful, if not actively vicious, failing entirely to take into account the gap between executives' estimates and engineers' realities.

When completed, the Full Service Network will be a remarkable achievement. It will offer a full range of video-on-demand offerings, interactive shopping, education, banking, and transactions with government agencies, like the Department of Motor Vehicles, municipal agencies, and the library. In its final incarnation, subscribers will be able to get both wired and wireless telephone service. Orlando businesses, hospitals, and schools will have high-speed, two-way communication facilities to exchange high-resolution video and graphics, as well.[12]

One impact of the complex configuration needed to deliver these sophisticated services is the equally complicated installation. It takes so long, Time Warner sends the members of the household out for a gourmet meal. While the residents are eating, " . . . installers and technicians check the home, install personal identification numbers, install the hardware, collect credit card information, and review the system," said Robert Benya, vice president of marketing for the project.[13]

Time Warner wants to recruit 80% of the homes passed by their cable into the test. To accomplish this ambitious goal, they have launched a massive recruiting effort. They begin with direct mail then follow up with a personal visit and telephone contact. The promotional giveaways and discounts include free movies, games, and advertiser-contributed products. The advertising effort involves promoting the service over attractive on-air "barker channels," that tell potential subscribers what's hot, newspaper ads, and messages over the in-home printer.

Time Warner has paid a steep price for pioneering. It is possible that their complicated design will never diffuse widely and be superseded by simpler, all-fiber systems. Nevertheless, in 1996, Time Warner will know more than any company in the world about how residential customers use an advanced television network.

Results of cable company testing

Since testing is so costly, it is probably not surprising that getting information about results is no easy task. The occasional story in the trade press provides only the bare outline of what researchers are discovering from the interactive television testbeds.

So far, few efforts have moved beyond technical testing, so this is the area where the most has been learned. Cable companies agree on several important technical issues. For example, they believe that the hybrid fiber-coax design, pioneered by Jim Chiddix and Dave Pangrac at Time Warner, is the most advantageous for them given their perspective of system evolution. HFC makes it possible for MSOs buildout from 550 MHz systems, gradually adding capability and services over time. The PC-over-cable rebuilds will probably be cable's first venture into interactivity, simultaneous with providing some relatively simple digital services. Telephony and video will come later. This evolutionary buildout allows cable companies to make incremental investments as they absorb the lessons of two-way communication. There is also general agreement that the MPEG-2 digital compression will be implemented.

Another area where much learning has taken place is with video servers and storage. Designs incorporating massively parallel processing have emerged as superior to any other approach, as companies like Time Warner have had to learn the hard way.

Software has also benefited from testing. MSOs like TCI and Time Warner have used trials to learn how to integrate software functions and how to incorporate dynamic network management, administration, and transactional billing into their software. Cable bills based on subscriber fees cost about 50 cents to process while trisection-based telephone bills cost $1.50 per month, so TCI considers lowering that cost a significant objective.

PC-over-cable trials

With few exceptions, cable companies only reluctantly consider the two-way communication capacity of their physical plant. Even the most sophisticated sites plan a rudimentary return loop that subscribers can use to send back messages, whether to the service provider or to others outside the system. One configuration for PC-over-cable is shown in Figure 10.3. Cable operators are much more comfortable seeing expanded capacity as a way to offer more programming downstream, rather than as a way to provide upstream service.

However, the rocketing popularity of online services, linking home computers through narrowband telephone lines, has caused cable companies to re-think their strategy. Computer users want high-speed capacity and many have the money to pay for it. However, providing PC access over cable requires an expensive rebuild so before proceeding, 23 MSOs are now testing to see if there is sufficient demand to justify the investment.[14]

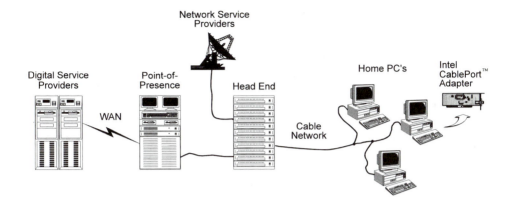

Figure 10.3 Intel's end-to-end architecture for PC networking over cable. Source: Intel Corporation.

Accessing online services and the Internet over cable requires a special modem, called a "cable modem," that runs from 15 times to 400 times faster than conventional modems.[15] They're not cheap, costing between $300 and $1,000. The higher end models will let subscribers download video clips, graphics, and audio nearly in real time.

Cable modems are either "symmetrical," meaning they send and receive at the same rate or they are "asymmetrical," meaning they receive data at a faster rate than they send it. The asymmetrical models are one-half to one-quarter less expensive. Intel, General Instrument, Scientific-Atlanta, and Hewlett-Packard are all developing asymmetrical modems that receive data at 30 Mbps and send it at 1.5 Mbps. Digital Equipment Corp. and LANcity plan symmetrical models that operate at 10 Mbps, while the Zenith modem will run at 4 Mbps.[16]

Each data-over-cable trial differs in the technical configuration and services it offers. The speed of the connection differs, as do the number and kinds of limits, such as the choice of which online service the consumer can access. Table 10.2 lists the announced PC-over-cable tests.

Focus on PC-over-cable: Cox/San Diego   One of the first PC-over-cable tests was undertaken by Cox Cable in northern San Diego, an area populated by computer-sophisticated high tech workers in the knowledge industries. The company found that the best customers for the online access service were those who already used telephone lines for their computer connections.[17]

Table 10.2 Cable Companies Testing PC Access/Cable Modem Tests

Co./Location	Service under test
Cablevision/ Johnson City, SC	Database access with high-speed cable modem
Cablevision/Long Island, Yonkers, NY & site in Mass.	Gauge consumer demand for PC access, using Zenith cable modems MA site uses cable modem for distance educ.
Columbia/ Portland, OR	Cable modem for environmentalist organization to transfer high-speed data
Comcast/Lower Merion, PA	PC network over cable
Continental, Performance Systems, Int'l/ Cambridge, MA	Technical, market, with a few beta testers Connect individuals and businesses with high-speed Internet access Charge $125 to indivs., $400–2,750 to businesses
Continental/sites in New Hampshire, Mass., Florida	Cable modem use for businesses needing high-speed data transfer; New Hampshire site is 300 HH, connected to CompuServe online svc.
Cox/San Diego	150 homes with high-speed access to Prodigy online service
Jones/Denver, CO	Connect homes with high-speed access to Internet, using cable modem
Media General/Fairfax City, VA	Connect homes with high-speed access to Prodigy online service
Scripps-Howard/ NW Georgia	High-speed data over cable: Links physicians at 3 sites for high-res imaging and video-conferencing, via cable modem
TCI/St. Louis, MO	High-speed access for governmental agencies via cable modem
TCI/Lansing, MI	Uses DEC's ChannelWorks to link MSU Radiology Dept. to 67 local medical practitioners and to the Internet
Time Warner/ Hawaii	Remote database access via high-speed cable modem

Table 10.2 *Continued*

Co./Location	Service under test
Time Warner/ Westmont, PA	Technical test with cable modem
Time Warner, Time, Inc./Elmira, NY	Link to Time, Inc.'s Internet "Pathfinder" site, and Gannett-owned Elmira Star Gazette, America Online, and Compuserve. Includes 500 homes, schools, libraries, and government offices. TW charges $14.95 for service, and $9.95 more for high-speed access to Internet.
Viacom/Castro Valley, CA	Connect subscribers to Prodigy online service, using cable modem

The Cox test began in November, 1993 and went to 150 customers, connecting them to the Prodigy online service. With symmetrical Zenith modems they could receive data at 1.5 Mbps. If they use a camera they can send and receive VHS quality video. Cox estimates subscribers will pay between $10 and $15 a month. Customers had three motivations for signing up: data speed, savings, and instant access. Cox plans to learn more about price points and additional service options through further testing.

Telephone Company Trials

Before the breakup of AT&T, telephone testing took place at Bell Laboratories, one of the preeminent research and development centers for communication technologies. AT&T conducted the first foray of telephone companies into a special form of interactive television: video telephony.

Bell Labs began experimenting with sending and receiving images over phone lines in the 1950s and unveiled the new technology at the World's Fair in New York in 1964.[18] AT&T tested the resulting PicturePhone in-house and introduced the product for the business market in the late 1960s. The PicturePhone never caught on. AT&T took it off the market in 1973, after investing between $130 and $500 million.[19]

Current tests

By the 1990s, telephone companies once again set their sights on delivering pictures. However, this time the telcos are on their own, cut loose from Ma Bell since the breakup of AT&T and they took this watchword seriously: test before you invest.

Every major telephone company has engaged in one or more trials. Table 10.3 lists trials underway in 1994 and 1995 sponsored by telephone companies. An example of the type of system the telephone companies are planning is shown in Figure 10.4.

There are some differences between the way cable companies and telephone companies approach testing. Unlike the scattered holdings of MSOs, telephone companies have developed by providing service to geographically-contiguous areas—that's why they're called "Regional Bell Operating Companies," or RBOCs. This growth pattern affected the way the telcos plan testbeds. They wire up a few homes, then rollout service to the neighborhood and region. They may conduct tests at some specified facility but they don't maintain relatively permanent test sites in some group of consumer homes like cable companies have.

Telcos also vary from one another in their approach to testing. Pacific Bell announced they would conduct trials, then canceled them and decided to simply roll out an advanced network throughout California on a commercial basis. Later, they modified even that plan, limiting it to northern California instead of the whole state.[20]

Similarly, Ameritech put together a handful of small-scale, carefully targeted test situations in 1993 and 1994, delivering home shopping to 40 households in Birmingham, Michigan, connecting healthcare facilities in Wisconsin, testing video servers on a few employees in Chicago, trying out distance learning in Ohio, and networking 150 school children in Warren, Michigan. Once Ameritech acquired a feel for these applications, they decided to forego further testing and, like PacBell, will roll out advanced networks without more testing.

By contrast, Bell Atlantic has been very active conducting trials of alternative technologies, including asymmetrical digital subscriber line (ADSL), hybrid fiber-coax (HFC), and fiber-to-the-curb (FTTC). B-A also built a digital production center where they developed the StarGazer navigation system.

In 1995, Bell Atlantic, NYNEX, and Pacific Bell formed an alliance called PlatCo. These three powerful companies provide telephone service to six of the biggest seven markets in the U.S.; only Chicago is missing. The purpose of the super-telco partnership is to standardize hardware for telco advanced networks.

Table 10.3 Telephone Company-Sponsored Trials

Co./Site	Network Description	Type test/Offerings
Ameritech/40 HH, Birmingham, MI	Varies; Vendors include IBM, AT&T, ADC	Use test: home shopping
Health agencies, Wisconsin		Use test: transfer high-res images, med records
125 5th graders, Warren, MI		Use test: "ThinkLink" connect students and teachers
Health agencies, Dayton, OH		Link hospitals for remote diagnosis and video-conferencing
Distance learning, NW Ohio		"SkillLink"—remote learning for workers
Bell Atlantic Video Services/ Reston VA/lab facility	Alternative tech: nCube, DEC, HP servers; Oracle, MicroWare, Apple, and EDS s/w	Technical test and integration lab
Bell Atlantic/Fairfax City, VA; 2000 subs	ADSL; nCube servers; DiviCom settops; Alcatel switches; database access; TV wavelet compr.; Microware software	Technical/market/VOD StarGazer navig.; dig. svcs.; shopping; programs on demand; test pricing, packaging & promo strategy
Bell Atlantic/ Tom's River, NJ	HFC	Tech. mkt./dig. info. svcs; TV on demand
BellSouth/Atlanta, GA (Chamblee)/pass 12,000 HH	Interactive/HP server; Sci-Atl settops; EDS pgm mgmt; Sybase s/w	Tech. and mkt./70 analog chs.; 240 dig. chs./VOD; IA advertising, online access
BellSouth/state of N. Carolina	Fiber/ATM switches	Tech. and use/links 106 schools for distance learning
GTE/Durham, NC	Fiber network	Market/link area businesses
GTE/Manassas, VA/pass 5,000 to get 1,000 subs	HFC	Market and pricing/ VOD; shopping; games, databases

Table 10.3 *Continued*

Co./Site	Network Description	Type test/Offerings
MCI, TASC/6 nodes in U.S./re-cruit companies	Fiber/Nortel ATM switches	Use and mkt./prgmg. user-provided: video conf.; med. imaging
NYNEX	Fiber; NetWindow software	Tech. and mkt./fast-packet network svcs; frame relay, ATM, SMDS
NYNEX/NYC/ 2500HH	SDV-FTTC/DEC server; StorageWorks storage; GIGA switches; Microware software	Mkt./Subs choose Liberty or TW cable; 160 chs.; CitiVision; VOD; databases; IA games; 2-way video; educ.; news on-demand
Pacific Bell/No. Calif./50 com-panies	SDV-FTTC/ISDN; T-1 and ATM	Technical and use/ Media Park; content user-supplied; shared video
Puerto Rico Telco/P.R./ 380HH	TBA	Technical
Rochester Tel./ Rochester NY/ 45 apts.	ADSL/ADS; USA Video servers and settops	Tech., use, mkt/NVOD shopping, music, educ.
Southern New England Tel. Co./ W. Hartford CT/ 1500HH, to 150,000HH	SDV-FTTC; HP servers; AT&T switches; ADC Homeworx; Sybase s/w	Tech., use, mkt/NVOD, local dig. svcs.; shopping
Southwestern Bell/Richardson TX 2500HH to 45000HH	SDV-FTTC/Lockheed; Broadband Tech. integrators; Iceberg software	Tech., mkt., diffusion/ b'cast & cable ch.; IA services; VOD; shopping; polls; advtsg; online
Sprint/Wake Forest, NC/ 1000HH	Fiber network analog	Tech., market/80 chs.; 30 dig.; NVOD; Sega games; online access

Table 10.3 *Continued*

Co./Site	Network Description	Type test/Offerings
Sprint/San Jose 7 companies, 2 univs., 8 elementary	Fiber network/SONET, ATM, SGI servers, 2.5 Gb/s	Technical and use/"Silicon Valley Test Track"—content user-provided; educational video-conferencing; collaborative work; online access
US West/Minnesota/healthcare agencies: Mayo Clinic; Univ. of Minnesota	Fiber network/ATM, ISDN, SMDS; Fujitsu switches; Compass s/w	Technical and use/ provides b'cast quality videoconf., high-res image transmission
US West/ Omaha/2,500HH to 9,000HH to 40,000HH	SDV-FTTC/DEC; Sci-Atl & 3DO set tops; AT&T telephony	Technical, market/76 analog chs.; 800 dig chs.; shopping; VOD; Yellow Pages; U.S. Avenue services

Results of telephone company testing

Important results came from the technical tests conducted by Bell Atlantic, whose experience influenced the network design which would be adopted by most of the telcos. In mid-1995, Bell Atlantic announced that it would deploy switch digital video (SDV) FTTC architecture, rather than the cable-favored HFC. This strategy makes sense for telcos because they are, in essence, starting from scratch while the current broadband configuration of cable companies makes the evolution-enabling HFC design more practical.

Once Bell Atlantic changed their construction plan, U.S. West announced a similar change, followed by Pacific Bell. This standard will be promulgated by PlatCo and its content twin, Tele-TV, whose programming will fully exploit the FTTC structure.

Bell Atlantic also has data on how interactive television must be presented to consumers.[21] For example, they believe that products must bridge from what people already know and are comfortable with, rather than launching headlong into the most advanced possibility. Bell Atlantic tried giving people immediate fast-forward and rewind service—and found they are accustomed to the "smear" of

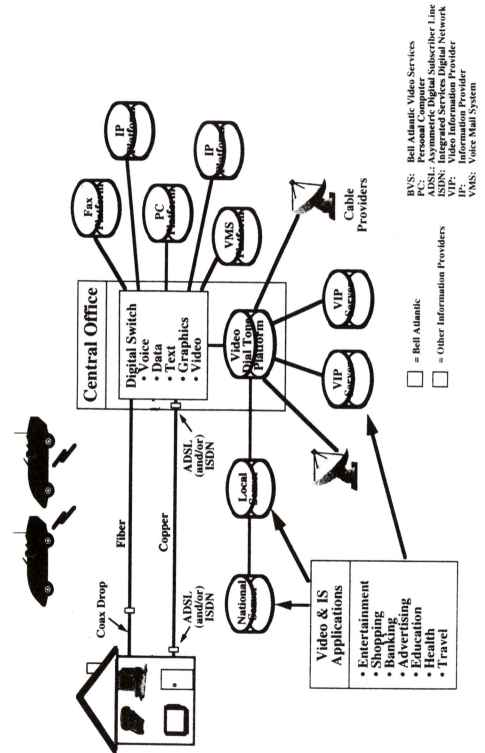

Figure 10.4 Bell Atlantic's future network architecture. Source: Bell Atlantic Video Services.

rolling video and still wanted it, even though digital random access makes it unnecessary. Also, people are used to paper program guides and B-A found that it was important to give consumers a printed schedule in addition to the electronic program guide.

At the same time, it is important to offer improved services. Thus, the best marketing model is to build out from the familiar to the better, much as CDs stuck with familiar branding and packaging of "albums" but delivered superior audio quality.

In the marketing arena, B-A results show that promotion and cross-promotion with products across the mix are the most important elements of raising buy rates. They also found that some concepts simply cannot be explained or marketed. For example, consumers didn't understand bit rates and their relationship to quality and didn't accept the idea of paying more for higher quality. They want a picture at least as good as they are now accustomed to. Subscribers also favor billing in a familiar form: They want to be billed by the movie, not the minute.

Focus on advanced television by a telco: SNET's PersonalVision

Southern New England Telephone began a 400-household diffusion test in West Hartford, Connecticut in mid-1994. The network, a 750 MHz system with both wired and wireless elements, passes 1,500 households. Subscribers are actual customers who pay for services.

PersonalVision offers several tiers of service: broadcast stations, basic cable networks, premium cable networks, and video-on-demand, with about 300 ever-changing choices (costing $2 to $4 to see a video). SNET plans to add electronic Yellow Pages, interactive shopping, and local programming in the future.

The SNET service attracted a 25% sign up rate of homes passed; most of the consumers had previously subscribed to the local TCI cable system. Overall, 40% of TCI's customers defected to the telco, according to SNET.[37] The telephone company-operated service offered three months of free service to get the initial sign-ups and 70% remained after that period. SNET undercuts TCI, charging $19.95 per month and offering 45 basic channels.[22]

These numbers are similar to the sign up rate of Bell Atlantic, which sent direct mail invitations to 16,000 consumers in Fairfax County, Virginia. Within 10 days, B-A received 1,100 positive responses, a 7% response rate, leading them to believe that they will reach about 25% before the test is over.[23]

Whether a 25% sign up rate should be considered a successful result depends on the perspective. These results would not thrill a cable company, accustomed to subscriber rates of 60% to 65%. For telephone companies in a new business, 25% is significant. When cable companies offer telephone service, they will consider 25% market penetration very exciting.

Focus on innovative networks developed by telcos

Telephone companies have engaged in some very exciting and unusual tests. Sprint is involved in two fascinating efforts: the Silicon Valley Test Track and the Starbright Pediatric Online Network.

The Silicon Valley Test Track is a 150-mile fiber-optic SONET ring that links 7 high-tech firms, 2 universities, and 8 elementary schools in the Laguna Salada School District, as shown in Figures 10.5 and 10.6.[24] The SVTT is a site for technical trials and use tests. Each participant has their own vision of how to use the technology, but the common thread is that they want to understand more about how to function in a networked broadband environment.[25]

Sprint is also providing the long distance lines for the Starbright Pediatric Online Network (SPON), covered in Chapter 3. Headed by director Steven Spielberg, Starbright is a nonprofit agency that provides high-tech entertainment to hospitalized children.

Another innovative project is underway by BellSouth. They built a statewide broadband network in the state of North Carolina, linking together 106 public sites, including many schools. The fiber optic network incorporates ATM switches for two-way interactive services with a focus on interactive distance education.

Power Utility Company Tests

Power companies may seem unlikely providers of interactive television, but before Sprint, few people thought of a railroad offering telephone service. (The name Sprint is an acronym based on the Southern Pacific INTernal communication system.) Power companies have thousands of miles of installed fiber optic cable and use it to meter power usage by large customers, like industrial plants. Energy management only requires 5% of the fiber optic bandwidth so they are considering offering advanced TV services over the remaining capacity.[26]

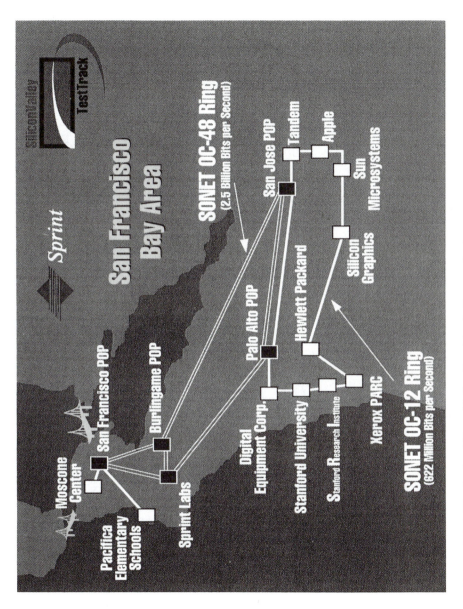

Figure 10.5 The Sprint Silicon Valley Test Track. Source: Sprint.

Figure 10.6 Teacher and students in one of the classrooms linked to the broadband Sprint SVTT. Source: Sprint.

There are several power utility-sponsored trials going on and they are listed in Table 10.4. Most of these tests involve hybrid-fiber coax networks, the same as deployed by cable companies.

On the energy side, customers get real-time information about their energy consumption and its cost. In some trials, appliances are monitored so the subscriber can see how much each appliance is contributing to overall usage. This allows residents to turn their air conditioning up or down, depending on the price they are paying to run it. It also lets the utility company read the meter remotely, so they don't have to send a truck and meter reader out to the home, a tremendous savings to the company. Also, the company can send a bill to the customer indicating the contribution of each appliance towards the total bill. In some configurations, consumers can control their lighting and security systems as well. Figure 10.7 shows the configuration of the tests involving First Pacific Network's PowerView system.

The remaining 95% capacity of the network can be used to offer television, information, and telephony services. In some tests, the utility packages the programming itself; in others, the power company contracts with a cable company to provide the content.

Table 10.4 Advanced Communication Tests Sponsored by Power Utility Companies

Co./Location	Tests/Applications
Central & Southwest Power/Laredo, TX/2500HH	Technical, use, market/750 MHz cable services; NVOD; energy mgmt.
DukeNet/Greensboro, NC/2,000HH	Technical, use, market/Cable TV; PPV; EtherNet wide area network, home security, energy mgmt
Entergy Corp./New Orleans LA, 1500HH	Technical, use, market/Energy mgmt., possible cable TV services; NVOD
Entergy Corp., Arkansas Power & Light/50HH	Technical, use, market/Energy mgmt., possible cable TV services; NVOD
Pacific Gas & Electric/Walnut Creek, CA/	Technical, use, market/Uses TCI cable; Diablo settops; Msoft s/w; provide auto meter reading; outage detect; usage info; energy mgmt; pricing, billing & payment info
Public Service Electric & Gas, w/AT&T/1000HH in Moorestown, NJ	Demonstration project for energy management HFC cable systems, metering, monitoring, remote on-off.
Scana Corp./Columbia, SC	Technical, use, market/Provide voice, video, and energy management services
Southern Calif. Edison Co./Palm Springs CA/25HH	Technical, use/Provide energy management for 5 appliances; shows usage on TV screens and on-screen energy mgmt.
Southern Co./Gulf Breeze FL/240HH	Technical, use/energy mgmt services; plans extensive future trials of TV and telephony services.
Virginia Power	Technical, over a Cox Cable system, using Nortel equipment

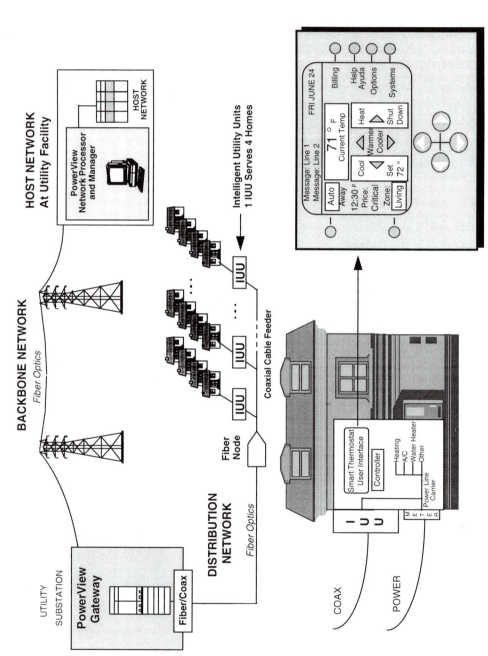

Figure 10.7 First Pacific's PowerView system. Source: First Pacific Network.

Results

Since most of these tests are only recently underway, there is only one instance of reported results. In Southern Company's trial in Gulf Breeze, Florida, which went to 240 homes, test households averaged about $8 less for electricity than control homes.[27]

However, it is clear that utilities, especially those that are publicly-owned, could offer cable companies potent competition. In Glasgow, Kentucky, the municipally-owned Glasgow Electric Plant Board built its own cable TV network in 1988. The utility was more concerned about delivering service to its customers, rather than making a profit, so it provided more channels at a lower cost than the incumbent cable company, TeleScripps. Within a short time, the power company accounted for 55% of cable customers. TeleScripps was forced to lower its basic cable fee from $20 per month to $5.95 and to upgrade its plant from 21 to 48 channels.[28]

Focus on a power utility test of an advanced TV system: DukeNet

Duke Power is a large conglomerate that owns the Washington Post and several small telephone companies, so it is well able to consider delivering broadband switched service.[29] An unregulated subsidiary of Duke Power, DukeNet, plans to build a fiber optic network in a Greensboro, North Carolina, neighborhood of 2,000 homes.

DukeNet does not plan to take a cable company partner to bundle content. The company will put together its own service to offer traditional cable, pay-per-view cable, an EtherNet local area network, high-speed Internet access, and security services, all for less than $100 per month. The parent company, Duke Power, will continue to provide power.

Focus on power management: PG&E, TCI, and Microsoft

In California, Pacific Gas & Electric has adopted the strategy of partnering with a cable company that bundles the television portion of the network service. PG&E uses TCI's bi-directional network in Walnut Creek, California, to provide what the company terms "energy information services." PG&E's goal is to build consumer loyalty in light of deregulation of the energy industry, which will bring competition to the utility.[30]

The test goes to 1,000 homes, with Microsoft providing an interface that will make it easy and convenient for customers to manage their energy consumption. Software capability will be built into the settop box that also serves up the television and information services, sending PG&E signals to an energy management unit. Customers will be able to manage individual appliances and monitor their energy usage on their television sets.

Programming Tests

We have now looked at tests by different industries. Despite their differences, they have one commonality: programming. There is a known demand for linear television services so placing over-the-air broadcast stations, basic and premium cable networks, and some kind of pay-per-view service for movies is a given. In addition, tests draw from the same small pool of interactive programming available and sometimes deliver content that the sponsoring company has developed in-house.

Content is an integral part of market testing and much attention is paid to it. In the next section, we will look at programming tests now underway; Table 10.5 is a list of those tests. We will also present the results that have started to come in about features interactive programming must have for success. Tests cover many types of content: suites of service, navigation and program guide tests, near-video-on-demand and video-on-demand, and YCTV, which provides on-demand television programs.

The most broadly-focused trial of programming was conducted by AT&T in a two-year test in Chicago, using 140 employees in 40 households. Although the sample is not representative and the findings are nearly three years old, the results remain the most comprehensive that are available. In addition, the study itself was scientifically methodical, covering devices, programming, and customer psychology.[31]

AT&T found that for programming to become popular, it must have 4 attributes: entertainment, transactions, information, and communication. In other words, people want to have fun, get something, learn something, and tell someone about it. According to AT&T, it matters less what the service is about (games, story-telling, information, etc.) than how the service is offered and what it lets people do.

Table 10.5 Programming Tests

Test/site/No.HH	Service/Apps
Use and market (attitudes, interests) Ventura County, CA, 1000HH	ACTV, Prime Ticket/Provides different camera angles; unique audio on multi-multiplexed channels; sports statistics and database access
Time Warner/Orlando test 4000HH	Andersen Consulting/virtual supermarket and drugstore for home shopping, stores deliver later
Market/Part of Interactive Channel trial/Denton, TX/150HH	Compton's New Media/search & retrieve technology for CD-ROM over cable, using interactive encyclopedia
Market/Viacom-Castro Valley; Time Warner-Orlando; GTE mainSt.	CUC International/Home shopping service with more than 50 catalog and merchandise partners
Technical and market Philadelphia, Los Angeles, Washington, D.C./several hundred HH	EON/Simulcast with b'cast programming over FM (IVDS); IA games, advertising, drama, and sports
Market/IA loop over phone	Family Channel/Interactive playalong with game shows; several hours of programming
Use and market/Continental, MA & NH; Daniels Cablevision, San Diego	GTE mainStreet/70 1 and 2-way IA svcs: educ., finance, games, sports, travel, shopping. Some full-motion video
Use and market/Denton, TX, 150HH; Birmingham, MI, 40HH	Interactive Channel (IT network)/IA advertising (Foote, Cone & Belding); on-demand news, travel. Yellow Pages; still color images w/ audio. Denton TX: sports, cinema & TV guides
Tech., use, mkt. pricing/ Hagerstown, MD, Coral Springs, FL, 2000 subs; 5 more FL communities	Interaxx TV Network/CD-ROM based service: TV shopping (100 retailers, games, edutainment, stock market ticker

Table 10.5 *Continued*

Test/site/No.HH	Service/Apps
Tech., use, mkt./Portland OR, 2000HH; Barcelona, Spain; Argentina; Australia	InTouch, by Interactive Services/Home shopping, playalong with b'cast TV, sports predictions, polling, local event schedules, games, news, sends dig. info to a thermal printer to targeted subs at home.
Market (determine if niche for music-related products/Nat'l test: 60 million HH	MTV, VH-1, Nick at Night/An IA entertainment superstore—shopping
Technical and market/sites TBA	Nordstrom's/Home shopping via video catalog of Nordstrom's merchandise; plan 15 catalogs per year
Technical and market/Time Warner/Orlando	Spiegel/Home shopping for upscale, full-price apparel
Market/Ventura County, CA, 1000 subs	Turner Educational Services, Inc. (TESI)/Uses footage from CNN Newsroom and Knowledge Learning
Market/Cablevision, Boston, passes 110,000HH	TVN/Provides 8 channels of movies and sports events for PPV, NVOD, or VOD channels
Market (pricing, pgm selection and timing)/Mt. Pleasant, IL, passes 8,900 subs	YCTV/24 channels of television programs for VOD or NVOD; includes ABC, CBS, NBC, HBO, TLC, Discovery Ch., and more

The study also found that the service must be very easy for consumers to use. They want television, not computers, and will not use anything that smacks of complex PC operation. Finally, AT&T notes that interactive services depend heavily on a variety of content. Subscribers wanted new material throughout the day, requiring as much footage as over-the-air network television and more new material than a cable service, such as CNN or rotation-scheduled movies.

The next sections will review the tests of specific types of programming and the results that have appeared.

Navigation programs

Navigation programs have been one of the earliest and most tested areas of content testing because they are acknowledged to be a critical part of delivering interactive television. The navigation program sets the parameters for program presentation. Its attractiveness, ease-of-use, and clarity all have an impact on how, or whether, subscribers will access the available products. Testing generally explores if consumers are able to get to each service area, how many steps it takes to get to a desired service, what design and colors attract and hold viewers.

Bell Atlantic Video Services created a navigation software program called "StarGazer," which helps subscribers access more than 700 choices, including traditional over-the-air and cable channels, and many on-demand video, movies, and music choices. B-A's findings reinforce those of AT&T: Navigation must be simple and intuitive. At the same time, it must be visually compelling. Consumers didn't like excessive branding, either. They wouldn't buy from a Bell Atlantic department store, although they would accept other versions, such as "Nordstrom's—brought to you by Bell Atlantic."

In their Castro Valley, CA trial, Viacom explored how much subscribers would pay for their electronic program guide (navigator) StarSight Telecast. They learned that 50% of the subcribers would pay $5 a month for the service. Lowering the price increased buy rates: 58% would buy at $4 a month and 68% would take it at $3 a month.

The Time Warner Full Service Network in Orlando will test two interface/navigator programs, each in 2,000 households. The first interface is based on a graphically-rendered carousel metaphor that turns showing 10 accessible services. When the viewer chooses one, they start with built-in promotions. The second interface, called "Omnia," employs a stationary 9-button approach with the titles of the services printed in text on them. The viewer uses the handheld remote to highlight a button and to choose the title.

Video-on-demand versus near-video-on-demand

Long touted as the killer app that would pay for interactive television systems, VOD has been tested alone and in combination with NVOD to assess the relative costs versus take-up rate of the two services. Some services are also testing movies-on-demand (MOD). Table 10.6 provides information on the results from these tests.[32]

Table 10.6 Results of VOD and NVOD Testing

Who/where	Service	Cost	Take-up rate
Cablevision/ Boston	12 ch. NVOD, 30 min.	$3.95	50%, w/60% penetration
Comcast/West Palm Beach, FL	5 hit movies, 30	$3.99	Happy w/ results
DirecTV/DBS	50–55 ch./NVOD, 15 and 30 min.	$2.99	150–200%
NYNEX/NYC	MOD, 6 films at a time, 25 title	new/$3.95 old/$2.95	+4–5 buys per HH, per month
SNET/W. Hartford, CT	NVOD and VOD	NVOD—$2.95 VOD—$4.	NVOD slightly higher
TCI, U.S. West, AT&T/Denver CO	VOD vs NVOD (15 min. start times)		250%+ (2.5 movies per HH, per month)
TCI/Mt. Prospect, IL	24 ch. NVOD, 19 hits chs., every 30 min. 2 action, 2 adult chs.		200%–250% + adult = 40% of all $.
Time Warner/ Queens, NY	57 NVOD channels, 15 min., 40 titles		191% above PPV
TVN/C-band satellite svc.	8 PPV ch., and NVOD chs.		NVOD 100%+
Viacom-CA	Cell 1 (Variety): 16 continuous run movies, over 16 channels		NVOD rate is 8 × PPV
	Cell 2 (Convenience): 3 titles every 30 minutes, then every 15 minutes.		No results published

Veteran journalist on the tech beat, Matt Stump, summarized what the testers have learned.[33]

- Overall buys increase when start times decrease, especially at the 15 minute break;
- People want variety, and are willing to wait for movies they want;
- In systems whose subscribers tend to be older, NVOD will perform as well as VOD;
- On average, subs with NVOD buy from 1 movie every 2 months, to 4 movies a month; PPV rates are about 1 movie every 5 months, or 1 of 5 subs.

YCTV: Television programs on demand

Your Choice TV is a special case of VOD and NVOD—instead of movies, the service offers television programs. ABC, CBS, and NBC have provided network-produced programming (mostly news and public affairs shows and daytime soap operas) to YCTV for testing.

In starting YCTV, founder and chairman John Hendricks, believed that viewers would pay for the convenience of being able to see time-shifted programs. YCTV has been tested on at least 10 systems, but the TCI/Mt. Prospect system has been the best venue because it was able to allocate 24 channels to YCTV.

In the Mt. Prospect test, subscribers spent an average of $4.40 per month on YCTV, at 49 cents to $1.50 per program. YCTV found that people didn't necessarily request the shows that had received the highest ratings. Rather, they wanted to watch programs they missed, like daytime soaps and high-quality titles, such as documentaries and children's programs. YCTV anticipates that popular selections will add 5% to 10% additional viewers to a program.

Behaviorally, the tests showed that people turn on their sets and cycle through their repertoire of six or seven channels a couple of times. If they don't find anything they like, they choose from movie and program on-demand selections, concentrating on high-quality branded network shows and the highest-quality cable originals.[34]

YCTV does not provide as much revenue to cable operators as on-demand movies do. And the delays in deployment of digital settop boxes has forced YCTV to operate in an analog world where there is little room for additional channels. (The amount of room on a cable system is called "shelf space.") This shortage is a disadvantage for cur-

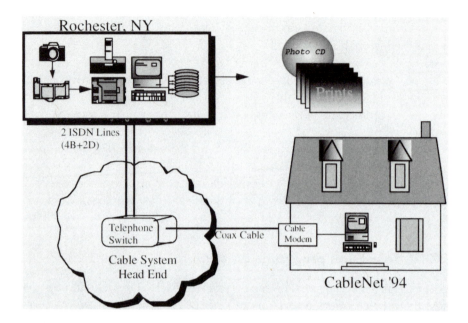

Figure 10.8 Eastman Kodak's interactive personal imaging demonstration. Source: Eastman Kodak.

rent deployment of the YCTV service because it cannot perform as well on a revenue/per MHz equation as movies-on-demand. Nevertheless, there is every indication that YCTV's founder, Hendricks, is in the video-in-demand business for the long-run and has plans to launch worldwide on both wired and wireless systems.

Tests of Technology

Manufacturers, both alone and in consortia, create facilities to test their products. One such test is shown in Figure 10.8. Table 10.7 lists important current tests that will have some impact on the future of advanced television systems.

International Trials

Advanced television network test sites are something new on the international scene. As few as three years ago, such tests were unheard of. Today there is a long list, as shown in Table 10.8. The types of tests, the questions asked, the systems' design and configuration, even the suppliers and vendors are the same as in the U.S.

Table 10.7 Technology Testbeds

Org./Location	*Descrip. of Tech.*	*Test/Applications*
Alcatel/Spain and U.K.	WDM over 930km fiber optic link/at 10 Gb/s (2.5 Gb/s in 4 channels)	Demonstration
CableLabs/Louisville CO/NA	1 GHZ HFC, permanent	Technical only
Eastman Kodak Company/Rochester NY cable system	2 ISDN phone lines	Consumers can receive digitized copies of their processed photos.
U.S. Dept. of Energy, Sandia Laboratory, Lockheed/San Francisco, Los Angeles, CA/1000 people; 54 projects 100+ institutions	Fiber optic network/ demonstration facility: ATM, SMDS, ISDN, Sw-56, frame relay	Use/Distributed interactive simulation; concurrent collaboration w/work; research; educ., civicnet for Cupertino
BBN Hark Systems/Time Warner, Orlando	BBN HARK	Technical, use/Offers speech recognition for sys. navigation
Texas Instruments/ in-house testbed	Digital micro mirror devices	Conduct eye-tracking studies of viewers while watching TV
UtiliCorp United Inc., Novell/Site TBA	Use home electrical wires as comm. network. Embed chips in appliances to make them power "smart."	Allows consumers to start and stop appliances
WavePhore/Advanced TV Evaluation Lab, in Canada	Data broadcasting in vertical blanking interval	Test effect of inserting data into TV signal

It will be some time before many of these testbeds are built and even longer before the results of testing are known. However, it will be fascinating to see how the different cultures and societies respond to this new technology. As covered earlier, there is a tradition of testing advanced television service in other countries. However, this new wave of testing is more extensive than ever before, both in terms of the

Table 10.8 International Trials

Org./Location	Network Description	Test/Apps
Europe:		
Cable & Wireless/ Global: Links NY, Hong Kong, London, Tokyo	Fiber, ATM switch/ StrataCom BPX is major vendor	Use/Content user-supplied
Canadian Network for Advancement of Research Industry and Education (CANARIE)	40-ATM switch network by Newbridge	Use/Distance educ.; real-time medical consulting
Cambridge, U.K./ 100HH	ATM/Online Media settops	Use and market/b'cast & cable chs.; VOD; IA advtsg; shop; banking
British Telecom/London, Ipswich, Colchester/2500HH	ADSL/Westell; nCube server; Alcatel Bell switch; Apple settop; Oracle, Sequent s/w	Tech., use, mkt/b'cast & cable TV; IA advtsg, shop, bank, music, dig svc
British Telecom/London/100HH, then 1000HH	Tech/use/mkt/Gen'l Instru. switch, settop; DEC server, "pure" VOD	
Bell Cablemedia, NYNEX TeleWest Comm./U.K./2000HH	TBA/will test several diff. vendors & equip.	Tech. in lab & field; market; navig/VOD; IA educ, advtsg, bank, shop
Videotron/W. London, Southhampton	TBA	Tech., use/IA advtsg by ad agencies
Belgacom/Brussels, Belgium/50HH	ADSL and ATM/DEC Alcatel Bell switch; Apple settop	Technical/VOD
France Telecom/3 tests, sites undisclosed; "Dora," 50,000, "Camille," and "Jasmin" TBA	Dora-fiber optic net; Camille-ADSL (Visiopass settops; Jasmin, not disclosed	Tech., and mkt/Dora-broadband Camille–news, VOD, shop, online access Jasmin—impulse shopping
VebaCom/Germany Ruhr Valley, (Cologne-Dusseldorf)	200km fiber optic ring Vendors: Apple; Nokia; Bertelsmann; DEC; Ericcson; Philips	Use, program/Cable & b'cast TV by Vox,Viva, and RTL; online InfoCity

Table 10.8 *Continued*

Org./Location	Network Description	Test/Apps
Swiss Telecom/Nyon, Grenchen/400 HH each site	ADSL, Grenchen; FTTC, ATM, 860 MHz, Philips is main vendor	Technical, mkt. VOD; shop; bank; educ; telemedicine; online access
Italy-SIP/Italy, site TBA	HFC/MicroWare s/w	TBA
Asia: Hong Kong Telecom/Hong Kong/ 400HH	ADSL, migrate to fiber optic ATM full svc network/Integration by Andersen Consulting; equip. by IBM, MicroWare	Tech.; program/cable TV, shop; bank; VOD
Korean Telecom/Seoul	TBA/MicroWare	TBA
MITI/Okazaki City	TBA/IBM server & s/w	TBA
Kajima/Sendai-Osaka/2 business sites	Fiber optic network for collaborative engineering work	Use/shared video in construction industry
NTT, Time Warner, U.S. West, Toshiba/ Japan site	HFC/Integration by Broadband Technologies, Msoft s/w	Tech., use, mkt
Taiwan	TBA/Integration by Broadband Technologies	TBA
Telecom Australia/ throughout Aus./300HH	ADSL/Switch NEC; Settops by Magnitude	Tech., use, mkt/7 b'cast chs.; VOD
Telstra/Centennial Park, Sydney	TBA/NTL major vendor	TBA
Singapore Telecom/ Singapore	Fiber optic network/ HP servers; Fujitsu switch; Philips settop	TBA/VOD
Astroguide, LaSalle IL IL/sites in Kuwait Bahrain	MMDS wireless, 11.7–12.2 GHz	Technical/28 chs.

Table 10.9 Commercial Rollouts of Advanced Television Networks

Organization/Time	Location/Technology
Ameritech/1997	Detroit, Columbus, Cleveland, Chicago, Milwaukee/HFC
Bell Atlantic/5 years	Northern New Jersey; Philadelphia and Pittsburgh, PA; Baltimore, MD; Norfolk, VA; Florham Park, NJ; Virginia Beach, VA; Washington, D.C./SDV-FTTC network/combined, passes 3,215,000 HH
BellSouth	Daniel Island, Charleston, SC/IA svcs. to 7,000 HH
Contel	Manassas, VA/passes 90,000 HH
GTE/1997	Thousand Oaks CA; St.Petersburg/Clearwater, FL; Honolulu, HI; northern Virginia; Ventura City, CA; combined, passes 1,500,000 HH/HFC
NYNEX/1995	Rhode Island, suburban Boston/SDV-FTTC/passes 394,000 homes and businesses
Pacific Telesis/1995	Los Angeles; Orange County; San Diego; San Jose/SDV-FTTC/total passes 1.2 million HH
SBC/1996	Montgomery County, MD/HFC
Southern New England Telephone (SNET)/1998; state by 2010	500,000 HH in Connecticut/HFC
US West (Southern Multimedia Comm.)/1998	Atlanta GA; HFC 750 MHz network/passes 500,000 HH
US West/1997	Denver; Minneapolis; Portland; Boise/SDV-FTTC networks/combined, about 1,000,000 HH
Viacom/1997	San Francisco Bay Area/Fiber optic ring, 2.4 Gb/s
YCTV/1995 or 1996	National rollout of 4 analog channels via satellite to headends

number of countries conducting trials and the scope of the technology and offerings.

Beyond Testing: Rollouts

As mentioned earlier, a number of system proponents are rolling out networks commercially, shortening the testing process—or eliminating it altogether. Table 10.9 shows the companies, mostly telephone companies, that plan commercial rollouts.

The next chapter will cover how companies building both wired and wireless systems use the results of trials to develop a viable business model that justifies and supports commercial ventures.

Notes

1. Carol Wilson, "Don't know much about history?" *Inter@ctive Week* (November 7, 1994):19.

2. William Dutton, "Driving into the future of communication? Check the rear view mirror," Paper delivered at 'POTS to PANS: Social issues in the Pictures and Network Services, the BT Hintlesham Hall Symposium, Suffolk (March 28–30, 1994):1–39.

3. Peter Krasilovsky, "Interactive TV: a slow revolution," *Telemedia Week* (December 1994).

4. Karl Weick, *The Social Psychology of Organizing* (New York: Addison-Wesley, 1969).

5. Susan Tyler Eastman, *Broadcast/Cable Programming,* 4th ed. (Belmont: Wadsworth, 1993).

6. Jim Cooper, "Your choice breaks ground" *Cablevision* (July 11, 1994):47.

7. Carol Davidge, "America's talk-back television experiment: Qube" *Wired Cities: shaping the future of communications,* eds. William Dutton, Jay Blumler and Kenneth Kraemer (Boston: G.K. Hall, 1981):75–101.

8. Dutton, op. cit., 10.

9. Ibid.:18–19.

10. Ibid.:8.

11. Joan Van Tassel, "Test patterns," *The Hollywood Reporter* (May 18, 1995):S-9.

12. Tom Steinert-Threlkeld and Carol Wilson, "Time Warner tunes in TV trial," *Inter@ctive Week* (December 12, 1994):6.

13. Leslie Ellis, "Tales of selling interactive: free meals, simplicity," *Multichannel News* (June 12, 1995):2A.

14. Peter Lambert, "The broadband PC connection," *On Demand* (March 1995):12.

15. Adam Bauman, "Alacazam! Welcome to the magical world of cable modems," *Los Angeles Times* (May 3, 1995):D4.

16. Lambert, op. cit., 10.

17. Ibid.

18. A. M. Noll, "Anatomy of a failure: PicturePhone revisited," *Telecommunications Policy*, 16:4 (May/June, 1992):307–316.

19. Edward Dickson and Raymond Bowers, *The VideoTelephone*, (New York: Prager, 1974).

20. Amy Harmon and Leslie Helm, "PacTel to cut $1 billion from telecom project," *Los Angeles Times* (September 2, 1995):D1, D4.

21. This information came from Kenneth Van Meter of Tele-TV, speaking at a seminar at Digital World, June 18, 1995, Los Angeles.

22. Peter Krasilovsky, "Interactive TV: a slow revolution" *Telemedia Week* (December 1994):7.

23. "Bell Atlantic Video Services connects first paying customers," *Interactive Television Association* 2:22 (June 2, 1995):2.

24. Christine Perey, "Perspectives on broadband network technologies and applications," *New Telecom Quarterly* (2nd Quarter, 1995):35–41.

25. Craig Kuhl, "Fast track?" *Convergence* (March 1995):34–39.

26. Steven Rivkin, "Positioning the electric utility to build information infrastructure," *New Telecom Quarterly* (2nd Quarter, 1995):30–34.

27. Harry Jessell, "Infohighway power play," *Telemedia Week* (December 1994):26–28.

28. George Lawton, "The utilities' role: building the ubiquinetwork, Pt. 5," *Communications Technology* (December 1994):80–85.

29. Tim Clark and Carol Wilson, "New power player: electric utilities," *Inter@ctive Week* (December 12, 1994):41.

30. Ibid.

31. John J. Keller, "AT&T's secret multimedia trials offer clues to capturing interactive audiences," *Wall Street Journal* (October 6, 1993):B1, B4.

32. Matt Stump, "NVOD: a how-to guide," *On-Demand* (April 1995):13–19.

33. Matt Stump, op. cit.,13. Also, see Matt Stump, "NVOD vs. VOD," *On Demand* (December 1994/January 1995):21–24.

34. Mark Berniker, "Pioneering on the digital highway," *Telemedia Week* (December 1994):34–39.

$$11$$

Business Development and Scenarios of Adoption

Introduction

The rule of thumb for the cost of building wired networks is "a billion per million." There are 92 million U.S. TV households, so for wiring the U.S. will cost nearly $100 billion. Wireless systems cost less: MMDS systems cost between $400 and $600 per household and LMDS systems run between $250 and $600, depending on the assumptions made about the population density. Thus, it would cost between $23 and $55 billion to provide advanced wireless service to all 92 million households.

In the face of such staggering investments, even the largest corporations must be sure there will be equally large returns. As we have seen, the demand for communication bandwidth has expanded steadily since the 18th century. Perhaps it will level off. However, even if growth continues, the useless metal of dead tech is scattered across the business landscape: Specific communications technologies have a high rate of failure. Examples include interactive videodiscs, videotext services, AT&T's PicturePhone, early facsimile machines, and countless others. An entire communications infrastructure is quite a different matter. With so many billions at stake, no one can afford to make a financial error of such magnitude.

How does a technology spread throughout a society? What factors make it more likely the technology will be adopted? Who are the adopters? These questions are addressed by a theory called the "diffusion of innovation," a dominant paradigm guiding much of the market research underlying business plans to provide advanced television. The chapter begins with this important theory, then turns to the industries providing advanced TV systems and the elements of their business plans: competition, costs, marketing research, opportunities, and lessons learned from the past.

The Diffusion of Innovation

The formal study of how new ideas, technologies, and practices spread throughout some given group began in geography by examining how agricultural products moved from one locale to another. Early diffusion studies retained a spatial bias until the work of scholar, Everett Rogers.[1] In essence, he moved the study of diffusion from a geographically-based discipline to a communication-based approach.

Rogers became familiar with diffusion processes at the University of Iowa, where he studied Agricultural Sociology. As a student, he became interested in the work of Wilbur Schramm and the then-new field of communication research. Rogers realized that the mass media and electronic communication changed the processes of adoption by overcoming the spatial limitations of earlier eras. In 1962, his compelling work, "The Diffusion of Innovation," brought these ideas to the forefront.

The theory of the diffusion of innovation is a stage-process model of change brought about by the introduction of new ideas, technologies, or practices. The stages are: 1) the introduction or invention of a new technology, idea, or practice, 2) the diffusion of that innovation through communication (or perhaps its rejection), and 3) the changes the innovation brings to individuals, societies, and cultures.

The second stage includes adoption, a process whereby individuals, organizations, or even nations, decide whether to accept or reject an innovation. If it is adopted, that process begins with awareness of the innovation, followed by interest in it, marked by information-seeking behavior. With more knowledge, the innovation is evaluated and, if it is seen to meet some need or desire, then the adopter tries it. After trial, the adopter may decide to continue using the product or practice. At any time, a decision can be made to discontinue using the innovation.

The diffusion S-curve in Figure 11.1 illustrates how innovations spread throughout a population. It shows a simple accumulation of adopters over time. Acceptance begins slowly, adding one adopter after another. At a certain point, adopters network, causing an exponential rise in the number of people accepting the innovation: This point is

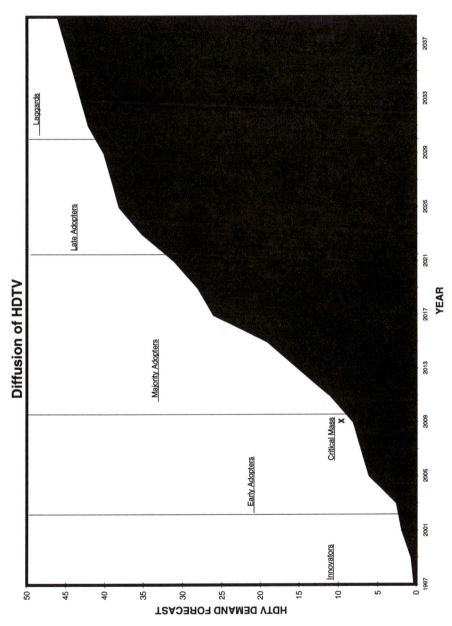

Figure 11.1 The forecast of the diffusion S-curve for HDTV.

called "critical mass." Critical mass marks when the innovation is used by enough people to make its adoption likely, at least for a time. Critical mass is especially important to communication because of the "sufficient pair problem": that is, the ability of a person to reach enough people who they want to reach to make adoption of the communication system worthwhile. For example, few people would pay for a telephone system that only reached people in a city 2,000 miles away.

The theory also presents a profile of adopters based on when they accept the innovation: *Innovators* are the first to adopt. They are willing to accept risks. As a group, on the average they are young and have high social status. They have close contact with scientific sources and other innovators and use the media to get information. They seek information from many sources and some act as opinion leaders within their social network.

Early adopters are often role models with high social status. They have contact with change agents, such as marketers and industry outreach efforts, and they tend to be secure within their local groups. Adopters in the *early majority* tend to be deliberate decision-makers, who accept an innovation only after peers have tried and liked it. They have middle social status and have considerable contact with change agents and early adopters. They may be leaders within their social group.

Adopters in the *late majority* are skeptical, adopting only after there is almost no alternative. They have below average social status and income. They get their ideas from like-minded others and occasionally from adopters in the early majority rather than from the mass media. There are few leaders within this group.

The last group of people to adopt an innovation are labeled by the unflattering term of *"laggards."* They are traditional and oriented to the past. They have the lowest social status and income and are often older. They may get information from interpersonal sources who share their values or they may be social isolates.

Attributes of the innovation itself, as well as the adopters, make a difference. The characteristics which have the most effect are: cost, complexity, compatibility, trialability, and substitutability. Costs include monetary outlays, opportunities lost through foregone options, social and personal costs, and other subtle or hidden penalties for accepting an innovation. Complexity refers to the number of different parts that have to be taken into account from the adopter's point of view regardless of the underlying technology. For example, automobiles and computers are complex; televisions and staplers are simple.

Compatibility includes product interoperability but also covers other aspects. For example, birth control pills may not be compatible with the cultural practices in Catholic countries; free speech may not be compatible with the political practices in a socially repressive regime. Thus, compatibility involves the way an innovation interacts with existing technologies, cultural and social practices, and individual preferences.

Trialability is the extent to which a person can try out an innovation before they adopt it, like test driving a car. One problem selling computers has been that until people reach a certain level of expertise, they can't even try it. Perhaps the new "plug and play" multimedia players will entice many new buyers who see how easily they can play graphic and video CD-ROMs.

Substitutability refers to the ability of an innovation to accomplish a task more efficiently than the old product or practice—a better way to skin a cat. Sometimes this quality is called "relative advantage." According to Peter Drucker, a product must be at least 10 times better than the one it replaces in order to gain a foothold in the market; that is, more efficient by an "order of magnitude." Five times better won't cut it, observes Drucker.[2]

Cost and complexity have an inverse relationship to adoption; to the extent that an innovation is costly and complex, its adoption will be hindered by those factors. Compatibility, trialability, and substitutability are positively related to adoption; the more compatible, the easier it is to try, and the more functions it will substitute for, then the more likely it is that the innovation will be accepted.

This clear and practical theory has been extraordinarily useful to scholars, national development specialists, technology transfer programs, and marketers. A careful study of the research in this tradition will guide those who wish to introduce innovations, from democracy to videophones.

The diffusion of interactive television

The rollout of IATV must occur successively: first by corporate conduit providers, then by consumers. Let us first consider IATV from the perspective of the companies who must invest billions of dollars to build the IATV physical plant, before interactive services can be offered to customers.

Corporations became aware of IATV as early as the 1960s, with the QUBE system in Cincinnati, OH. Two technical advances have increased their interest: reductions in the costs of installing high-bandwidth fiber optic cable and increases in the processing power of computers that are sufficient to digitize video and to perform other system tasks. Conduit providers have been carefully evaluating the technical and economic viability of IATV since the late 1980s.

As we have seen, the cost of advanced television systems is enormous and their complexity intimidating, slowing their diffusion. Figure 11.2 shows how Carl Podlesney of Scientific-Atlanta analyzes the cost impact of network improvements to add advanced services. Compatibility issues, especially the lack of standards that would insure the interoperability of equipment and software, continue to inhibit the spread of IATV. Without them, the overall project can become mired in a hopeless tangle of incompatible components. The Time Warner testsite in Orlando, delayed for more than a year and a half, is an example of such problems.

In the face of the problems of cost, complexity, and compatibility, we have seen that all the large cable operators and most of the telcos

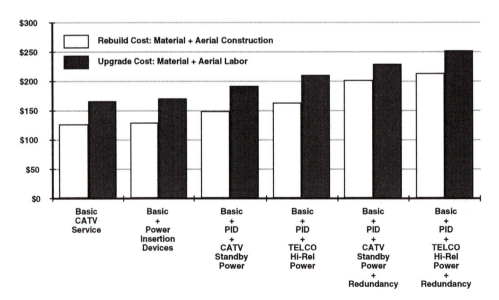

550 Plant Mile Hub/200 to 300 Home Serving Areas/135 Passings per Mile

Figure 11.2 Cost impact of network improvements for advanced services. Source: Carl Podlesney, Scientific-Atlanta.

have launched trials of their own and watched the tests of others very carefully. The trialability of IATV is an important element in the final configuration these systems will assume and their viability as businesses.

Finally, for the companies building advanced television systems, this new broadband technology substitutes for their old infrastructures. The telcos will be freed from the bandwidth constraints of narrowband copper wire plant. Cablers will have two-way networks that will allow them to expand and improve their core business—and to enter new markets.

Companies, as well as individuals, have adopter profiles. AT&T, Time Warner, GTE, Silicon Graphics, General Instrument, and many others are innovators. Some companies are early adopters, majority and late majority adopters, and even laggards. It is not unusual for one division of a company to innovate while other departments fit into another category. However, it often pays off to be a player at certain crucial points in time.

There are distinct advantages to being in the early adopter and early majority categories. Innovators often suffer the consequences of a lack of standards, product and service characteristics, and consumer awareness. A good example is the Japanese HDTV effort which lost out because of the development of digital technology. By contrast, at least some of these deficiencies have improved for those entering the market a little later. On the other hand, late adopters and laggards may suffer, as well. They have little influence on the development of the market and may have lost business because they failed to provide a service that their customers wanted and sought elsewhere. However, late entry in the high tech arena isn't always penalized. The U.S. success with HDTV is a good example of the rewards for the industry that waits.

This cursory evaluation of IATV using the diffusion model demonstrates the power and clarity the theory brings to the understanding of an innovation. From this base of knowledge, it is possible to make rational assessments about the likely future of an innovation. This aid to prognostication accounts for the popularity of the theory of diffusion of innovation in many different areas of research and business. Now we turn to the industries most likely to establish a broadband infrastructure.

A Tale of Four Industries

Four industries are poised to deliver interactive television: the cable, telephone, utility power, and computer industries. For the past few years they have enacted the most intricate corporate choreography, jockeying for competitive advantage and joining for cooperative benefit.

When it comes to the financial ability to pay for the expensive infrastructure advanced television systems, the power companies have the most capital with $200 billion in annual revenues, followed by the telephone companies that bring in $100 billion annually, and the cable companies, with $50 billion yearly income.[3] For the most part, the computer industry doesn't build infrastructure. Computer networks are either provided by the telephone companies (and perhaps by cable and power companies in the future), or by individual companies who communicate on LANs (local area networks) and WANs (wide area networks), using some combination of proprietary infrastructure and leased lines.

Each of these powerful industries not only has a substantial installed base of equipment, they also work within a sometimes invisible culture that guides the way they conduct their business. Their experiences and practices, as well as their existing technological configurations, shape the strategies and investments of these major players.

One way this influence can be seen is in the failure of both cable and telephone companies to address broadband interactivity in a serious way—neither plans to provide much two-way functionality. For the cable companies, it is an aspect of organizational culture: They are, essentially, broadcasters and have little interest or expertise in the communicative aspects of the systems they are planning and building. For telephone companies, it's more of an economic issue. They want to add the income of cable-casters to their revenue stream and don't consider two-way broadband communication an important aspect of their new systems.

Both these industries believe that the key to profits lies in sending information downstream to people, rather than creating infrastructure for people to communicate upstream. Thus, they limit their provision for interactivity to creating sufficient return bandwidth for a "clickstream" of button pushes to let customers order films, buy merchandise, and conduct banking transactions such as paying bills and moving money from one account to another. No systems have enough upstream bandwidth to allow customers to send complex graphics or full-motion video at high speed. If interactivity functionality comes to consumers, it will arrive through the demands of computer users at a price high enough to entice a conduit provider to build two-way capability into their systems.

In spite of the installed base of both cablecos and telcos, their physical plants have formidable strengths and significant vulnerabilities in a competitive marketplace. Cable networks have large capacity, but they are fragmented into discrete systems and information flows only one way. They are not switched nor entirely compatible with other networks. Telephone companies have compatible, switched, global two-way networks, but they will not handle video or other broadband services.

Although power, telephone, cable, and computer companies all bring some influence to bear on the eventual outcome, no one industry can entirely predict or control the evolution of the global infrastructure. Starting with their unique technological configuration, these industries face different network design and construction challenges, yet must end up with the same capabilities. This final product is a network capable of delivering and receiving compatible, switched analog and digital signals for voice, data, and video to anyone in the nation and possibly the world.

Cable companies

Around the country about 75% of cablers have upgraded to fiber optic cable in the backbones of their networks; about 30% have put it in their feeder lines as well. Paul Kagan estimates they will spend $5.6 billion on the total rebuild.[4] This investment, necessary in the face of increasing competition, is one cause of the increasing consolidation of cable system ownership. In 1994 and 1995, more than half of the 25 top companies either joined others or announced intentions to do so. As a result, 16 MSOs account for 80% of cable subscribers and only two of them, Telecommunications, Inc. and Time Warner combined, have 37% of all subscribers.[5]

The cable industry has adopted hybrid fiber/coax (HFC) as a network structure, one that allows them to make a smooth migration to full service networks. In 1995, there were three important network issues facing MSOs: creating a national, interconnected network; providing telephone service; and allowing subscribers to link computers to the cable via cable modems.

Interconnecting disparate, discrete cable systems MSOs are connecting the systems they own into "superheadends" or "master headends" that feed smaller headends eliminating the cost of redundant equipment. There are many examples of this trend: TCI and Viacom have joined to build a fiber ring around San Francisco; Continental Cablevision is linking their northeast systems; and in Dallas, Sammons Communication is collapsing 12 headends into one.[6]

Once cable systems are consolidated, MSOs in the region will jointly construct "metro area networks" ("MANs") to carry voice, data, and video. This plan will allow them to enter the lucrative local telephone business. They will hand off their telephone traffic to long distance carriers, avoiding local exchange carriers' expensive carriage and access charges.

Cable MSOs are experimenting with both wired and wireless telephony. Time Warner Communications plans to offer telephone service in 1995 to users in Manhattan via a 22-mile fiber optic ring, challenging NYNEX in its primary service area. Cox Communications will market wireless personal communication services (PCS) throughout Southern California where they have several systems and in Omaha.

Connecting PCs to cable   The cable plant requires an upgrade similar to that required for telephony in order to offer high-speed connection for PCs to the Internet and other online services. Specifically, two-way capability and switching are necessary modifications that are both quite expensive.

Cable companies need a high-speed cable modem that will attach to the new generation of settop boxes. At least eight companies are developing them, both for the business (LANcity, Digital Equipment Corp.) and residential markets (GI, Scientific-Atlanta, Intel, Motorola, H-P, and Zenith). Business cable modems cost about $5,000 in 1995, while consumer modems run between $200 and $2,000. At the high end, these prices are far too high to attract the large number of subscribers required to pay for the upgrade.[7] In spite of the barriers, in 1995, MSOs Cox Cable, Comcast, Cablevision, TCI, Viacom, and Rogers are all testing high-speed PC hookups to some of their subscribers.

Telcos

The copper wire plant of the telephone companies has taken 100 years to build at a cost of $1,500 per person![8] Naturally, telcos have been interested in technologies which would extend the use of this extensive, expensive infrastructure.

One new technology may bring moving video to households, although it is not sufficient for video-on-demand or broadcast quality service. AT&T Paradyne's GlobeSpan technology uses ordinary copper telephone wires to bring voice, video, and data into the home or office. "Anyone with standard telephone service will have access to a multitude of multimedia services," says a company brochure. Consumers will be able to attach a telephone line to a converter box connected to either the TV or the household PC. The new technology will eliminate the need for ISDN or fiber optic lines and could be available in most parts of the country within 12 months.[9]

"ISDN," the integrated services digital network, is an interim medium-band solution that telephone companies have been trying to deploy for a decade. Basic rate ISDN provides two 64 Kbp/s digital channels over one circuit and another channel for signaling to residential customers. For larger businesses, primary rate ISDN delivers 1.544 Mb/s at a higher cost. This technology has been available for 15 years. The requirement for the consumer to buy expensive terminal adapters and transmission devices, incompatibility between ISDN offerings, and bottlenecks that prevented customers from using the full bitrate of ISDN all worked to discourage adoption of the technology. ISDN is now being widely deployed by Regional Bell Operating Companies but many observers feel it is too little, too late and say ISDN stands for "It Still Does Nothing."

Still another way telephone companies thought to extend their copper plant was through the use of two related technologies, "asymmetrical digital subscriber line" (ADSL) and "high-speed digital subscriber line" (HDSL). However, recognition that these technologies would be nearly as expensive as more advanced networks, while failing to offer a smooth migration path to future services, has resulted in telcos abandoning ADSL/HDSL technology.[10]

The larger telcos will build fiber-to-the-curb, switched digital video systems, overlaying the current narrowband telephone lines. Figure 11.3 shows Bell Atlantic's vision of what advanced TV systems will look like. They are adopting the SDV configuration because it will offer more robust interactive services at the same installation cost as HFC. To make this decision, U.S. West assumes that 25% of households will subscribe to video dialtone. Seventy percent of these households in the 25%, about 19% of the total, will use digital services. Given the high cost of installing this dual-mode system, the breakeven point will be reached in 7 to 8 years.[11]

Computer companies

Computer networks are far from ready for interactive television. Businesses and institutions have built "local area networks" (LANs) and

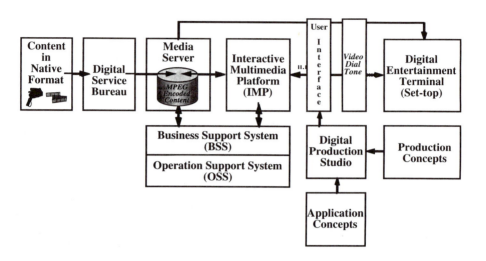

Figure 11.3 Bell Atlantic's view of the complete advanced TV provider. Source: Bell Atlantic Video Services.

"wide area networks" (WANs) to handle organizational needs. They are mainly private, dependent on the telephone companies for inter-connection and public access.

Two types of networks prevail: 10 Mbp/s EtherNet and 4 Mbp/s or 16 Mbp/s Token Ring. These networks are "duplex" (two-way), switched, connected, and digital. However, few are designed for video delivery; the quality is not high enough nor is their packet-switching method practicable to deliver full-motion images. Intel is leading a 40-member consortium to develop a Fast EtherNet standard to operate at 100 Mbp/s which would alleviate the quality problems. However, Fast EtherNet would do nothing to help the fact that LANs and WANs are fragmented and lacking interconnection, except via the "public, switched telephone network" (PSTN) with its lack of broadband capacity.

Yet we see computer networks moving slowly towards providing full-motion video. In the past 3 years, still photographs, complex graphics, and audio have all become available on computer networks. The data download for photos may be slow and the audio may be relatively poor—but the trend towards information-richer material over computer networks is unmistakable.

Powercos

Power utility companies are potentially major players in the telecommunications delivery business. Like railroads, power utilities are granted governmental right-of-way which allows them to put up poles, dig ditches, and string cable wherever they need to without paying fees like cable companies must.

Since the 1970s, the power companies have been laying fiber optic cable to meet their own internal communication needs and have about 18,000 miles of installed fiber cable. In the next three years, they plan to add an additional 12,000 miles.[12] Fiber optic cable offers another advantage to the utilities, which explains why they adopted it so early, before almost anyone. Utilities couldn't communicate usage information along the same lines that were used to transmit power because electricity interferes with the information passing along copper wire (or any other metal). However, silicon can carry data right next to a power cable, with no interference, consuming no more than 5% for energy management. And more, silicon fiber is an excellent insulator.[13]

Power companies have several attributes that make entry into the broadband conduit business attractive for them. They are already connected to 95% of households in America and proponents of power company provision of telecommunication services note that they are more likely than any other provider to be able to offer universal service. In addition, utilities have particularly good relationships with their customers.[14]

To provide telecommunication services, however, power companies have a big investment in front of them. Their networks are not powered, "lighted," two-way, or switched. Nor does this industry have any experience providing communications, information, or entertainment services.

The emergent convergence

In spite of their differences, these industries have a lot in common. Each has a legacy system—outdated, expensive to maintain and operate, and unsatisfying to the customer base. Each faces major competition in its core business, has difficulty cooperating with the other industries with the expertise it needs, and must adopt some version of an advanced communications network.

They even share the same technology. Table 11.1 shows how they are converging on the same configuration.

Table 11.1 Network Infrastructure by Industry

Industry	1980s	1990s	2000s
Telephone	DS-3 access IDSN	SMDS BISDN	Sonet/ATM
Cable	Analog coax/fiber trunks	Analog & digital fiber	Sonet/ATM
Computer	Bridges & routers	100 Mbps LANs	Sonet/ATM
Power	Fiber for internal communication	Remote metering via fiber	Unknown

Chapter 4 introduced the theory that all the communications industries are converging into a single, seamless system and there is much evidence that this notion is now occurring. While the advantages of convergence at the hardware/software level are easy to see in the abstract, its virtues are less obvious to the thousands of companies actually pursuing businesses in various industries. For example, the union of computing and television suggests that the display unit (TV screen and computer monitor) should be compatible. As we saw in Chapter 1, there is considerable research to perfect a standardized reception device for moving images. However, for companies in the television business, converting their production and distribution facilities from interlaced-mode to progressive-scan mode will cost hundreds of millions of dollars.

When the cost of convergence is multiplied across the telephony, data processing, broadcasting, cable, microwave, satellite, xerography, and facsimile machine industries, it isn't hard to see why enthusiasts are often observers on the sidelines rather than business people in the trenches. Nevertheless, the efficiency, benefits, and opportunities offered by convergence are likely to overcome the barriers—at least in the long run.

The High Cost of Advanced Television Systems

Earlier, we introduced the rule of thumb to estimate the cost of building two-way broadband networks: "$1 billion per million"; that is, it costs about $1 billion to install a wired advanced television system to one million residential customers. The $1,000 per household includes $750–800 or so for the network and $200–250 for the settop box. These figures are hardly stable. For example, the relative costs of installing hybrid-fiber/coax (HFC) and fiber-to-the-curb switched digital video (FTTC/SDV) designs are changing rapidly, in favor of the SDV architecture. In 1992, HFC cost $400 per home passed; by 1995, that amount had risen to $800. By contrast, in 1992, SDV/fiber-to-the-curb cost $1,200 per household; in 1995, costs had declined to $800. In only 3 years, SDV went from costing 200% more than HFC, to cost parity, encouraging deployment of SDV.

According to Chris Nolan, writing in Cablevision, costs will decline even further in the very near future. He assumes that by the end of 1995, digital cable networks will cost about $850, including the set-top box. However, activating the return path for interactive services will cost $600–$1,200 per mile; assuming average density of 75 homes per mile, this would add $8–$16 to the cost.[15]

Jim Chiddix, Senior VP for Time Warner, disputes these estimates—and he probably has more knowledge about it than just about anybody. He told cable operators that a bare-bones fiber/coax architecture will cost $125–150 per home. Switching will run an additional $700–800 per home and $15–20 is needed to power telephony. Finally, operators should expect to pay $600 per subscriber to deliver personal computer services. Chiddix estimates it will cost operators $1,440–1,570 to provide interactive communication services.[16] High as these costs are, CableLabs President, Dr. Richard Green, estimates that the cable industry can upgrade its plant " . . . to provide two-way interactive multimedia services for about $20 billion, which is just a fraction of the $400 billion required for the telephone companies to comparably rebuild their local networks."[17]

By contrast, wireless systems cost far less to install than wired systems because they do not require the same rights-of-way, digging, and beneath-the-earth maintenance, between $200 and $600 per household.[18] More and more, wireless systems offer the same functionality as the more expensive wired networks.

However, it wasn't until almost mid-1995 that MMDS companies had the wherewithal to finance the upgrade to digital and it came from a novel source: the telephone companies.[19] With wireless cable, NYNEX, and Pacific Telesis and other telcos all see a way to get into the video delivery business quickly, migrating to wired architectures as they come online.[20]

Pacific Telesis has committed to investing $5 billion for video dialtone by the year 2000, reaching 5 million households. Their investment in Cross Country Wireless (MMDS) will enable PacTel to provide video service to 5 million customers in Southern California by the end of 1996.

The initial investment was $175 million and the telco will spend $20 million to upgrade the analog system to digital. The company must pay for antennas, settops, and possible in-home wiring at an average cost of $600 per home. These numbers mean PacTel will be in the video delivery business over wireless cable for about $200 million, paying out the $600 per home only for those that subscribe. By contrast, wired systems pay to pass households whether or not they buy the service.[21]

The cost of running a wireless system, as well as entry costs, is also lower. MMDS operating costs are in the 25% of income range while wired cable operating costs are 36% to 39% of gross revenue. There are no broken cables to fix, no rights of way to lease, or franchise fees to pay. Further, advertisers are impressed by their ability to target viewers within a 35 mile radius, a far greater area than covered by wired systems.[22]

The cost per customer for LMDS service falls between $250 and $600, depending on the assumptions about the number of subscribers in the cell area. The 2–3 mile radius makes LMDS an excellent choice for very densely populated urban areas, while MMDS's 35 mile radius is more efficient for suburban regions. Rural and other low-density regions are better served by some form of satellite service.[23] Table 11.2 compares the systems under discussion.

The above discussion provides a sense of the real costs to conduit providers. To conceptualize their cost/income projections and develop a business model, they begin with a brute force approach. For example, an organization that plans to deliver interactive television over a wired system will spend about $1,000 per home passed. Of these, an optimistic projection would have about 60% actually sign up, making the cost to wire each subscriber home about $1,666. If the aver-

Table 11.2 Comparison of Operating Costs of Wired Versus Wireless Infrastructures

System Type	Construction Cost (per HH)	Operating Costs, as % of Gross Revenue	Area of Coverage
HFC	$800–1,000	36–39%	Few miles
SDV	$800–1,000	36–39%	Few miles
MMDS	$400–600	25%	35 miles
LMDS	$250–600	Not avail.	2–3 miles
DBS	$275–$25 (depends on #)	41%	Thousands of sq.mi.

age cable bill is currently about $33 per month, the yearly income from that home is about $400 a year. Assume that the new services will add $12 to the bill, on the average, to reach $45/month and $336/year (some subscribers will take only basic digital services for $7.00, while others will order VOD movies, games, and so forth). At that rate, allowing for $17 (about 37.5%) in expenses for product and maintenance, it will take almost 5 years to recoup the investment.

At the point of construction, the cost of a system is fixed. If for some reason the system attracts fewer subs, or subs sign up for fewer services, it would take even longer to get the service into the black. If the company can show it can successfully market more services, it makes the prospect of this massive spending much more appealing to the board of directors and investors.

One incentive to delay implementation is the rapid decline in costs as time passes. A good example is the cost of storing a library of video offerings. The server, switching, and storage are a major expense in upgrading both wired and wireless systems for video-on-demand, about $500 to $700 of the total costs. Just five years ago the cost of storage was about $1.00 per megabyte; today that cost is 25 cents. The disk drives to store 1,000 movies, assuming 25 cents/megabyte, would cost about $800,000 to feed 10,000 simultaneous streams—enough for 40,000 to 50,000 subscribers, assuming 20% utilization. This would place storage costs at $80 per bitstream and $16–20 per subscriber, figures not out of line when the total costs for the system are considered—except that next year they'll be lower yet.

Competition and Cooperation

The tempting revenues from communication products and services and the emergence of alternative delivery technologies leads to increased competition. The high cost of providing them encourages cooperation resulting in mergers and alliances.

In 1975, there was one source of TV; in 2005, there will be 11. Cable, telephone companies, power companies, and computer networks can bring TV over wired networks. Broadcasting, MMDS, LMDS, low power TV, DBS, DSS, and LEOS are all potential wireless technologies.

The battle for the settop box and navigation software

Every aspect of advanced television systems is fiercely competitive— every industry, design, hardware component, software element, program, and service. The markets for navigation software and settop boxes are particularly contentious. Settop boxes drive the decisions about equipment to provision the network; navigation software controls consumer access to product.

Starting with the settop market, the success of the manufacturers of many network components—chips and microprocessors, servers, switches, modulators, multiplexers, and software—depend on the success of the settop box company they have chosen to support. The need for interoperability to the settop has driven many of the various alliances between the interactive television vendors.

Settops are a lucrative market, as they are needed by cablers, telcos, and DSS companies to place in 50 million homes. At an average cost of between $175 and $250, the stake ranges from $875 million to $1.25 billion. In 1993, the market leaders in analog settops were General Instrument, with the most market share (56%), followed by Scientific-Atlanta (15%), Pioneer (7.5%), and Zenith (7.5%); the rest share the remaining 14%.[24]

At present, it is impossible to provide a settop box with advanced two-way features at the $250–$300 price point where they will be cost-effective. However, over the next few years improvements in processing capacity and increased volume may bring prices near this range. C-Cube, Micro-systems, and Hyundai Electronics America introduced smaller integrated chips at the 1995 National Association of Broadcasters' convention. The benefits are increased processing, lower power usage, and, ultimately, lower cost, which will be seen sometime in 1997.

Today's settop box belongs to the cable system and subscribers pay a monthly fee for its use. However, there are moves afoot to allow consumers to buy their own settop boxes at retail. For example, Senator William Cohen (R-Maine) introduced an amendment to the 1995 telecommunications reform bill to provide for retail sale of the units, and similar legislation is favored by powerful leaders in the House of Representatives.[25]

The next generation settop is a hybrid analog/digital model. Three telephone companies alone, Bell Atlantic, Pacific Telesis, and NYNEX, have ordered 4 million settop units worth about $800 million.[26]

Scientific-Atlanta's 8600X allows for graphic displays of electronic program guides, accepts different software downloads that change what the settop box can do, and displays information quickly. S-A has sold more than 1 million 8600Xs for between $135 and $200.[27]

General Instrument's CFT is modular, flexible, and upgradable. It also has a security port, expandable memory, and a radio device for wireless upstream pay-per-view orders. It costs between $140 and $180, depending on the quantity ordered.

Analyst Carl Lehmann, with BIS Strategic Decisions (Norwell, MA), projects that digital settop boxes won't reach 1 million homes until 1997. By the turn of the century, he believes that of 51 million settop boxes, 22.5 million will still be analog, 2.5 million will be hybrid analog/digital, and only 7 million will be fully digital.[28]

There will be an even more advanced all-digital settop in the future. It will be nothing more than a powerful computer that can decompress and descramble downstream video, and compress, scramble, multiplex, and address material the customer sends upstream.[29]

Intimately tied to the settop box is the navigation system which is the software that helps the viewer move through the offered programming and services. The reason it is such a competitively pursued product category is that it plays an important role in capturing a market in advanced television systems. In the online world, a World Wide Web browser called Netscape is a recent example that demonstrates how a proprietary customer interface can build market share with prominent gateways and tailored menus.

Netscape markets server software for companies that want to establish a website. The company also gives away software to Internet users to access the web. Every time a person starts Netscape on their computer, the software logs into Netscape Corporation's own site—an attractive, exciting gateway that touts other Netscape wares, carries advertising, and directs users to other sites that use Netscape software. This feature cannot easily be disabled.

Even after the user links out of Netscape's site, it remains the most accessible site through one click on the permanent button on the screen. Only Netscape software can access the most advanced features on Netscape sites, including secure communications for enabling financial transactions. Using this system, Netscape has come to dominate navigation of the web portion of the Internet almost overnight. In fact, the second most popular website is Yahoo—prominently featured on Netscape's first level menu.

Netscape's strategy is extremely successful. They have captured significant market share in a very short time. If a company wants a lot of people to check out their website, who do they talk to? If a marketer wants users to be able to send secure credit card information to their server, whose software do they buy? It's important to recognize that all this is accomplished without Netscape owning the conduit in the presence of many other browser products. Netscape was simply first to market with an excellent product.

What Netscape is doing now is in many ways a tiny scale model of the kind of system Bell Atlantic plans. What will happen when this strategy is translated to the "monster model" of an IATV conduit-content provider?

Under B-A's system, it would be impossible for a customer to opt out, when B-A controls the pipeline, the server, the settop box, the software, and the content. This monopoly power offers great potential to manipulate viewing habits. B-A's marketing strategy is like Netscape's. They will sell their software to content providers for implementation so even when the content migrates to another system, B-A will still get a piece of the action through software royalties. Also like Netscape, B-A's proprietary settop box will give content providers using B-A's software access to a richer set of features than to those not using B-A's software.

Consumers like competitive markets and in many cases prices do decline. For example, in the Glasgow, KY market where the local electric company also provides cable services, prices are significantly lower. This pattern prevails in other two-provider markets, as well.[30]

Competitive strategy: synergy through alliance and merger

The *Wall Street Journal* headline read: "Mergermania in commtech sector," noting that acquisition activity in the information technology sector soared 76% in the first half of 1995, with 649 transactions.[31] One way for organizations to cope with a tumultuously competitive environment is to merge or ally with others whose core business is complementary to their own. If it adds to horizontal or vertical control of a market, the deal is said to "leverage" the company's efforts or to provide "synergy."

The majority of the blockbuster alliances that have taken place are between content providers and conduit owners. Examples are the agreement between Disney and ABC; the partnership between 3 baby bells headed by former CBS executive, Howard Stringer; Paramount and Viacom; and Time Warner and U.S. West (although these last two companies also make a good conduit/conduit partnership with little system overlap).

There are two models for the relationship of conduits to content. The oldest is the "common carrier" model, developed out of the regulation of telephone companies. In this view, the provider of the conduit is prohibited from providing content and must interconnect their conduit with others in an international infrastructure. The other model is what Gilder terms the "Malone model," derived from John Malone's control over the Telecommunications Inc. conduit and its content as well.[32]

One way to look at these agreements is to examine whether the partners are content or conduit providers. Table 11.3 indicates that there are content-content alliances and conduit-plus-content partners, but few conduit-conduit ventures.

One industry observer doesn't believe this current crop of alliances will ultimately prove to be viable. Futurist George Gilder argues that such partnerships work against both parties in a bandwidth-rich market. He points out that it doesn't make sense for content providers to limit their material to a single outlet; they need their products to appear on as many conduits as possible. And conduit providers who want to maximize the business opportunities afforded by plentiful capacity should seek additional offerings to attract new customers and to entice existing customers to maximize usage.[33]

Media economist David Waterman of Indiana University provided a thoughtful response to the Disney acquisition of ABC on the Internet: "I am surprised at the wild enthusiasm for this merger on Wall Street, with all the discussion about new synergies, guaranteed access to TV viewers through the ABC network for Disney products, etc. To follow, I am sure, will be articles and discussion about the threat to the public from media conglomerates being too big, etc.

"These are smart people making these decisions, and it illustrates the signficance of strategy in successful competition, so perhaps it's a good move for them. But the enthusiasm, as well as general alarm, also are indicative of the comic book view that most people seem to have of how companies compete with each other.

"There is a vision of huge giants with many arms slugging it out in the ring until one succumbs and the other stands up on top grunting and snorting. In reality, only the different particular segments of these big companies (e.g., record distribution, movie distribution, cable networking) really compete with each other directly, and this competition is essentially unchanged by conglomeration.

"Vertical integration creates strategic and cost advantages, but in my view they are much more limited than people seem to imagine. ABC Network has approximately the same incentive to carry Disney

Table 11.3 Deals Between Content and Conduit Providers

Alliance	Content	Conduit
Apple with:	s/w	h/w
IBM		h/w, system s/w
Scientific-Atlanta		h/w
Kaleida Labs	authoring s/w	
TCI with:		cable systems
Acclaim	video games	
Prodigy with:	online service	
BellSouth Corp.		telco
Nynex		telco
Pacific Bell		telco
Residential Broadband Services Forum:	all content providers	
America Online, Bellcore, CNN, Media General Cable, MicroMall, 6 others.		
Broderbund with:	edutainment	
Learning Co.	educational software	
HBO with:	TV production	
Warner music	music and videos	
Tele-TV: Headed by executives from CBS and FOX	broadcast TV	
NYNEX		telco
Pacific Bell		telco
Bell-Atlantic		telco
Disney with:	movie/TV studio	
CapCities/ABC		TV network
Ameritech		telco
BellSouth Corp.		telco
Southwestern Bell		telco
Time Warner	movie/TV studio/ interactive content	cable systems
U.S. West		telco
Dreamworks	TV, films, IA content	
Microsoft	s/w	
Paul Allen	s/w	
IBM	digital library	
SGI		h/w

Table 11.3 *Continued*

Alliance	Content	Conduit
Microsoft with:	s/w, games, services, CD-ROMs	online network
NBC	News and public aff. content	TV network
Southwest Bell		telco
Sony	films, music	h/w
Hewlett-Packard		h/w
U.S. West		telco
Alcatel Cable		h/w
Viacom with:	TV, films, music	cable company
Paramount	TV/movie studio	
MCI		LD telco
News Corp (FOX)	TV/movie studio	

programs that they did in the past, and from an organizational standpoint, the risk is that a decision to carry a Disney program gets made without the discipline of the market really putting the program's value in the ABC schedule to the test. And ABC network is hardly the only distribution outlet for Disney programs.

"All these visions of giant slugfests between monsters carry right over into the public interest arena. If there are competitive problems with this merger, it will have to do with particular segments of the media industries, such as record distribution, cable networking, broadcast networking, etc., in which market shares become too large. At a glance this is hard to imagine in this case. The vertical integration could create a public interest problem conceivably, but it is highly unlikely that the substitution of an ownership relationship for what was previously a contractual relationship in which there are the same audiences and the same incentives to satisfy them, will make any serious difference to the public and what they get for their cash."[34]

Competitive strategy: camp-building alliances

Some deals that purport to be made for developing product are actually exercises in "camp-building." Companies engage in building camps when they need to share technical and marketing information to make their products available and compatible with one another.[35]

Thus, camp-building describes how companies form alliances to set standards with the objective of building a market. For example, Prodigy joined with telcos BellSouth, NYNEX, and Pacific Bell to make sure its subscribers would have access to ISDN technology and the high-speed services it allows.

Chris Halliwell studied the standards-setting process over time and her work suggests that standards may play a different role, depending on the stage of the market for the technology. She used the categories of the theory of diffusion of innovation to describe how camp-building takes place.

Table 11.4, adapted from Chris Halliwell's model, shows why each type of adopter buys (or does not buy) a new product, the role that standards play in the buying decision, what producers need to do to increase the size of the market, and how to evaluate standards.

This fascinating analysis suggests that standards are important throughout the life of a product, not just at its beginning. In the early stages, producers need to get together to develop the standards needed for interoperability so there can be a market. As the Early Ma-

Table 11.4 Diffusion and the Standards-Setting Process

	Innovators	Early Adopters	Early Majority	Late Majority	Laggards
Attitude towards tech.	Tech for its own sake	Compet. advantage or personal advantage	Practical use	Use, not tech-oriented	Tech averse
Attitude towards standards	Don't care	Don't care	Want stable, accepted stds.	Demand end-to-end solution	Cling to old stds.
Role of stds.	Verify technology	Deploy stds.	Partially accept stds.	Industry-wide stds.	Extend prev. stds.
Develop market-	Mfgrs agree to grow mkt	Cultivate early users	Get partners for pkg. solutions	Develop stds. and products	Invest in educ.
Eval. of stds.	Meets needs of users	Meets needs of users	If no de facto stds. need them	Multiple stds. required	Use stds. to grow mkt

All: Address cost of adoption to users.

jority begins to buy, the market enters a mainstream stage and standards are essential if it is to grow. The Late Majoritarians demand end-to-end solutions and application-specific products, which probably require multiple products that comply with their own standards.

An example of camp-building is the Microsoft effort to build alliances to promote its IATV operating system. The company has brought in U.S. West, Deutsche Telekon, Telstra Corp. of Australia, Rogers Cablesystems, SBC Communications, and TCI Cable to support its network management software, "Microsoft's Media Server" (MMS). It provides the applications environment, acts as spectrum manager, and handles billing and other operating functions all in one package.[36]

An important new element of camp-building is the emergence of the "gorilla in the tornado." Author Geoffrey Moore describes high tech markets as bursting upon the marketplace like a tornado. Inside the tornado is a gorilla, a company that grows at an extraordinary rate and quickly dominates the market. Writes Moore: "All of a sudden, you go from 'We're not doing it,' to 'We do it all at once.' What it does is it creates phenomenal market demand that wildly outstrips supply . . . As it turns out, the tornado will always have a gorilla. In fact it must have a gorilla because the gorilla becomes the basis for de facto standards. Technological deployment can't happen without standards." In this way, a new company becomes the center of a larger camp-building exercise when it is joined by existing companies that move in to develop the new market created in the tornado.

Competitive strategy: branding

Driving media conglomerates to control their own content creation facilities is the belief that the key to survival in a multi-channel environment is to build a network identity, a brand name, around the content itself.[38] The necessity of branding has established an environment that encourages "conduit-content" or "pipeline-programming synergy." Examples of this type of synergistic consolidation is the 1994 Viacom acquisition of Paramount Pictures and the 1995 merger of Disney and Cap Cities/ABC.

There are several ways to execute a branding strategy, all based on content. Broadcast networks use logos, advertising slogans ("ABC's the One!"), and "signature shows" (for example, "Seinfeld" on NBC) to build their brand. Cable networks title their service to identify their offerings, such as Comedy Central, The Family Channel, The Game Channel, the Cartoon Network, and so forth.

"You have to control your programming," said Bob Jones, vice president of programming for the Mind Extension University. "And one of the best ways to control that brand and look is to create it yourself." Thus, the purpose of branding is to offer a service that is readily identifiable by consumers through controlling the service's image and content.

Identifying Markets and Estimating Demand

The first step in approaching IATV is to identify what market will be important. Hewlett-Packard has released data from a pair of studies that will illuminate the discussion in this section. The first study covered market segments that companies should prepare themselves to enter, as shown in Figure 11.4.

The problem is more complex when it comes to forecasting consumer demand, even when a company is certain about the product being offered. An idea of the difficulty can be gleaned from the following anonymous message that was posted to an Internet discussion group

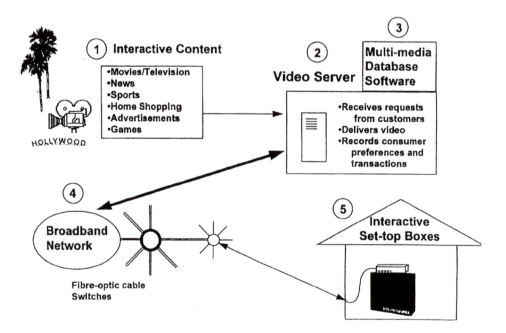

Figure 11.4 Hewlett-Packard's view of important markets in IATV.

in August of 1995: "Many of my friends on the Web say they can't afford to subscribe to cable, a newspaper, a couple of magazines, an online service like AOL, a monthly web browser charge, and then . . . on top of all that . . . an additional service like USA Today OnLine or the Washington Post's Digital Ink. To draw those extra few bucks from people's pockets is quite a challenge and it had better be something very unique, personally valuable, or extremely entertaining (or X-rated)."

To figure out this difficult problem, companies marketing TV systems, programs, and services do market research to estimate potential demand. In the previous chapter, the many projects underway show how important media companies believe this iterative testing is to their ability to formulate realistic business plans. The bulk of the results from these trials will not be available until 1996, and many may remain proprietary even after that time.

In the meantime, until there is robust market research from actual experience, consulting firms, investment houses, and research companies conduct surveys to try and estimate the market for advanced television. In fact, so many studies of the demand for IATV have been undertaken, there has even been a study of the studies.[39] The results are somewhat disparate, largely due to methodological differences. However, there were some similarities and they confirmed the findings of one of the best studies, the Hewlett-Packard omnibus study, so we will consider it in detail.

HP sponsored a year-long study that was conducted between early 1992 through early 1993 by American LIVES, Inc., a San Francisco-based research service. The study used surveys, focus groups, personal interviews, and home visits to learn how large the potential market for IATV services is, what consumers wanted, what would motivate them to purchase services, and what the barriers to purchase might be.[40]

The HP study began by identifying a qualified U.S. population of 37 million households, defined as those where the head of the household is between 25 and 65 years old and has an annual income greater than $25,000. The researchers found they could divide the qualified households into three segments:

Core market	14,000,000
Next market	12,000,000
Out of market	11,000,000

People in the Core market were further divided by their motivations to purchase interactive services. One group just liked the idea of them; researchers dubbed them the Interactive Television Fan Club. The second group were called Utilitarian Time Savers who liked applications that save them time and make chores more convenient.

The people in the Next Market also fall into two groups. The Ambivalents are interested in the benefits of IATV but they have several concerns that are barriers to acceptance, including risks to personal privacy, fears over security of their data, bank or credit accounts, impulse buying, and uncontrolled use of the system by the children in the household. The Lukewarms just aren't that interested in IATV, period.

Finally, the Out-of-the-Market segment includes Anti-TV Rejecters, who don't like the current television system and don't like IATV either. A second group, the Laggards, are very cautious in their buying habits and wait to see how others use services before purchasing themselves. Laggards may never purchase.

The study found that about 1 million households are Innovators, who are likely to purchase services soon after launch. These consumers are younger than other groups. The services that interest them are:

90%	Information and educational services
88%	Education resources
88%	Information from advertisers
83%	Comparison shopping information
82%	Order from customized catalogs
78%	Bill-paying

Chapter 10 considered the problem of "virtuous" responses and the researchers who conducted this study encountered this problem. They noted in the study that they believe the desire for entertainment was probably vastly understated. While educational applications may motivate the purchase decision, entertainment applications appear to motivate actual use.

The results indicate how important it is to market to the entire family. The research found that sales to households depend on applications that enable families to learn and play together. Children play a big role in motivating a purchase. They intuitively know how to use it and why it is fun. If parents believe their children will benefit, they are much more likely to purchase. However, ease-of-use, including a simplified remote, is essential to family purchase.

Estimates of what people will pay for IATV service range from $5 to $25 per month above the current cable bill. Several studies agree that movies on demand, home shopping, video games, living room gambling, and education will be highly popular, but government services and infomercials would not be successful.

Advertisers will be pleased to learn that advertising on IATV networks appears to be successful. Videoway customers in Quebec chose a Molson beer commercial to view, from among several, and their recall was 40% higher than those who didn't have the opportunity to choose. In the Netherlands, 38% of customers who received a McDonald's coupon printed out by their settop box redeemed them, compared with the average 1% redemption rate.

Companies planning to market communication conduits and content must conduct extensive research before they can offer products. However, the task is easier, with greater certainty of success, if the company chooses to enter an existing market.

Moving in on existing markets   Media economist Michael O. Wirth, University of Denver, analyzed this "least risky" approach to developing additional revenue streams. Table 11.5 shows the markets he identified: 1) Video dial tone, telco provided cable service; 2) NVOD/VOD; 3) Home shopping service; 4) Advertising; 5) DSS TV delivery; and 6) Interactive games.[41]

Wirth also writes about "riskier alternative revenue streams," where consumer demand does not yet exist. These include information-on-demand channels, video telephony, distance learning, and HDTV. Businesses need to exercise caution about investing in these riskier alternatives, says Wirth, citing Tryg Myrhen, president of the Providence Journal Company: "It's great to be on the cutting edge, as long as you stay behind the blade."

Business Models

How a business conceptualizes its product depends on its model of the marketing situation. Each industry brings a set of practices which influence how they go about developing their business plan. One insider summed up the various industry models: "deliver eyeballs, get ads; deliver value, get subscriptions." Table 11.6 shows the various industry business models in a more formal manner.

Table 11.5 Existing Markets for Additional Revenue Streams

Market	Existing revenues	Projected revenues
Video dial tone	Cable industry: $25.7B in 1995	27% share—6.2% (Arthur D. Little) 15% share—3.5% (Malarkey-Taylor)
NVOD/VOD	1995 Home video rent.: $11.5B 1995 Home video pur.: $8.6B	$10B/yr. by 2005 Cable: 75%, $7.5B Telcos: 15%, $1.5B (Malarkey-Taylor) $3B/yr by 1997 $9B/yr by 2003 (Bernstein Res.)
Home shopping	Catalog shopping: $46B Home TV shopping: $2.2B (Donaldson, Lufkin & Jenrette)	
Advertising	B'cast TV: $26.6B Radio: $9.2B Cable TV: $10B (23% viewers, 7.6% ad revenue) (Donaldson, Lufkin & Jenrette & CableVision)	
DSS	Estimated breakeven: 3 mil. subscribers (Sky Report)	6% TVHH by 2000 (Morgan Stanley) 10% market share by 2005 (Malarkey-Taylor) 10% TVHH by 2000 (industry belief)
Interactive games	1993: $6.5B	
Online services	$344.9 mil. in 1994 (The Electronic Marketplace)	$3.3B in 1997 (Dataquest)

Lessons Learned

In spite of a quarter of a century of investment, testing, hype, commitment, and effort, IATV has not yet been a successful innovation. While corporate publicists and presidents wax eloquently about this new

Table 11.6 Business Models by Industry

Industry	Business model/description
Cable	*Subscription:* subscribers pay a flat monthly fee *Value added:* pay extra for really neat stuff, like an important boxing match or special
Telcos	*Flat service fee:* A monthly basic phone bill *Usage fees/metered service:* charges for calling within an area, long distance access, and long distance fees *Network utilization:* charges incurred when your phone is "off-hook," like an 800 number or cellular phone
Power companies	*Usage fees* *Flat service fee:* some have introduced to let customers pay the same every month, but it's adjusted at the end of the year anyway.
Computer online	*Subscription:* monthly flat fee *Usage:* most charge extra for chat rooms and information services
Film & video	*Transactional:* you pay, you get, you go--every rental deal is independent and unique

technology that will "change the way we live, work, and play," couch potatoes recline on their sofas and scarf up "Roseanne" and the disease-of-the-week (or murder-of-the-week) Sunday night movie. Life goes on.

For Professor William Dutton, today's IATV trials are nothing new.[42] He has been studying such tests since the 1970s, compiling the results from earlier testing efforts, as well as the current generation. Dutton finds there are 6 causes of failure—failure in the sense that the technology does not diffuse for widespread use. The 6 barriers to success are:

1. The lack of killer apps (trigger services, in Dutton's parlance) to spur demand;
2. a wide gap between the image and promise of a technology and its actual implementation;
3. failure to reach a critical mass of users;
4. inability to change habits of media usage;
5. concerns about privacy and security; and
6. high costs.

Many of the causes of failure pointed out by Professor Dutton exist in today's generation of IATV systems. For example, despite a concentrated search over the past 5 years, no one has uncovered a clear killer app. In the absence of a single service or program that would attract users, some developers argue that the killer app is actually the totality of offered services. Maybe so, but Dutton's work suggests there may be a problem.

Then there's the hype. Virtually every site, except possibly the Time Warner testbed, fails to live up to its PR and the reality is almost sure to disappoint consumers.

Nor are providers addressing consumer desires. Nearly every demand study shows that consumers want communication services. Companies are leery of attempting to put together these services because they require a critical mass to become viable (enough people have to be on the system to reach desirable communication partners). However, reaching a critical mass is quite difficult, expensive, and prone to failure. After all, unless a communication technology improves and enhances the life of consumers, they will not alter their habits to use it.

Recent examples of technologies that caused people to change their habits are the fax machine, the CD player, and e-mail. By contrast, the teletext and the CD-I multimedia player were apparently insufficiently useful or attractive to influence people to change their media habits.

The points Dutton raises about privacy and security worries on the part of consumers are also important. These issues continue to weigh heavily on people's minds and they are prominent among their concerns, as shown in the Hewlett-Packard study summarized earlier. Finally, cost is important to everyone.

Unless system planners address these consumer issues, it will be impossible for them to succeed. Dutton's research on both past and current makes it clear that builders, designers, developers, and marketers must deal decisively with each of these elements.

At the many trade shows and expositions that cover advanced television, companies often create videotapes of their systems. A cursory viewing will show the most casual observer that today's crop of interactive television test sites often suffer from one or more of the very problems that Dutton points out.

Media companies are focused on downstream entertainment and information, not the communication services their customers tell them they want. Construction costs are high and providers want to charge equally high fees to recoup their investments quickly. In their efforts to promote IATV, they overpromise. At the same time, consumer worries about security and privacy are inadequately addressed and, in any case, it isn't clear that solutions are even available. Given these problems, it is likely that a critical mass of users will not change their media habits. Thus, even if they build it, people will not come unless planners and providers change their ways.

Nevertheless, advanced television systems are on their way—over the wire and over the air. Too many technologies are ready, too much money has already been spent, and there's plenty of profits for the companies that figure it all out.

As mentioned in the previous chapter, several telephone companies have suspended trials and have commenced building. They are taking a page out of Thomas Edison's book when the Edison Electric Light Company started up its first generating station in New York in 1882. In what was probably the fastest diffusion of any wired technology on record, by 1887, there were more than 120 generating stations serving 325,000 lamps! What Edison didn't do was to set up a 5-year test facility with white-coated researchers running around with clipboards, asking people, "Now what are you going to do with all that light and energy?"

The philosophy of control that pervades modern American culture demands that media companies control the content, control the conduit, and ultimately control the audience. There is some irony that in doing so, they run the risk of controlling IATV right out of business before it ever gets started. The willingness of the telephone companies to simply get on with it is welcome, as is their provision of video dial tone—if they can actually resist the temptation to monopolize their networks.

The end of the 20th century in the U.S. is not a pretty picture. We are witnesses to the first generation of children to have less opportunity than their parents, to homelessness, hollowed out inner cities, crime, taxation, and the failure of public education. A broadband two-way network is a generational dream, a dazzling, thrilling, and positive legacy that the largest-ever demographic cohort, the Baby Boomers, can leave. Go forth and communicate!

Notes

1. Everett M. Rogers, *The Diffusion of Innovation* (New York: Free Press of Glencoe, 1962). Also see Everett M. Rogers, *Diffusion of Innovation* 3rd ed. (New York: Free Press, 1983).

2. Peter Drucker, *Innovation and Entrepreneurship* (New York: Harper & Row, 1985).

3. Revenues for utility industry came from Peter Lambert, "Utility networking percolates," *On Demand* (December 1994/January 1995):16. Revenues for the cable industry were reported in Tom Coyle, "1994 local telco revenue per average switched line," *America's Network* (June 15, 1995):14.

4. Don West and Harry Jessell, "The way uphill: How cable's making it on the infohighway," *Broadcasting & Cable* (November 28, 1994):34–44.

5. Jim McConville, "Cable business: a tale of two operators," *Broadcasting & Cable* (May 15, 1995):24.

6. Leslie Ellis, "National cable interconnect activity heats up," *CED* (September 1994):28–32.

7. Peter Lambert, "PC Connection," *On Demand* (March 1995):9–12. See also Fred Dawson, "My kingdom for a modem," *CED* (June 1995):127–134.

8. Sanjay Kapoor, speaking at Digital World, April 1995, Los Angeles.

9. AT&T Globespan product brochure.

10. Lawrence K. Vanston, "ADSL/HDSL: Dangerous temptation," *New Telecom Quarterly* (3rd Quarter, 1994):53–61.

11. John McConnell and Jane Lehar, "HFC or SDV architecture? Economics drives the choice," *Communications Technology* (April 1995):34–40. See also "FTTC costs come down," *Investor's Business Daily* (May 25, 1995):A7.

12. Harry Jessell, "Infohighway power play," *Telemedia Week* (December 1994):26–28.

13. Thanks to James Bromley for this observation and leading me to explore this idea further.

14. S. R. Rivkin, "Electric utilities will build telecom infrastructure," *New Telecom Quarterly* (2nd Quarter, 1994):15–19. See also S.R. Rivkin, "Positioning the electric utility to build information infrastructure," *New Telecom Quarterly* (3rd Quarter, 1995):30–33.

15. Chris Nolan, "Reality check: What's here? What's now? What's never?" *Cablevision* (August 22, 1994):22–32.

16. Mark Berniker, "Experts bullish on cable's migration to digital services," *Broadcasting & Cable* (May 15, 1995):42, 46.

17. Leslie Ellis, "Technology debate obscures MSOs digital agenda," *Multichannel News* (May 17, 1995):1A, 16A-17.

18. "DBS Competition," *Specs Technology* (December 1994):6.

19. Joel A. Strasser, "Telco power," *Wireless Cable* (June 1995):10–13.

20. Mark Berniker, "Bell Atlantic, Nynex purchase CAI wireless systems," *Broadcasting & Cable* (April 3, 1995):40.

21. Strasser, op. cit.

22. David Tobenkin, "The wireless system that could," *Broadcasting & Cable* (May 1, 1995):20.

23. Barry W. Phillips, "Broadband in the local loop," *Telecommunications* 26:11 (November 1994):37–44.

24. Joe McGarvey, "Competition heats up early digital set-top market," *Inter@ctive Week* (January 16, 1995):26.

25. Ted Hearn, "Maine Sen. introduces retail set-top box bill," *Multichannel News* (April 10, 1995):10.

26. Chris Nolan, "The telcos' set-tops," *Cablevision* (April 3, 1995):56.

27. Chris Nolan, "It's analog for now," *Cablevision* (April 3, 1995):54.

28. "Digital delay," *Inter@ctive Week* (January 16, 1995):27.

29. Joe McGarvey, op. cit.

30. George Lawton, "The utilities' role: Building the ubiquinetwork—Part 5," *Communications Technology* (December 1994):80–85.

31. "Mergermania in commtech sector," *Wall Street Journal* (July 7, 1995):B-8.

32. George Gilder, "Washington's bogeyman," *Forbes ASAP* (June 6, 1994):115.

33. George Gilder, "Mike Milken and the two trillion dollar opportunity," *Forbes ASAP* (April 10, 1995):104.

34. David Waterman, Professor at Indiana University, electronic document, posted to discussion group on telecom regulation, August 10, 1995.

35. Chris Halliwell, "Camp development: The art of building a market through standards," *IEEE Micro* (December 1993):10–18.

36. Microsoft promotional literature. See also Chris Nolan, "Slow and steady?" *Cablevision* (March 20, 1995):22.

37. Geoffrey A. Moore, *Inside the Tornado* (New York: Harper Collins, 1995).

38. Linda Haugsted, "Brand or bust is cable operator's latest motto," *Multichannel News* (June 12, 1995):36.

39. Tim Clark, "Interactive attitude survey roundup," *Inter@ctive Week* (May 8, 1995):24.

40. Hewlett-Packard, "Is there a market for interactive television?" Proprietary research available from Casey Lemus, Interactive Television Appliances, Hewlett-Packard Company, Palo Alto, CA, 408/553-2948.

41. Michael O. Wirth, "The emergence of the alternative media revenue streams," paper delivered at the Broadcast Education Annual Convention, Las Vegas, NV, April 7, 1995. Michael Wirth is professor and chair, Department of Mass Communications, University of Denver, Denver, CO.

42. William H. Dutton, "Driving into the future of communications? Check the rear view mirror," paper delivered at POTS to PANS: Social issues in the multimedia evolution from Plain Old Telephony Services to Pictures and Network Services, the BT Hintlesham Hall Symposium, Hintlesham, Suffolk, March 28–30, 1994.

Glossary

Entries marked with (*) were taken from *Newton's Telecom Dictionary*, written by Harry Newton, published by Flatiron Publishing, Inc., New York. You can get his book from 800-542-7279 or by fax at 212-691-8215 or orders@flatironpublishing.com. This 1,400 page compendium covers telecommunications, computer telephony, data communications, voice processing, networking, and the Internet. It is the largest-selling telecom dictionary in the world. I recommend it highly for its thoroughness, ease of use, and witty writing, a rare quality in writing about technical topics.

active matrix liquid crystal display (AMLCD)   Most likely candidate to replace the current television screen.

addressibility   The capacity of a wired system to send material to a particular terminal or household.

adopters   One group exisiting in the Theory of Diffusion of Innovation. This is someone who tries out an innovation and either continues or discontinues its use.

ADSL (asymmetrical digital subscriber line)   A system that is designed to carry video to the home by way of standard telephone lines.

affiliate,   An affiliate is a contractual partner of a television network. The contract specifies when and how much programming will be supplied by the network, at what cost, and under specified conditions.

aliasing   In sampling, the impairment produced when the input signal contains components higher than half of the sampling rate. Outcome will produce a diagonal step graph.

AMLCD   See **active matrix liquid crystal display**

*amplitude   The distance between the arc and the trough of a radio wave, sometimes called "height."

*amplitude modulation (AM)   A method of adding information to an electronic signal in which the signal is varied by its height to impose information on it. "Modulation" is the term given to imposing information on an electrical signal. The information being carried causes the amplitude (height of the sine wave) to vary.

*analog   Comes from the word "analogous," which means "similar to." In telephone transmission, the signal being transmitted—voice, video, or image—is "analogous" to the original signal. In other words, if you speak into a microphone and see your voice on an oscilloscope and you take the same voice as it is transmitted on the phone line and ran the signal into the oscilloscope, the two signals would look essentially the same. The only difference is that the electrically transmitted signal (the one over the phone line) is at a higher frequency.

API (Application Programming Interface)   In advanced television networks, APIs are used to translate between applications and the network. APIs will include standardized interrupts, calls, and formats that allow all parts of the network to work together. Control of the API dictates which content providers can use the system.

*asynchronous transfer mode (ATM)   ATM is the fast, cell-switched technology based on a fixed-length 53-byte cell. All broadband transmission (whether audio, data, imaging, or video) are divided into a series of cells and routed across an ATM network consisting of links connected by ATM switches. Each ATM link comprises a constant stream of ATM cell slots into which transmissions are placed or left idle, if unused. The most significant benefit of ATM is its uniform handling of services, allowing one network to meet the needs of many broadband services.

*attenuation   The decrease in power of a signal, light beam, or light wave either absolutely or as a fraction of a reference value. The decrease usually occurs as a result of absorption, reflection, diffusion, scattering, deflection, or dispersion from an original level and usually not as a result of geometric spreading. In other words, attenuation is the loss of volume during transmission and is measured in decibels.

avatars   In computer games and online services, an avatar is the graphic representative of a user. Users choose their avatar from a library or construct it in a draw program. The avatar is seen by all the other users.

*bandwidth   The range of electrical frequencies a device can handle. The amount of bandwidth a channel is capable of carrying tells you what kinds of communications can be carried on it. A broadband circuit, for example, can carry a TV channel. A broadband circuit capable of providing one video channel can also provide 1,200 voice telephone channels.

*bit   Bit is a contraction of the term BInary digiT. It is the smallest unit of information (data) a computer can process, representing either high or low, yes or no, or 0 or 1. It is the basic unit in data communications.

bitstream   A continuous stream of data bits transported over a communications network, or transmitted over the air.

bpp (bits per pixel)   The number of bits required to represent the color of each pixel in a digitized video image.

bps (bits per second)   The number of bits that transmitted or transmitted over some channel in a second of time. BPS measures the speed of the data flow or bitstream.

byte   Eight bits of digital information, made up of zeros and ones. Sometimes a byte is called a "digital word."

*cache   A high-speed memory designed to hold upcoming to-be-accessed or recently-accessed data. A cache speeds up a computer's operation because high-speed memory sends the information to and from the computer's central microprocessor much faster than a hard disk could.

*carrier signal   A continuous waveform (usually electrical) whose properties are capable of being modulated or impressed with a second information-carrying signal. The carrier itself conveys no information until altered in some fashion, such as having its amplitude changed, its frequency changed, or its phase changed. These changes convey the information.

*cascaded   To connect the output of a device into the input of another device, often the same type as the first.

cathode ray tube (CRT)   A regular television screen.

cell   In switching, when Asynchronous Transfer Mode is used, the data is reformatted into cells, also called octets. An octet is actually the same as a byte—8 bits of data. Each ATM packet is composed of 53 cells (or octets or bytes). Seven cells are the "header" which contains the addressing and destination information; the remaining cells are the "payload" which is the actual message.

chrominance   The part of the television signal that provides information about color. Chrominance has two aspects: the color itself, or hue, and the strength of the color, or saturation. Low chroma makes the color picture look washed out or muddy. High chroma makes the color too intense. Chroma is not brightness; that characteristic is called "luminance."

*cladding   A layer of material of lower refractive index, in intimate contact with a core material of higher refractive index.

Clarke orbit   An orbit, or circling of the earth, by a satellite 22,300 miles above the equator. This positioning—also referred to as a geostationary arc—creates the illusion that the satellite remains in a fixed position. More important, this also allows line-of-sight transmission to anyplace on earth (except the poles) using only three satellites. It is named after science fiction author, Arthur C. Clarke, who first suggested the use of satellites for communication purposes and suggested this positioning for maximum coverage.

*coaxial cable A cable composed of an insulated central conducting wire wrapped in another cylindrical conducting wire. The whole thing is usually wrapped in another insulating layer and an outer protective layer. A coaxial cable has great capacity to carry great quantities of information, usually high-speed data.

codec CODer-DECoder. Equipment that converts analog signals to digital signals and vice versa. The term is used to refer to both audio and video material. When the equipment only translates one way, it is usually called an A (analog)-to-D (digital) converter (ADC) or a D (digital)-to-A (analog) converter (DAC).

COFDM (coded orthogonal frequency division multiplexing) The basis for all transmission designs under consideration in Europe. The cheapest of all transmission schemes, it also lends itself to transmission from small, geographically-dispersed cells.

compaction encoding One of the last steps in digital video compression where the reduced information is finally expressed in as parsimonious a way as possible.

*compatible In the computer world, two computers are said to be compatible when they will produce the identical result if they run identical programs. Another meaning is whether the equipment can be used interchangeably in place of another or in conjunction with a system.

*compression Reducing the representation of the information, but not the information itself. Reducing the bandwidth or number of bits needed to encode information or encode a signal, typically by eliminating long strings of identical bits or bits that do not change in successive sampling intervals. Compression saves transmission time or capacity. It also saves space on storage devices such as hard disks, tape drives, and floppy drives.

contouring In a digital system, the appearance of patterns in a digitized image because quantization did not have enough levels.

convergence The concept that all the different communication structures, such as the telephone, television, computer, and image processing industries are converging into a single, seamless digital structure.

*circuit switching (CS) The process of setting up and keeping a circuit open between two or more users, such that the users have exclusive and full use of the circuit until the connection is released.

cybernetic systems Cybernetic systems have a goal that is established externally, such as a thermostat where the homeowner sets the temperature (the goal) and the heating system and its parts try to keep the room temperature at the goal state. In homeostatic systems, the goal is set internally.

*cycle One complete sequence of an event or an activity. Often refers to electrical phenomena. One electrical cycle is a complete sine wave.

D2-MAC   A hybrid digital/analog format for European advanced television.

dark fiber   Unused fiber optic cable that is not "lighted," meaning that no light is transmitted through it.

DBS   See **direct broadcast satellite.**

dBw   A measure of the strength of satellite signals that stands for "decibels relative to one watt."

DCT (discrete cosine transform)   A transform is a way of manipulating information from one domain to another. The DCT translates spatial pixel information into temporal frequency information. DCT is the most common form of video compression. It arranges information carried by television signals into blocks. A single coefficient describes how often similar blocks appear. This type of transform is the basis for both the JPEG and the MPEG schemes.

demodulation   See **modulation.**

digital tier   Marketers of cable system services divide up their offerings into "tiers." The basic tier gives over-the-air broadcast channels and a few basic cable channels. The satellite tier offers the cable channels people want, like CNN and MTV. The premium tier includes HBO, Showtime, Disney, Cinemax, and The Movie Channel, for which subscribers pay extra. The digital tier is a new tier that will offer all the digital services available on advanced TV systems.

direct broadcast satellite (DBS)   Any service which delivers television programming from a satellite. Includes both TVRO (TV Receive Only) dishes and DSS (Direct-Satellite-System) dishes. TVRO needs a bigger dish because it is on the lower frequency C-Band, while DSS (sometimes called DTH for direct-to-home) is on the higher frequency Ku or Ka bands that can beam to smaller, less expensive backyard dishes.

DSS (direct satellite service)   See explanation under **direct broadcast satellite.**

direct-to-home (DTH)   See explanation under **direct broadcast satellite.**

dubs   Short for "duplicated." Now used to describe copies of videotapes ("It's a dub."), it was originally adopted to refer to copies of films that were "dubbed" for distribution.

*EDFA (erbium-doped fiberoptical amplifier)   A form of fiber optical amplification where the transmitted light signal passes through a section of erbium-doped fiber and is amplified by means of a laser pump diode. EDFA is used in transmitter booster amplifiers, in-line repeating amplifiers, and receiver preamplifiers.

electromagnetic spectrum (EMS)   Waves of light energy, or electricity, that move through the air.

extensible   In advanced television, extensible means that the technology can be extended to work with other communication systems. For example, it is considered important that the final digital television format be extensible to computers.

fiber optics   Hair-thin, flexible glass fibers that carry light signals over long distances. Fiber can carry more signals over a longer distance for less cost and use less power than other transport media such as copper wire and co-axial cable.

fiber to the curb (FTTC)   Fiber has been placed first in the long distance net-work, then the local distribution plant. In this case, the fiber then proceeds through the local backbone, feeders, and neighborhood nodes to the curb, with copper going from the curb to the home itself.

field   An NTSC television picture has 525 lines and is made up of frames. To create a moving picture, 30 frames per second are scanned. Each frame is composed of two fields. One field begins with the scan of the even lines, 2–524; the second field begins with the scan of the uneven lines, 1–525. Thus the NTSC picture has 60 fields per second.

*filtering   A process used in both analog and digital image processing to re-duce bandwidth. Filters can be designed to remove information content such as high or low frequencies, for example, or to average adjacent pixels, creating a new value from two or more pixels.

*firmware   Software kept in semipermanent memory. Firmware is used in conjunction with hardware and software and shares the characteristics of both. Its unique feature is that it will not be "forgotten" when the power is shut off.

flame   A term to describe angry, nasty, or abusive electronic messages that appear on online services. Penners of such messages are called "flamers." Non-flamers who are temporarily vitriolic signal their intentions by typing <FLAME ON> Angry message typed <FLAME OFF>.

*footprint   The area on the earth's surface where the signals from a specific satellite can be received. A footprint is shown as a series of concentric con-tour lines that show the area covered and the decreasing power of the sig-nal as it speads out from the center.

fractal compression   A compression technique that reduces the information in a video image substantially by storing it as mathematical equations, rather than in pixel format. Iterated Systems in Norcross, Georgia, is the most well-known of the companies that is developing fractal compression. The algorithm identifies similar patterns within the image and reduces the pattern to a formula. While the methods result in very small files, the pic-ture quality is merely adequate and contains noticeable artifacts.

frame   See **field**. The basic unit of an NTSC television picture. A frame is a complete scan of all 525 lines, which occurs 30 times each second.

*frequency   The rate at which an electrical current alternates, usually measured in Hertz. Hertz is a unit of measure which means "Cycles per second." So, frequency equals the number of complete cycles of current occurring in one second.

*frequency modulation   A modulation technique in which the carrier frequency is shifted by an amount proportional to the value of the modulating signal. The amplitude of the carrier signals remains constant. The deviation of the carrier frequency determines the signal content of the message. Commercial TV and FM radio use this technique, which is much less sensitive to noise and interference than AM modulation.

FTTC   See **fiber to the curb.**

game players   In the the consumer study by Hewlett-Packard, game players are a market segment composed of people who would buy digital services that let them play games.

*gateway   A gateway is an entrance and exit into the communications network. In data communications, they are typically referred to as a node on a network that connects two otherwise incompatible networks.

GEOS (geostationary earth-orbiting satellite)   A satellite that orbits the earth at 22,247 miles in space.

gigahertz (GHz)   One billion cycles of a wave. See **hertz.**

global information infrastructure (GII)   A plan to bring integrated, seamless information services to the entire world.

*guard band   A narrow bandwidth between adjacent channels which serve to reduce interference between those adjacent channels. That interference might be crosstalk. Guard bands are typically used in frequency division multiplexing. They are not needed in time division multiplexing.

H.261   This standard was set by the ITU for videoconferencing over telephone lines by the International Telecommunications Union. By setting a standard for the coding of the video, the digital format, and forward error correction, equipment from different manufacturers will interoperate.

hardware   Equipment. The physical "stuff" that makes up communications technology: metal, plastic, and silicon.

HD-MAC   A high definition TV standard that originated with the Independent Broadcasting Authority in the U.K. It is compatible with the television sets currently in use in Europe.

HDTV   High Definition Television. High definition TV describes an array of technologies that were developed to improve the television picture, including digitizing the TV picture and compressing the signal so it can be transmitted by local TV stations. The picture is rectangular and contains about 5 times as much information as the current TV picture. The result is an image that has more resolution, richer color, and far superior digital audio quality.

headend The electronic equipment located at a cable television company's facility. This equipment could include antennas, earth stations, preamplifiers, frequency converters, demodulators, and related cable transport equipment.

*header The portion of a message that carries the information to guide the message to the correct destination. This information contains such data as the sender's and receiver's addresses, precedence level, routing instructions, and synchronization pulses.

hedonists In the Hewlett-Packard study of potential customers of interactive television, hedonists are motivated by pleasure.

*hertz Abbreviated Hz. A measurement of frequency in cycles per second. A hertz is one cycle per second. A kilohertz (KHz) is a thousand cycles per second, a Megahertz (MHz) is a million cycles per second, and a Gigahertz (GHz) is a billion cycles.

HFC See **hybrid fiber coaxial.**

*horizontal blanking interval The period of time during which an electron gun shuts off to return from the right side of the monitor or TV screen to the other side in order to start painting a new line of video.

HTML (hypertext markup language) A standardized format for writing material for sites on the World Wide Web. The format allows users to read material and link to other parts of the material and to other sites on the Web.

hybrid fiber coaxial (HFC) A combination of fiber optic and coaxial cable in a cable TV system. Fiber provides high capacity, distortion-free transmission over longer distances and coaxial provides low-cost, multi-channel capacity from the neighborhood node to the home.

IATV The merging of television, computer, and multimedia technology. IATV is delivered through a settop or personal computer and will include such services as media-on-demand, home shopping, home banking, interactive game playing, and learning services for home and schools.

innovators A category of adopters in Everett Rogers' theory of the diffusion of innovation. These people are the first to use the new technology or practice.

*Integrated Services Digital Network (ISDN) A concept of what the world's telephone system should be according to AT&T. The proposal is to overcome the problems with the world's telephone service in 4 ways: 1) By providing an internationally accepted standard for voice, data, and signalling. 2) By making all transmission circuits end-to-end digital. 3) By adopting a standard out-of-band signalling system. 4) By bringing significantly more bandwidth to the desktop.

Internet   An outgrowth of the Arpanet, which was a government-sponsored computer network that linked research facilities, government agencies and departments, and defense contractors. Since the late 1980s, it has expanded to include other types of organizations. In 1992, the Net, as it is called, was revolutionized by the creation of the World Wide Web, which made it easy for people to access a variety of different organizational and personal information sites.

interoperability   An aspect of standards that specify that different pieces of equipment will function as part of a whole system. For example, VCRs must interoperate with any TV set, regardless of who manufactures the components.

ISDN   See **Integrated Services Digital Network.**

*iteration   The process of repeatedly processing a bunch of instructions. Each repetition, theoretically, comes progressively closer to the desired result, the "correct" answer.

IVDS (Interactive Video & Digital Service)   The FCC auctioned a portion of the electromagnetic spectrum for IVDS services. An FM radio signal is used to complement television images. Some systems use FM as the return path as well, while other systems use the telephone.

JPEG   Pronounced jay-peg, JPEG is an international panel named the Joint Photographic Experts' Group. The group was responsible for considering and developing worldwide standards for still image compression. The system removes redundancies within a single image, a technique called spatial compression.

*killer app   Short for Killer Application. The high-tech industry's lifelong dream to discover a new application that is so useful and so persuasive that millions of customers will rush to purchase it. The term derives from the PC industry where a killer app was so powerful that it alone justifies the purchase of a new computer.

kilohertz (KHz)   See **hertz.**

laggards   In Everett Rogers' theory of diffusion of innovation, laggards are the last people to adopt a technology or practice.

*latency   The time interval between when a network station seeks access to a transmission channel and when access is granted or received.

LEOS (low earth-orbiting satellite)   Low earth-orbiting satellites are being proposed for world-wide wireless communications systems. These satellites operate at a low altitude of between 300 and 600 miles—much lower than the 22,300 miles of a typical satellite. They travel across the sky rather than remaining stationary over a single point.

LMDS (local multi-point distribution service)   A proposed cell-based distribution system for video distribution. It is hoped that someday it will play a role in two-way voice, video, and data transmissions.

*lossless   Image- and data-compression applications and algorithms that reduce the number of bits a picture would normally take up without losing any data. In this way, no information is lost or altered in the compression and/or transmission process.

*lossy   Methods of image compression, such as JPEG, that reduce the size of an image by disregarding some information.

LPTV (Low Power TV)   An FCC designation for low power broadcast television at VHF or UHF frequencies.

lukewarms   In the Hewlett-Packard study of potential consumers of interactive television, Lukewarms are people who don't much care whether they are offered the service, but are not actively opposed to receiving it.

luminance   The black and white, or brightness, information in the video signal. Chrominance must be added to provide color.

*MANs (metro area networks)   A loosely defined term generally understood to describe a broadband network covering an area larger than a local area network. It typically interconnects two or more local area networks, may operate at a higher speed, may cross administrative boundaries, and may use multiple access methods. It may carry data, voice, video, and high resolution still images.

MDS   See **multi-point distribution service.**

media server   A file server that provides voice, video, graphics, still images, and data upon request, then saves information about the transaction for metering and billing purposes. The media server is necessary to deliver video-on-demand, a potential "killer app" for interactive television systems.

megahertz (MHz)   See **hertz.**

MEOS (Middle Earth-Orbiting Satellite)   Satellites positioned more than 1600 kilometers above the earth.

*mesh   A data network that provides multiple paths between points. Internet working devices choose the most efficient paths in moving data from one point to another. A mesh network might be constructed for greater reliability or it might be constructed because all the points on the network need to be connected.

*microwave   Electromagnetic waves in a radio frequency spectrum above 890 Megahertz and below 20 Gigahertz. Microwave is a common form of transmitting telephone, facsimile, video, and data conversations used by common carriers as well as by private networks. Microwave signals only travel in straight lines. In terrestrial microwave systems, they're typically good for 30 miles, at which point you need a repeater. Microwave is the frequency for communicating to and from satellites.

*modulation   The process of varying some characteristic of the electrical carrier wave as the information to be transmitted on the carrier wave varies. Three types of modulation are commonly used for communications: Amplitude (AM), Frequency (FM), and Phase (PSK). Demodulation is the reverse of modulation—the process of retrieving an electrical signal from a carrier wave or signal.

MPEG-1 (Moving Picture Experts' Group)   Pronounced em-peg, it is the international standard for compressing and decompressing of audio and video signals and synchronizing audio and video signals during playback. MPEG-1 was adopted to set the standards for the storage and playback of video and audio for CD-ROMs. In addition to reducing the information of a single image, MPEG also reduces the data from one image to another, using temporal compression algorithms.

MPEG-2 (Moving Picture Experts' Group)   The quality of MPEG-1 is not sufficient for broadcast-quality images. Like MPEG-1, MPEG-2 employs spatial and temporal compression techniques but produces a much higher quality of video images.

MSO (Multi-System Operator)   A cable operator that owns and operates several cable systems, usually a corporation or holding company.

multi-channel   In programming, refers to a package of program services or networks. In communications engineering, it means a common channel that contains two or more channels on different frequencies (FDM) or broken up into different time allotments (TDM).

multiplexed services   1) Two programs transmitted simultaneously over one channel or 2) one program transmitted over several channels at set intervals.

multi-point distribution service (MDS) and multi-point, multi-channel distribution service (MMDS)   Established in the 1970s, MDS is a common carrier service that is used primarily for pay television programming services. In the beginning, MDS systems carried only 2 channels; today, channels are grouped together allowing distributors to send up to 30 channels of programming. When several channels are delivered, this service is called MMDS, or wireless cable.

Near Video-on-Demand (NVOD)   A form of video-on-demand, except that the customer does not control start times of programs. Movies are placed on multiple channels with start times at designated intervals. The customer must tune into the channel that plays the movie of his choice at the nearest start time available.

Net   See Internet.

NII (Network Information Infrastructure)   A vast system of communications networks, computers and their databases, and consumer electronics that will provide large amounts of information to its users.

NIM (network interface module) Telephone companies conceptualize the next generation of settop boxes as being composed of separate modules so they can be upgraded piece by piece, rather than all at once. The NIM is the portion of the settop that would connect the subscriber to the network.

nodes A branching or exchange point for networks.

NTSC (National Television Standards Committee) A group of experts on radio signals that set the North American standard specifications for television signals. The TV signal has 525 lines and interlaced scan at a rate of 30 frames per second.

NVOD See **Near Video-on-Demand.**

octet A term used in ATM (asynchronous transfer mode) systems to describe a byte.

*OSI (Open Systems Interconnection) The only internationally accepted framework of standards for communication between different systems made by different vendors. The International Standards Organization's major goal is to create an open system networking environment where any vendor's computer system, connected to any network, can freely share data with any other computer system on that network or a linked network.

*packet switching Sending blocks of digital data in packets through a network to some remote location. The data to be sent is subdivided into individual packets of data, each packet having a unique identification and each packet carrying its own destination address. The packets are reassembled into their proper sequence after they reach their destination.

PAL (Phase Alternation Line) A European standard for color television accepted in 1967. The system is 625 lines with interlaced scan 25 times a second.

payload The portion of a message that carries the message proper, as distinguished from addressing and error correction messages (sometimes called overhead information or header).

photodiode In a fiber optic system, the photodiode senses the incoming light and changes its signals into an electrical current that matches the intensity of the light signals.

pixel (contraction of picture element) The smallest unit of a television screen. It is made of a phosphor that lights up when electric current passes through it, giving the screen its color and brightness.

pixellation When the pixels are large enough in a digital image to become individually visible.

point-to-multipoint A characteristic of a communication system that allows a single source to send messages to multiple receivers at the same time. Examples of point-to-multipoint systems are broadcast television and existing cable systems.

point-to-point   A characteristic of a communication system that opens up a link between two terminal points. A good example is a telephone call from one person to another.

*polarization   Characteristic of electromagnetic radiation where the electric-field vector of the wave energy is perpendicular to the main direction, or vector, of the electromagnetic beam.

PS   See **packet switching.**

QAM (Quadrature Amplitude Modulation)   An elegant modulation technique that uses variations in signal amplitude to represent either 16 or 32 different data states. The number of states it supports makes it efficient in achieving high data rates. This method was championed by General Instrument to serve as the modulation for digital HDTV but it lost out in field tests to Zenith's VSB-16 technique.

quantization   The second step in digitizing an analog signal. First, the signal is sampled. Then some number of states of the sample are determined. The number of states is the level of quantization the digitizing scheme allows.

radio frequency (RF)   Electromagnetic waves between 10 kHz and 3 MHz that are propagated and transmitted over the air.

RAM (random access memory)   In a computer, RAM is memory that is immediately available to the user and does not have to be accessed on a peripheral device. It is also the most expensive type of memory and subject to loss if the system goes down.

rejectors   In the Hewlett-Packard study of potential consumers of interactive television, these people rejected the idea of receiving interactive services entirely.

RF   See **radio frequency.**

RISC (Reduced Instruction Set Chip)   This is a way of making computer chips faster. By organizing its instructions into hierarchically arranged data, RISC chips are more efficient than SISCs (standard instruction set chips).

sampling   The first step in digitizing analog signals. At specified intervals, a measure of the continuous signal is taken. Sampling consists of making decisions about the appropriate interval and devising measurements that capture the variation in the underlying analog signal.

sampling rate   The measurement of an analog signal at specified intervals is called the sampling rate. It defines how often the underlying analog signal must be measured in order for the least amount of data to be lost.

scalability   An important aspect of the standards-setting process, scalability refers to the ability of a technology to remain the same regardless of the size of the surrounding system. For example, a video server may operate well in a small network of 100 users but crash when 500 users are hooked up. If the server is scalable it can be enlarged to handle the additional users by adding additional processors and ports. If the company must buy a new server altogether, then the old one was not scalable.

scanning line   In television, a scanning line refers to a single pass of a sensor from left to right across a given image.

SDV   See **fiber to the curb (FTTC).**

SECAM (Systeme Electronique Couleur Avec Memoire)   A standard for the television signal standard in France, eastern European countries, Russia, and some African countries. SECAM has 819 lines with an interlaced scan picture. It is the highest resolution picture in current use.

semiconductor   A semiconductor is an electronic component that has a conductor of electricity on one side, usually germanium, and a resistor to electricity on the other side, usually silicon. Hence, its name comes from these two opposite and complementary functions.

server   A server is a computer processor and an array of memory devices which store digital data that can be accessed by users. A server can be used by one person to store material, but it is commonly used by several, even thousands of users. A media server is one that keeps a transactional record for the purpose of charging for material. A video server provides broadband material to the users. See **media server.**

sidebands   Frequencies on either side of the carrier signal in a television signal.

*sine wave   A complete set of one cycle and one negative alternation of current.

SMCR model   In communication theory, this refers to the model that specifies Source–Message–Channel–Receiver.

software   A detailed set of operating instructions that specify and control the actions of a computer. The term differentiates between instructions and hardware.

spectrum flexibility   A term coined by broadcasters to press Congress to let them use the 6 MHz allotted for HDTV for other high definition service, such as datacasting or multiplexed NTSC channels.

stagger-casting   Another term for near-video-on-demand. It refers to the practice of using multiple TV channels on a cable system to stagger the same movie every half-hour, fifteen, ten, or five minutes.

substitutability   A term from the theory of the diffusion of innovation. It describes the ability of an innovation to perform the same function of the older technology or practice. The innovation must offer "relative advantage" in the substitution in order for it to be adopted.

switched digital video networks (SDV)   See **fiber to the curb (FTTC).**

switching   In a network, switching is the capacity for a signal to go to one or more terminals but not others.

symmetric multiprocessing (SMP)   A way of designing large computers that combines processors. They share input and output functions and memory and have a shared operating system.

system integrator   A person, team, or company that is responsible for organizing and putting together all the needed vendors to construct an advanced television system. Ultimately, their task is to make sure the resulting system operates smoothly from end-to-end.

tap   In cable systems, a tap is where the drop cable to the subscriber household connects to the feeder cable delivering signals downstream.

TDM   See **time-division multiplex.**

telecine   Equipment that transfers motion picture film, photographs, and slides to video.

telcos   Abbreviation for telephone companies.

telephony   Telephone services.

temporal compression   Temporal means time. Temporal compression occurs by eliminating redundancies between successive frames of a television picture.

terrestrial transmission   Sending out a radio or TV signal over-the-air from a transmitter that is situated on the ground.

tiering   A marketing technique used by cable operators to create price differentiation among their programming offerings. See **digital tier.**

tiling   The effect seen when a digital signal loses enough signal to create a pixellation effect on the screen. In other words, when a portion of the intended picture is shown along with partial digitized information not appearing as a picture.

time-division multiplex (TDM)   Allowing multiple users to share the same transmission channel by assigning each user a dedicated segment of each transmission cycle.

traditionalists   In the Hewlett-Packard study of potential consumers of interactive television, these are people who aren't very interested in interactive services because they are loyal to traditional broadcast television viewing.

*transponders   A radio relay equipment on board a communications satellite. Transponders will receive a signal, amplify it, change its frequency, and then send it back to earth.

trialability   In the theory of the diffusion of innovation, the trialability of a product or technique is the extent to which potential adopters can try it out. The test drive before buying a car is a good example of a complex product that must be tried before the consumer is willing to put down the cash.

Turing test   Alan Turing, a British researcher who worked on early computers, defined as the ultimate test of a computer that it would be able to fool a human user that he or she was talking to another human.

UHF (Ultra High Frequency)   Frequencies of the electromagnetic spectrum that range from 300 megahertz to 3 gigahertz.

utilitarian time savers   In the Hewlett-Packard study of potential interactive television consumers, this group of people liked interactive TV a lot because they could see how it would allow them to save time in day to day tasks.

vector   A line. In video, it is the expression of a vector which typically includes the direction and magnitude of movement of some on-screen object.

*vertical blanking interval   The interval between television frames in which the picture is blanked to enable the trace (which "paints" the screen) to return to the upper left hand corner of the screen, from where the trace starts, once again, to paint a new screen.

VHF (Very High Frequency)   Frequencies of the electromagnetic spectrum that range between 30 and 300 megahertz.

video server   See **media server.**

video-on-demand (VOD)   An interactive television service that allows customers to receive requested video material and view it within a few seconds. Usually VOD service offers "VCR functionality": fast-forward, rewind, and pause.

*wavelength   The length of a wave measured from any point on one wave, to a corresponding point on the next wave, such as from crest to crest. In other words, a wavelength is the distance that an electromagnetic wave travels in the time that it takes to oscillate through a complete cycle.

wavelength division multiplexing (WDM)   A way to increase the capacity of an optical fiber by transmitting different signals at different wavelengths at the same time.

wavelet compression   Mathematically-based compression scheme that deconstructs and transforms the waves that carry television signals into wavelets.

WDM   See **wavelength division multiplexing.**

wetware   A literary term for human beings. Taken from Cold War spy terminology, the former KGB called killing "wet work." In cyberpunk fiction, the term wetware was adopted to refer to people.

World Wide Web (WWW)   An Internet service that lets users access the holdings of providers at what are called "websites." Although the underlying engineering is quite complex, the WWW is easy to use with software programs that describe the contents. These programs are called "browsers" or "spiders."

Glossary of Organizations

ACATS (Advisory Committee on Advanced Television Service) The advisory committee chartered by the FCC to resolve and recommend a standard for advanced television broadcasting in the United States. Composed of volunteers from the television industry.

ATSC (Advanced Television Standards Committee) Established in 1983, the ATSC is a group composed of executives from the U.S. television industry. They come from broadcast networks, stations, cable companies, producers, electronics equipment manufacturers, and satellite companies. This influential group makes recommendations to the U.S. State Department on pending issues before international standards-setting organizations, such as the CCIR and the ITU.

CCIR (Consultive Committee on International Radio) A department of the International Telecommunications Union that reviews recommendations on broadcast policies.

Council on Competitiveness A coalition of chief executives from business, labor, and higher education.

DAVIC (Digital Audio/Visual Council) An influential private international group that is attempting to set worldwide standards for advanced television systems. Composed primarily of equipment manufacturers.

EC (European Community) Known in the U.S. as the Common Market, the EC is composed of 12 nations in Western Europe. Headquartered in Brussels, Belgium, the EC has a political structure and decision-making process. The organization is primarily economic with the aim of reducing barriers and promoting trade between member states. The members are Belgium, Denmark, France, Germany, Greece, Ireland, Italy, Luxembourg, the Netherlands, Portugal, Spain, and the United Kingdom.

Federal Communications Commission (FCC) Established by Congress in the Communications Act of 1934, the FCC is part of the executive branch of government. It regulates spectrum allocation, licensing of spectrum use rights, assigns call letters, and interstate communication.

FSN (Full Service Network)   Time Warner's advanced TV system in Orlando, Florida. The most technologically (and financially) ambitious of all the IATV test sites.

G7 nations   G7 is an abbreviation for the Group of Seven, an economic roundtable whose members are the U.S., Canada, the U.K., Germany, Italy, Japan, and France.

Grand Alliance   Former FCC Chairman Richard Wiley put together this group of companies who had developed proposals for U.S. HDTV separately. He suggested, urged, and finally prodded the former competitors to ally so they could all be winners in the race to advanced television.

INTELSAT (International Telecommunications Satellite Organization)   Operates a worldwide satellite network which provides members and non-members alike with telecommunications services. Earth stations used by members must conform to standards established by this organization.

ITU (International Telecommunications Union)   An intergovernmental agency whose membership is comprised of soverign states. The ITU develops worldwide telecommunication standards in the form of recommendations. The standards cover all technical aspects of systems and equipment. The ITU also allocates spectrum frequencies.

MMCF (MultiMedia Communications Forum)   Founded in 1993 to act as a common forum for consumers, business users, customer premise equipment vendors, software vendors, program and content providers, and network communications service providers. The MMCF acts to define an architectural template for multimedia communications and to guide and influence standards by issuing requirements papers. It also works to promote industry and user acceptance of multimedia solutions.

NHK (Nippon Hoso Kyokai)   Japan's powerful state broadcasting system. As well as acting as a creator and distributor of television programming, NHK is instrumental in the development of new TV technologies, such as HDTV.

NIST (National Institute of Standards and Technology)   An agency of the U.S. Department of Commerce, NIST has tried to involve itself in the standards-setting process. These attempts have met with stiff resistance from private industry. In 1995, the Congress tried repeatedly to defund this agency and was successful in cutting its budget substantially.

NTIA (National Telecommunications and Information Administration)   An agency of the U.S. Department of Commerce concerned with the development of communications standards.

SMPTE (Society of Motion Picture and Television Engineers)   The members of this organization work at television stations and technical facilities that process material for the television and motion picture industry. They exercise considerable influence as they are involved in the daily processing of programming.

SPON (Starbright Pediatric On-Line Network)   Starbright is a charity headed by Steven Spielberg that delivers high tech entertainment to sick kids. In 1995, Starbright began operating an innovative network that links hospital-ized children in 5 U.S. hospitals, providing an electronic playground for them.

USSB   A direct-to-home satellite service that delivers its programming to small, inexpensive backyard dishes, using the high power Ku-band part of the spectrum.

YCTV (Your Choice TV)   A programming service for TV systems that offer near-video-on-demand and video-on-demand to their customers. YCTV specialized in allowing customers to see TV programs on demand for about $1.00.

References

"1990: pivotal year for DBS hopefuls." *Multichannel News* (Jan. 7,1991):4.

"A virtual Library of Congress." *New York Times* (September 12, 1994):B-1.

Abel, M., Bell, R., Perey, C., and Zakowski, W. "The MMCF transport services interface." *New Telecom Quarterly* (4th Quarter, 1994):38–44.

Acker, Stephen R. "Designing communication systems for human systems: values and assumptions of 'socially open architecture'." *Communication Yearbook* 12 (Newbury Park: Sage, 1988):498–532.

Akhavan-Majid, R. *"Public service broadcasting and the challenge of new technology: A case study of Japan's NHK."* Paper presented at International Communication Association, Miami, FL, May 21–29, 1992.

Akwule, Raymond. *Global Telecommunications: The technology, administration, and policies.* Stoneham, MA: Focal Press, 1992.

Alexander, George A. *Seybold Report on Desktop Publishing* 8:8. (April 4, 1995):3–5.

Altman, I. and Taylor, D. *Social penetration: The development of interpersonal relationships.* New York: Holt, Rinehart & Winston, 1973.

Amdur, Meredith. "Adventure a strong U.S. export." *Broadcasting & Cable* (Aug. 29, 1994):39.

American Electronics Association, *"High Definition Television (HDTV): Economic Analysis on Impact."* Prepared by the ATV Task Force Economic Impact Team of the AEA, Santa Clara, CA, (November 1988).

Apodaca, P. and Shiver, J. "Southland firm spotlights FCC auction woes." *Los Angeles Times* (Aug. 20, 1994):D-1, 2.

Argyle, Michael and Dean, J. "Eye contact, distance and affiliation." *Sociometry* 28 (1965):289–304.

Associated Press. "Cable arrives to revolutionize South Korea for better or worse." *Los Angeles Times* (March 2, 1995):D4.

Associated Press. "Chinese being given greater remote control." *Los Angeles Times* (April 17, 1995):D5.

Associated Press. "Europeans seek quotas to blunt U.S. dominance of cyberspace." *Los Angeles Times* (Feb. 23, 1995):D-8.

"AT&T's Africa cable." *Telecommunication Reports*, 6:7 (July 1995):15.

Bandura, Albert. "Self-efficacy mechanism in human agency." *American Psychologist* 37 (1982):122–147.

Banks, Mark J. "Low power television," in August E. Grant (ed.), *Communication Technology Update*, 3rd ed. (Boston, MA: Butterworth-Heinemann, 1994):107–115.

Banks, Mark J. and Havice, M. *Low power television 1990 industry survey*. Unpublished report of the Community Broadcasters Association (Dec. 14, 1990).

Barish, Charles. "Superman's now super digital." *Videography* (Oct., 1993):30–32, 101–102.

Bateson, Gregory. *Steps to an Ecology of Mind*. New York: Ballantine Books, 1972.

Bauman, Adam. "Alacazam! Welcome to the magical world of cable modems." *Los Angeles Times* (May 3, 1995):D4.

Baylin, Frank. *Miniature Satellite Dishes: the New Digital Television*. Boulder, CO: Baylin Publications, 1994.

Bayus, Barry L. "High-Definition Television: Assessing Demand Forecasts for a Next Generation Consumer Durable." *Management Science* 39, no. 11 (November 1993):1319–1333.

Beaulieu, Mark, and Okon, Chris. *Multimedia Demystified*. New York: Random House, 1994.

"Bell-Atlantic Video Services connects first paying customers." *Interactive Television Association* 2:22 (June 2, 1995):2.

Berniker, Mark. "PacTel joins wireless migration." *Broadcasting & Cable* (April 10, 1995):35.

Berniker, Mark. "Bell Atlantic, Nynex purchase CAI wireless systems." *Broadcasting & Cable* (April 3, 1995):40.

Berniker, Mark. "Experts bullish on cable's migration to digital services." *Broadcasting & Cable* (May 15, 1995):42, 46.

Berniker, Mark. "'Star cam' player stats among ACTV sports features." *Broadcasting & Cable* (Oct. 31, 1994):21.

Berniker, Mark. "ABC signs NTN to create interactive services." *Broadcasting & Cable* (Oct. 24, 1994):30.

Berniker, Mark. "Eon mines for some IVDS gold." *Broadcasting & Cable* (Aug. 8, 1994):29.

Berniker, Mark. "GI picks Microware's OS for digital set-tops." *Broadcasting & Cable* (May 1, 1995):35.

Berniker, Mark. "GTE Interactive playing games." *Broadcasting & Cable* (June 27, 1994):33–34.

Berniker, Mark. "Interaxx plans trials in Florida, Washington." *Broadcasting & Cable* (May 7, 1994):29.

Berniker, Mark. "NBC Desktop Video to deliver news to PCs." *Broadcasting & Cable* (July 8, 1994):26.

Berniker, Mark. "Online with Steve Case." *Broadcasting & Cable* (Oct. 24, 1994):33–36.

Berniker, Mark. "Pioneering on the digital highway." *Telemedia Week* (Dec., 1994):34–39.

Berniker, Mark. "Preiss bringing TV shows to CD-ROM." *Broadcasting & Cable* (Feb. 13, 1995):28.

Berniker, Michael. "Microware creates de facto operating system for interactive tv." *Broadcasting & Cable* 124, no. 30 (July 25, 1995):30.

Besen, S. M. and Farrell, J. "The role of the ITU in standardization." *Telecommunications Policy* (April 1991):311–321.

Bhasin, Roberta. "Heaven sent." *Convergence* (March 1995):27–32.

Blank, Christine. "ShopperVision starts grocery shoot." *On Demand* (Dec., 1994/Jan., 1995):42.

Blankenhorn, Dana. "Wireless cable operators form alliance." *Newsbytes* (June 22, 1994).

"Blood and Gore." *Harper's* (December 1994):18.

Bolick, Brooks. "Piracy tab $1 bil in 94; data 'cannot be ignored'." *The Hollywood Reporter* (Feb. 17–19, 1995):1, 35.

Bowling, Tom. "A new utility." *New Telecom Quarterly* (4th Quarter, 1994):14–17.

Brown, Eric. "The Edutainers: thrills for skills." *NewMedia* (Dec. 1994):50–56.

Brown, Ralph. "Video Server Architecture." *1995 NCTA Technical Papers* (Washington, D.C.: National Cable Television Association, 1995):125–131.

Brown, Rich. "MSOs take direct approach." *Broadcasting & Cable* (June 27, 1994):26.

Brown, Rich. "TCI invests in $500 million Japanese venture." *Broadcasting & Cable* (May 30, 1994):28.

Brown, Roger. "Does SONET play in cable's future?" *CED: Communications Engineering & Design* (Sep. 1994):34–38.

Brown, Roger. "The return band: Open for business?" *CED: Communications Engineering & Design* (December 1994):40–43.

Buchman, Caitlin. "Back to the future: the art of interactive storytelling." *Film-Maker* (Summer 1994):34–39.

Burgess, John. "U.S. withdraws support for studio HDTV standard; Japanese suffer setback in global effort." *The Washington Post* (May 6, 1989):D-12.

Burgoon, Judee K. "Nonverbal Signals." *Handbook of interpersonal communication*, eds. M. L. Knapp and G. R. Miller (Beverly Hills: Sage, 1985):344–390.

Burgoon, Judee K. and Hale, J.L. "The fundamental topoi of relational communcation." *Communication Monographs* 51 (1984):193–214.

Burgoon, Judee K., et. al. "Adaptation in dyadic interaction: Defining and operationalizing patterns of reciprocity and compensation." *Communication Theory* 4 (1993):295–316.

Burnish, Christine. "HBO Studio Productions' Digital Dreams Come True." *AV Communications* (June 1989):26–29.

"Business leaders schedule meeting on G7 issues." *Telecommunication Reports,* 6:7 (March 17, 1995):6

"Buy rates skyrocket for DBS pay-per-view." *Interactive Video News* (April 3, 1995):1–2.

"C-Cor to provide line extenders for Chilean multimedia project." *Wireless International* (April 1995):20.

Carballes, J. C. "The impact of optical communications." *Alcatel Research* (1st Quarter, 1992):4–10.

Careless, James. "Interactivity in the here and now." *TV Technology* (March 1995):11.

Carlton, Jim. "Nintendo sues Samsung unit." *Wall Street Journal* (Jan. 16, 1995):B4, B6.

Carter, N. M. and Cullen, J. B. *Computerization of newspaper organizations: The impact of technology on organizational structuring.* Lanham: University Press of America, 1983.

Casti, John. *Complexification.* New York: HarperCollins, 1994.

Cegala, D. J. "A study of selected linguistic components of involvement in interaction." *Western Journal of Speech Communication* 53:3, (Summer 1989):311–326.

Clark, Tim and Wilson, Carol. "New power player: electric utilities." *Inter@ctive Week* (Dec. 12, 1994):41.

Clark, Tim. "Interactive attitude survey roundup." *Inter@ctive Week* (May 8, 1995):24.

Clark, Tim. "TV Channels go online." *Inter@ctive Week* (Sep. 18, 1994):10.

Colker, David. "Cyber stars of the next frontier." *Los Angeles Times Calendar* (Dec. 8, 1993):4–5, 82–86.

Colker, David. "Evolution Revolution." *Los Angeles Times* (May 13, 1995):F1, F14.

"Commission background note on digital tv." *Reuters News Service* (December 6, 1993).

Commission of the European Communities. "Wide-screen television lifts off." *RAPID* (press release IP: 94-21, January 14, 1994), both archived on NEXIS, an electronic information service of Mead Data Central.

Comsat World Systems press release. "Africa and Middle East Regions forecasted as next growth areas for international broadcasting." April 20, 1995, 6560 Rock Spring Drive, Bethesda, MD, 20817.

Cook, William J. "There's a battle ahead sparked by a revolutionary technology—even a death star." *U.S. News & WorldReport* (September 10, 1990):75.

Cooper, Jim. "Your choice breaks ground." *Cablevision* (July 11, 1994):47.

Coran, Stephen. "Low power subscription television." *Wireless Broadcasting Magazine* (April 1995):14–16.

Coyle, Tom. "1994 local telco revenue per average switched line." *America's Network* (June 15, 1995):14.

Cringely, Robert X. "Thanks for sharing." *Forbes ASAP* (February 1995):49–52.

Crutchfield, E. B., ed. *Engineering Handbook,* 7th ed. Washington, D.C.: National Association of Broadcasters, 1985.

Curtin, Michael. "Beyond the vast wasteland: The policy discourse of global television and the politics of American empire." *Journal of Broadcasting and Electronic Media* (Spring, 1993):127.

D'Amico, Marie. "Home music network." *Digital Media,* 4:2 (Jan., 1995):18–19.

Davidge, Carol. "America's talk-back television experiment: Qube." *Wired Cities: shaping the future of communications,* eds. William Dutton, Jay Blumler, and Kenneth Kraemer. Boston: G.K. Hall, 1981:75–101.

Davis, Frederic E. "A CD-R in every studio, a CD-ROM on every desktop." *Multimedia Producer* (April 1995):39–45.

Dawson, Fred. "My kingdom for a modem." *CED* (June 1995):127–134.

Dawson, Fred. "The state of the display: flat-panel screens coming soon to a PDA or computer near you." *Digital Media* 3, no. 9/10, (February 1994):11.

"DBS Competition." *Specs Technology* A Cable Labs publication (December 1994):6.

Dean, Richard. "Media revolution revolves around video servers." *World Broadcast News* (April 1995):90.

"DEC, SGI court content developers." *On Demand* (December 1994/January 1995):36.

Denison, D. C. "My excellent interactive adventure." *Boston Globe Magazine* (July 11, 1993).

Depp, Steven W. and Howard, Webster E. "Flat-Panel Displays." *Scientific American* (March 1993):90–97.

Development of a U.S.-based ATV industry. Washington, D.C.: American Electronics Association, May 9, 1989.

Dickinson, John. "Financial crash on the digital highway: lack of home banking standards." *Computer Shopper* 14, no. 4 (April 1994):68.

Dickson, Edward and Bowers, Raymond. *The VideoTelephone.* New York: Prager, 1974.

"Digital delay." *Inter@ctive Week* (Jan. 16, 1995):27.

Digital Equipment Corp. press release, "Digital announced three-tiered strategy for Digital Media Studio interactive content and application development business," June 5, 1995, contact (508) 841-2609.

"DMX Music Service first DBS user of AC-3 audio." *PR Newswire* (January 17, 1994).

Doherty, Richard. "Ultimate CD? The digital video disc." *Digital Video (DV)* (May 1995):74–77.

Dolnick, David. "Digital game delivery systems." *Broadband Systems & Design* (June 1995):23–24.

Donohue, William A. "Interaction Goals in Negotiation: A Critique." *Communication Yearbook* 13 (Newbury Park: Sage, 1989):417–427.

Doyle, Bob. "Crunch Time for Digital Video." *NewMedia* 3, no. 3 (March 1994):43–50.

Doyle, Bob. "How Codecs Work." *NewMedia* 3, no. 3 (March 1994).

Doyle, Marc. *The Future of Television.* Lincolnwood, IL: NTC Business Books, 1993.

Drucker, Peter. *Innovation and Entrepreneurship.* New York: Harper & Row, 1985.

Dupagne, M. "High-definition television: A policy framework to revive U.S. leadership in consumer electronics." *Information Society* 7, no. 1 (1990):53–76.

Dutton, William H. "Driving into the future of communications? Check the rear view mirror," paper delivered at POTS to PANS: Social issues in the multimedia evolution from Plain Old Telephony Services to Pictures and Network Services', the BT Hintlesham Hall Symposium, Hintlesham, Suffolk, March 28–30, 1994.

Dutton, William H., Blumler, Jay G., and Kraemer, Kenneth L., eds. *Wired Cities: Shaping the Future of Communications.* Boston: G.K. Hall, 1987.

Dvorak, John C. "Blow their heads off!" *PC Magazine* 13:10 (Nov. 8, 1994):93.

Eastman, Susan Tyler. *Broadcast/Cable Programming: Strategies and Practices,* 4th ed. Belmont: Wadsworth, 1993:29–30.

Ellis, Leslie. "National cable interconnect activity heats up." CED (Sep. 1994):28–32.

Ellis, Leslie. "Tales of selling interactive: free meals, simplicity." *Multichannel News* (June 12, 1995):2A.

Ellis, Leslie. "Technology debate obscures MSOs digital agenda," *Multichannel News* (May 17, 1995):1A, 16A–17.

Engineering Report. Washington, D.C.: National Association of Broadcasters, September 4, 1989:1.

"European utilities eye telephone business." *Wall Street Journal* (Dec. 2, 1994):B4.

Facts at a Glance: International Cable. Washington D.C.: National Cable Television Association, Spring, 1995:1.

Fantel, Hans. "HDTV faces its future." *New York Times* (February 2, 1992):H17.

Farnoux-Toporgoff, Sylviane. "The European Union postures the information society." *New Telecom Quarterly* (3rd Quarter, 1994):3–6.

Farrell, J. and Shapiro C. *Brookings Papers: Microelectronic 199.* Brookings Press, 1992.

Featherstone, Mike. "An Introduction." *Global Culture: Nationalism, globalization and modernity,* ed. Mike Featherstone. London: SAGE Publications, 1990:6.

Federal Communications Commission Advisory Committee on Advanced Television Service. *ATV System Recommendation, draft-SP Version.* Washington: FCC.10-11.

"Fiber-optics that work." *Los Angeles Times* (Jan. 7, 1995):D-5.

Flanigan, James. "TV networks evolve from dinosaurs to darlings." *Los Angeles Times* (Oct. 5, 1994):D1, 2.

Fleeter, Rick. "The smallsat invasion." *Satellite Communications* 18, no. 11 (November 1994):27.

Fredin, Eric S. "Interactive Communication Systems, Values and the Requirement of Self-Reflection." *Communication Yearbook* 12 (Newbury Park: Sage, 1988):533–546.

Freedman, Jonathan. "Fried green writers at the Viacom Cafe." *Los Angeles Times* (May 14, 1994):F2.

Freeman, Michael. "It's the hour of the hour." *Media Week,* 4:23 (Feb. 1, 1995):18.

"FTTC costs come down." *Investor's Business Daily* (May 25, 1995):A7.

"G-7 nations hop on info superhighway." *Los Angeles Times* (Feb., 27, 1995):D5.

"G7 countries agree to 11 'GII testbed networks'." *NextNet,* 4:5 (March 13, 1995):13.

Gagnon, Diana. "Toward an Open Architecture and User-Centered Approach to Media Design." *Communication Yearbook* 12 (Newbury Park: Sage, 1988):547–555.

Galbraith, J. R. "Designing the innovating organization." *Organizational Dynamics,* 10 (1982):5–25.

"Game Show Network goes interactive at NCTA." Game Show Network advertisement in *NCTA Show Daily.* May 19, 1995.

Gelman, Morrie. "Monte Carlo: Special Issue." *The Hollywood Reporter* (Feb. 1, 1994):S3–S16.

Gergen, Kenneth. *The Saturated Self.* New York: Harper Collins, 1991.

Gerwig, Kate. "Government grapples with privacy issue." *Interactive Age* (Nov. 14, 1994):27, 32.

Gilder, George. "Life After Television (revisited)." *Forbes ASAP.*

Gilder, George. "Mike Milken and the two trillion dollar opportunity." *Forbes ASAP* (April 10, 1995):104.

Gilder, George. "Washington's bogeyman," *Forbes ASAP* (June 6, 1994):115.

Gilder, George. "The new rule of the wireless." *Forbes ASAP* (April 11, 1994):99–110.

Ginsberg, Shane. "The digital revolution will be televised." *Digital Media* 5, no. 1 (June 5, 1995):3–12.

Golding, Peter. "The communications paradox: inequality at the national and international levels." *Media Development* 41 (April 1994).

Graf, James E. "Global Information Infrastructure: First Principles." *Telecommunications* (May 1994):72–73.

Gross, Lynne S. *The New TV Technologies,* 3rd ed. Dubuque: Wm. C. Brown, 1990.

Haddad, Charles. "Turner's epic expansion overseas." *International Cable* (April 1995):30–41.

Hall, Jane. "Cyberspace for sale." *Los Angeles Times* (May 21, 1995):D1, D-3.

Halliwell, Chris. "Camp development: the art of building a market through standards." *IEEE Micro,* (Dec., 1993):10–18.

Halonen, Doug. "FCC: Who pays for advanced TV?" *Electronic Media* (March 1995):1, 75.

Halpin, Mikki. "Detours to Utopia." *FilmMaker* (Summer, 1994):40–41.

"Hamamatsu City designated HiVision City by Ministry of Posts." *COMLine Daily News Telecommunication* (September 6, 1994), archived on NEXIS.

Hanson, Jarice. *Connections: technologies of communication.* New York: Harper-Collins College Publishers, 1994.

Harmon, Amy. "A Digital Visionary Scans the Info Horizon." *Los Angeles Times* (June 1, 1994):D6, D8.

Harmon, Amy. "Fun and games—and gore." *Los Angeles Times* (May 12, 1995):A1, A22, A29.

Harmon, Amy. "Invasion of the Film Computers." *Los Angeles Times* (Aug. 15, 1993):A1, 22–23.

Harmon, Amy. "Joining the multimedia Gold Rush." *Los Angeles Times* (Sept. 30, 1994):A1, A24.

Harmon, Amy. "Software giants Broderbund, Learning Co. agree to merge." *Los Angeles Times* (Aug. 1, 1995):D-2, D-11.

Harmon, Amy and Helm, Leslie. "PacTel to cut $1 billion from telecom project." *Los Angeles Times* (Sept. 2, 1995):D1, D4.

Haugsted, Linda. "Brand or bust is cable operator's latest motto." *Multichannel News* (June 12, 1995):36.

Hayes, Mary. "Working online, or wasting time?" *Information Week*, no. 525 (May 1, 1995):38.

"HDTV cooperation asked." *Television Digest* 29 (May 22, 1989):9.

"HDTV Developments in Japan." *Financial Times* (May 9, 1989), found on NEXIS, an electronic information service of Mead Data Central.

"HDTV Live Broadcasts." *Japan Economic Newswire* (February 23, 1988), found on NEXIS.

"HDTV production standard debated at NTIA." *Broadcasting* 116 (March 13, 1989):67.

"HDTV transmission tests set to begin next April." *Broadcasting* 119 (November 19, 1990):52–53.

"HDTV: Broadcasters look before they leapfrog." *Broadcasting* 117, no. 11 (September 11, 1989).

Hearn, Ted. "Maine Sen. introduces retail set-top box bill." *Multichannel News* (April 10, 1995):10.

Hewlett-Packard. "Is there a market for interactive television?" Proprietary study of potential interactive television consumers. Interactive Television Appliances, Palo Alto, CA, 408/553-2948.

Hift, Fred. "EBU adapting to compete with commercial TV." *Electronic Media* (April 10, 1995):72–73.

Hiltz, S. R. *Communication via computer.* Reading: Addison-Wesley, 1978.

"Hitch-hikers' guide to Asia's cable and satellite markets." *Cable and Satellite Europe* (Nov., 1994):22–28.

Holsendorph, Ernest L. "CBS Cable Bid Cleared by FCC." *New York Times* (August 5, 1981):D-1.

Holsinger, Erik. "Kai's Power Tools." *Digital Video (DV)* (April 1995):37–40.

Hong, Junhao. "High Definition Television." In August E. Grant and Kenton T. Wilkinson, eds. *Communication Technology Update: 1993–1994*, 3rd ed. Austin: Technology Futures, Inc., 1993:19–3.

Hontz, Jenny. "Infohighway bill passes Senate panel." *Electronic Media* (August 15, 1994):1.

Horton, Bob. "Standardization and the challenge of global consensus." *Pacific Telecommunications Review* (Sep., 1993):16–22.

Hugh, Annette K. "MVS Multivision pushes ahead despite peso's plunge." Wireless International, 2:5 (April 1995):6–7.

"Hypertext bowls of spaghetti," *EduTech Report* (May 1995):1.

Ingrassia, Joanne. "Canada's infohighway war: Cable, telcos fight to enter each other's businesses." *Electronic Media* (March 20, 1994):23, 52.

"International Scene." *Radio Communication Report*, 14:5 (March 13, 1995):26.

IT Network press release. "ASI to offer interactive television service in United States through new Cableshare licensing agreement." June 15, 1992. ITNetwork, 8140 Walnut Hill Lane, Suite 1000, Dallas, TX 75231.

Jacobi, Fritz. "High definition television: at the starting gate or still an expensive dream?" *Television Quarterly* 16:3 (Winter, 1993):5–16.

"Japan's Cable Growth." Forbes (November 6, 1995):44.

Jessell, Harry A. "Broadcasters come together over HDTV." *Broadcasting & Cable* (April 17, 1995):6.

Jessell, Harry. "Infohighway power play." *Telemedia Week* (Dec., 1994):26–28.

Jessell, Harry. "Level playing field on program guides." *Broadcasting & Cable* (April 10, 1995):58.

Kao, K. C. & Hockham, G. A. "Dielectric-fibre surface waveguide for optical frequencies." *Proceedings of the IEEE* 133, no. 7 (July 1966):1151–1158.

Kaplan, Karen. "Banks seek to branch into homes." *Los Angeles Times* (June 14, 1995):D-4, D-7.

Kapoor, Sanjay. Speech at Digital World, April 1995, Los Angeles.

Karpinski, Richard. "It's in the air: broadband goes wireless." *Telephony* 225, n. 9 (Aug. 30, 1993):16.

Karpinski, Richard. "Up close: U.S. West's new video strategy." *Interactive Age* (Nov. 14, 1994):53–54.

Kayton, Brad. "Very Small Aperture Terminals." in August E. Grant ed., *Communication Technology Update* 4th Ed. (Boston: Focal Press, 1995):307–317.

Keller, John J. "AT&T's secret multimedia trials offer clues to capturing interactive audiences." *Wall Street Journal* (Oct. 6, 1993):B1, B4.

Kellermann, Kathy. "Extrapolating Beyond: Processes of Uncertainty Reduction." *Communication Yearbook 16* (Newbury Park: Sage, 1992):503–514.

Kelly, Lindsey. "Group delaying interactive goal." *Electronic Media* (March 20, 1995):22, 36.

Kerver, Tom. "Turning TCI around." *Cablevision* (March 20,1995):36–42.

Kiesler, S., Siegel, J., and McGuire, T.W. "Social, psychological aspects of computer-mediated communication." *American Psychologist,* 39:10 (1984):1123–1134.

Kline, David. "Interview: Jeff Berg." *WiReD,* 2:3 (March 1994):99.

Krasilovsky, Peter. "Interactive TV: a slow revolution." *Telemedia Week* (Dec., 1994):7.

Krasilovsky, Peter. "Program guides face off." *Broadcasting & Cable* (February 20, 1995):51–54.

Krauss, Jeffrey. "Security issues and the NII." CED (Sep., 1994):24.

Kuhl, Craig. "Test track, race track or fast track?" *Convergence* (March 1995):34–39.

Kupfer, Andrew. "The U.S. wins one in high-tech TV." *Fortune* 60:4 (April 8, 1991):123.

Lahiri, Indrajit. "Star struck in India." *World Broadcast News* (April 1995):86–88.

Lambert, Peter. "PC Connection." *On Demand* (March 1995):9–12.

Lambert, Peter. "Utility networking percolates." *On Demand* (Dec., 1994/Jan., 1995):16.

Lambert, Peter. "Abel says multichannel options could pay for HDTV." *Broadcasting* 122, no. 44 (October 26, 1992):44–45.

Lambert, Peter. "ACATS orders issue of broadcast multichannels." *Multichannel News* 15, no. 9 (February 28, 1994):3.

Lambert, Peter. "All the news that's fit to digitize." *On Demand* (Dec., 1994/Jan., 1995):44.

Lambert, Peter. "BVS' video vision." *On Demand* (Dec., 1994/Jan., 1995):6–14, 50–51.

Lambert, Peter. "First ever HDTV transmission." *Broadcasting* 122:10 (March 2, 1992):8.

Lambert, Peter. "Interactive TV content moves to on-deck circle." *On Demand* (June 1995):4–8, 14

Lambert, Peter. "Software Roundtable: six players show their hands." *On Demand* (March 1995):15–22.

Lambert, Peter. "The broadband PC connection." *On Demand* (March 1995):12.

Landauer, Thomas. *The Trouble with Computers: Usefulness, Usability and Productivity.* Cambridge: MIT Press, 1995.

Lane, Earl. "The Next Generation of TV." *Newsday* (April 5, 1988): Discovery section, 6.

Lawton, George. "The utilities' role: Building the ubiquinetwork—Part 5," *Communications Technology* (Dec. 1994):80–85.

Lawton, George. "Deploying VSATs for specialized applications." *Telecommunications* 28:6 (June 1994):27–30.

Lehman, Tom. "Doing business on the Internet." *New Telecom Quarterly* (1st Quarter, 1995):46–54.

Levine, Marty. "Critical Time Nears For Setting HDTV Standard." *Multichannel News* (June 12, 1995):12A.

Lewyn, Mark, Thierren, Lois, and Coy, Peter. "Sweating out the HDTV contest." *Business Week* 33, no. 6 (February 22, 1993):92–93.

Lipmann, Andrew. "HDTV sparks a digital revolution." *BYTE* (December 1990):297–305.

Lippman, John. "Prime-time targets." *Los Angeles Times* (Oct. 12, 1994):D1–4.

Lippman, John. "Networks push for cheaper shows." *Los Angeles Times* (February 19, 1991):D-1.

Liska, Jo. "Dominance-seeking Language Strategies: Please Eat the Floor, Dogbreath, or I'll Rip Your Lungs Out, Okay?" *Communication Yearbook* 15 (Newbury Park: Sage, 1991):427–456.

Liu, Yu-li. "The growth of cable television in China." *Telecommunications Policy,* 18:3 (1994):216–228.

"Looking for that little extra." *Cable and Satellite Europe* (Nov., 1994):68–70.

Lowndes, Jay C. "14 seek direct broadcast rights," *Aviation Week and Space Technology* (August 10, 1981):60.

Luff, Bob. "Why take an interest in DAVIC?" *Communications Technology* (April 1995):18–19.

Luther, Arch C. *Digital Video in the PC Environment.* New York: McGraw-Hill, 1991.

Magel, Mark. "Friendly, inviting, informative: building a successful kiosk." *Desktop Video World* (March 1995):68–71.

Magel, Mark. "The box that will open up interactive TV." *Multimedia Producer* (April 1995):30–36.

Magyar Chiefs. *Cable and Satellite Europe* (Nov., 1994):38, 40.

Marshall, Tyler. "EU panel urges tighter TV import quotas." *Los Angeles Times* (Feb. 9, 1995):D-1.

Marshall, Tyler. "Hollywood faces a new fight with Europe on quota issue." *Los Angeles Times* (Feb. 10, 1995):A2.

Marx, Andy. "Agencies bank on brave new business." *Inter@ctive Week* (Nov. 7, 1994):78.

Marx, Andy. "PacBell & Hollywood's fiber freeway." *Inter@ctive Week* (Nov. 7, 1994):25.

Marx, Andy. "Where have all the numbers gone?" *Inter@ctive Week* (Jan. 16, 1995):20.

McConnell, Chris. "ABC Television on a digital spending spree." *Broadcasting & Cable* (July 18, 1994):61.

McConnell, Chris. "Broadcasters fending off spectrum grabs." *Broadcasting & Cable* (July 25, 1994):82.

McConnell, Chris. "On-line services blossoming." *Broadcasting & Cable* (June 6, 1994):60.

McConnell, Chris. "Primestar's miles-high problem." *Broadcasting & Cable* (May 8, 1995):92–94.

McConnell, Chris. "Turning data streams into revenue streams." *Broadcasting & Cable* (April 10, 1995):32.

McConnell, John and Lehar, Jane. "HFC or SDV architecture? Economics drives the choice." *Communications Technology* (April 1995):34–40.

McConville, James A. "Taking the CD-ROM plunge." *Computer Merchandising* (June 1994):34.

McConville, Jim. "Cable business: a tale of two operators." *Broadcasting & Cable* (May 15, 1995):24.

McElvogue, Louise. "Cannes now a prime-time player." *Los Angeles Times* (Oct. 13, 1995):D4, D5.

McGarvey, Joe. "Competition heats up early digital set-top market." *Inter@ctive Week* (Jan. 16, 1995):26.

McLaughlin, Laurianne. "Pentium flaw: a wake-up call?" *PC World* 13:3 (March 1995):50–51.

Meherabian, A. *Silent messages.* Belmont, CA: Wadsworth, 1971.

Mehler, Mark. "Network overload, unused capacity: fixing the optical fiber paradox." *Investor's Business Daily* (Jan. 16, 1995):A4.

"Member states ready to impose EU norm on digital tv." *European Insight* (June 10, 1994):615.

Mendelbaum, Jenny. "Communication Phenomena as Solutions to Interactional Problems." *Communication Yearbook 13* (Newbury Park: Sage, 1989):255–267.

"Mergermania in commtech sector." *Wall Street Journal* (July 7, 1995):B-8.

"Microsoft envisions nonconsumer interactive TV applications." *The Cable-Telco Report* (March 13, 1995):13.

Miller, Stuart. "Soft soil sifting in webs' rose bed." *Weekly Variety* (April 4, 1995):30.

Mitchell, William. "When Is Seeing Believing." *Scientific American* (February 1994):68–73.

Moore, Geoffrey A. *Inside the Tornado.* (New York: Harper Collins, 1995.)

Mountford, Joy. "Essential interface design." *Interactivity* (May/June 1995):60–64.

Mowlana, H. and Wilson, Laurie J. *The Passing of Modernity: Communication and the Transformation of Society.* (New York: Longman, 1990).

"Multimedia Communications Forum establishes workgroup to establish MIB." *OSINetter Newsletter*, 9 (Jan., 1994).

Nass, C. and Steuer, J. "Agency & Ethopoeia: Computers as Social Actors." *Human Communication Research* 19:4 (June 1993):504–527.

Nass, Clifford I. and Steuer, Jonathan. "Voices, Boxes, and Sources of Messages." *Human Communication Research* 19:4 (June 1993):504–527.

Nass, Clifford I. et al. "Anthropocentrism and computers." *Behavior and Information Technology* 14:4 (April 1995):229–238.

Nass, Clifford I. et al. "Are respondents polite to computers? Social desirability and direct responses to computers." Submitted to Public Opinion Quarterly (1995).

Nass, Clifford, Steuer, Jonathan, and Tauber, Ellen R. "Computers are social actors." Unpublished paper.

National TeleConsultants. "TV4. Sweden Goes Server Route from Get-Go." *Broadcasting & Cable* (April 3, 1995):S-2.

"Networks migrate to multimedia." *Broadcasting & Cable*, Telemedia Week (April 1995):14–19.

"New cable channels." *Broadcasting & Cable* (May 8, 1995):28–30.

Newton, Harry. *Newton's Telecom Dictionary.* New York: Flatiron Publishing, 1994:981.

"NHK and 9 private stations for trials of EDTV." *COMLine Daily News Telecommunication* (September 1, 1994), archived on NEXIS.

Nickelson, Richard I. "The Evolution of HDTV in the Work of the CCIR." *IEEE Transactions on Broadcasting* 35, no. 3 (September 1989):250–258.

Nicolis, G. and Prigogine, I. *Self-organization in non-equilibrium systems.* New York: Wiley, 1977.

Nolan, Chris. "It's analog for now." *Cablevision* (April 3, 1995):54.

Nolan, Chris. "Reality check: What's here? What's now? What's never?" *Cablevision* (Aug. 22, 1994):22–32.

Nolan, Chris. "Slow and steady?" *Cablevision* (March 20, 1995):22.

Nolan, Chris. "The telcos' set-top." *Cablevision* (April 3, 1995):56.

Noll, A. M. "Anatomy of a failure: PicturePhone revisited." *Telecommunications Policy*, 16:4 (May/June 1992):307–316.

NVOD technology: commitment's the thing." *On Demand* (April 1995):35.

"Online news predictions." *WiReD*, 2, no. 9 (September 1994):50.

"Orbital free-for-all." *Information Week* (Oct. 16, 1995):16.

Owen, Bruce M. and Wildman, Steven. *Video Economics.* Cambridge: Harvard University Press, 1992.

Papyrus Design Group press release. "Papyrus announces new multiplayer online service, code named 'Hawaii'." PDG, 35 Medford St., Somerville, MA, 617/868.5440.

Paskowski, Marianne. "New media blues." *Multichannel News* (June 2, 1995):50.

Patton, Carl. "Digital HDTV: on-the-air!" *ATM* 1 (Advanced Television Markets), no. 5 (April 1992):1–2.

Pelton, Joseph N. "Geosynchronous satellites at 14 miles altitude?" *New Telecom Quarterly* (2nd Quarter, 1995):11.

Peltz, James F. "Hughes DirecTV already a rival to cable." *Los Angeles Times* (Nov. 8, 1995):D-1, 11.

Peltz, James F. "ITT plans commercial satellite 'spaceport'." *Los Angeles Times* (January 27, 1995):D1, 6.

Perey, Christine. "Perspectives on broadband network technologies and applications." *New Telecom Quarterly* (2nd Quarter, 1995):35–41.

Perin, Constance. "Electronic social fields in bureaucracies." *Communications of the ACM*, 34:12(December 1991):74–79.

Phillips, Barry W. "Broadband in the local loop." *Telecommunications* 26:11 (Nov. 1994):37–44.

Pierce, John R. *An Introduction to Information Theory: Symbols, Signals & Noise*, 2nd ed. New York: Dover Publications, Inc., 1980.

Pioneer New Media Technologies, Inc. press release, "*Pioneer's PLUS: first year update.*" June 6, 1993, 600 E. Crescent Avenue, Upper Saddle River, NJ 07458-1827.

Pitta, Julie. "Let the video games begin." *Los Angeles Times* (May 9, 1995):D1, D5.

Pitta, Julie. "Microsoft lures NBC away from rival on-line services." *Los Angeles Times* (May 12, 1995):D5.

Platt, Charles. "Satellite pirates." *WiReD* (Aug., 1994):78, 122.

Podlesny, Carl. "Hybrid fiber coax: a solution for broadband information services." *New Telecom Quarterly* (First Quarter, 1995):16–25.

Press release. "Thomson, Sun select Broadvision to provide interactive electronic commerce management software." Contact: Thomson-Sun Interactive Alliance, Phyllis Scargle, 415/336-0514.

Price, Derek et al., "Collaboration in an Invisible College." *American Psychologist* 21 (1966):1011–1018.

Quittner, Joshua. "500 TV channels? Make it 500 million." *Los Angeles Times* (June 29, 1995):D2, D12.

Quittner, Joshua. "Johnny Manhattan meets the Furrymuckers." *WiReD* (March 1994):92–95, 138.

Rast, Robert. *Statement of Robert M. Rast, Vice President, HDTV Business Development, General Instrument Corporation Communications Division.* Washington, D.C.: FDCH Congressional Testimony, March 17, 1994, archived on NEXIS, an electronic information service of Mead Data Central.

Ratcliffe, Mitch. "Real Progress: the Internet as information utility." *Digital Media*, 4:12 (May 10, 1995):19–22.

Rathbun, Elizabeth. "MCI funds PBS new media ventures." *Broadcasting & Cable* (March 27, 1995):14.

"Reader's Digest books in multimedia format." *AtlantaJournal-Constitution* (Sep. 22, 1994):K-2.

"Refined HDTV cost estimates less daunting." *Broadcasting* 118 (April 9, 1990):40–41.

Reuters New Service Report 3. June 17, 1994, found on NEXIS, an electronic information service of Mead Data Central.

Rivkin, S. R. "Electric utilities will build telecom infrastructure." *New Telecom Quarterly* (2nd Quarter, 1994):15–19.

Rivkin, S. R. "Positioning the electric utility to build information infrastructure." *New Telecom Quarterly* (3rd Quarter, 1994):30–34.

Robinson, G. "Yo! MTV . . . shops?" *Los Angeles Times* (October 27, 1994):E1, E6.

Rogers, Everett and Kincaid, D. L. *Communication networks: toward a new paradigm for research.* New York: The Free Press, 1981.

Rogers, Everett M. *Communication Technology: The new media in society.* New York: The Free Press, 1986.

Rogers, Everett M. *Diffusion of Innovation,* 3rd ed. New York: Free Press, 1983.

Rogers, Everett M. *The Diffusion of Innovation.* New York: Free Press of Glencoe, 1969.

Rose, Steve. "Media server system overview." For a copy of the full report, contact Steve Rose at P.O. Box 100, Haiku, HI, 96708-0100.

Rosenstein, Aviva. "Interactive television." in August E. Grant, ed., *Communication Technology Update,* 3rd ed. (Boston, MA: Butterworth-Heinemann, 1994):55–66.

Rosenthal, Edmond. "FBC studies multiplexing strategies." *Electronic Media* (February 28, 1994):26.

Rutter, Derek. *Looking & Seeing: The role of visual communication in social interaction.* New York: Wiley, 1984.

Sablatash, Mike. "Transmission of all-digital advanced television: State of the art and future directions." *IEEE Transactions on Broadcasting* 40 (June 1994):2.

Samuelson, Pamela. "Digital media and the law." *Communications of the ACM,* 34:10 (Oct., 1991):23–29.

Sanchez, Jesus. "Two phone giants open cable TV's door." *Los Angeles Times* (Feb. 12, 1995):D1, 2.

Santo, Brian and Yoshida, Junko. "Grand alliance near?" HDTV grand alliance faces tough road." *Electronic Engineering Times* (May 31, 1993):1, 8.

Schlossstein, Steven. "Intelligent user interface design for interactive television applications." *1995 NCTA Technical Papers.* Washington, D.C.: National Cable Television Association, 1995.

Schmitz, Joseph, Rogers, Everett, Phillips, Ken, and Paschal, Donald. "The Public Electronic Network (PEN) and the homeless in Santa Monica." *Journal of Applied Communication Research,* 23:1 (Feb., 1995):26–43.

Schrage, Michael. "Humble pie: Japanese food for thought." *Los Angeles Times* (Nov. 4, 1994):D-1, 4.

Schrage, Michael. "Why Sonic the Hedgehog needs to jump onto the Info Highway." *Los Angeles Times* (Nov. 3, 1994):D1–D3.

Schwartz, Frank. "Set-top standards." *Electronic Design* 42, no. 19 (Sep. 19, 1994):151.

Scully, S. "Primestar buys compression for $250 million." *Broadcasting & Cable* 123:32 (Aug. 4, 1993):49.

Scully, Sean. "Satellite business looking up." *Broadcasting & Cable* 123:26 (July 12, 1993):48.

Shackelford, John. "Miami: Launch pad for Latin America." *International Cable* (April 1995):20–28.

Shannon, Claude and Weaver, Warren. *A Mathematical Theory of Information.* Urbana: University of Illinois Press, 1949.

Shiver, Jube Jr. "Congress urged to make broadcasters pay license fees for new technology." *Los Angeles Times* (Nov. 8, 1995)D1, D4.

Short, J., Williams, E., and Christie, B. *The Social Psychology of Telecommunications.* New York: Wiley, 1976.

Shrage, Michael. *Shared Minds.* New York: Random House, 1990.

"SMPTE: Seeking a universal, digital language." *Broadcasting* 121 (November 4, 1991):62.

"Soliton waves double fiber-optic capacity." *New Telecom Quarterly* (2nd Quarter, 1993):6.

Somheil, Timothy. "Real life in a box." *Appliance* 50, no. 8 (Aug., 1993):41.

Southwick, Thomas P. "Down Mexico Way." *Cable World* (April 10, 1995):8.

Stallings, William. *Data and Computer Communications,* 4th ed. New York: Macmillan Publishing Company, 1992.

Stefanac, Suzanne. "Sex and the New Media." *NewMedia* (April 1993):38–45.

Stein, Mark. "Satellites: Companies, nations fight for spots in space." *Los Angeles Times* (September 20, 1993):A1, A16.

Steinberg, Steven. "Spread-spectrum technology." *WiReD* (April 1995):72.

Steinert-Threlkeld, Tom and Wilson, Carol. "Time Warner tunes in TV trial." *Inter@ctive Week* (Dec. 12, 1994):6.

Stern, Christopher. "FCC moves to strengthen wireless cable." *Broadcasting & Cable* (June 13, 1994):11.

Stern, Christopher. "Telcos hedge bets with wireless wagers." *Broadcasting & Cable* (May 1, 1995):14.

Stilson, J. and P. Pagano, "May the best HDTV system win [an interview with R. Wiley]." *Channels* 10 (August 13, 1990):54–55.

Stockler, Bruce. "Water, Water Everywhere." *Millimeter* (Oct., 1994):43–4.

Strasser, Joel A. "Telco power." *Wireless Cable* (June 1995):10–13.

Stratimirovich, C. and Mallalieu, B. "Polsat license bid giant-killer." *Hollywood Reporter* (Feb. 1, 1994):1–5.

Stump, Matt. "Multimedia Diplomacy." *On Demand* (March 1995):5.

Stump, Matt. "NVOD vs. VOD," *On Demand* (Dec., 1994/Jan., 1995):21–24.

Stump, Matt. "NVOD: a how-to guide." *On Demand* (Apr. 1995):13–19.

Stump, Matt. "Program nets go interactive." *On Demand* (March 1995):24–29.

Switzer, Israel. "Methods of two-way service." *TV Technology* (September, 1994):31.

System Subcommittee Working Party 2, Document SS/WP2-1354. Washington: Advisory Committee on Advanced Television Service, September 1994.

Tadjer, Rivka. "Low-orbit satellites to fill the skies by 1997." *Computer Shopper* 15, no. 3 (March 1995):49.

Takashi, Fujio. "High-Definition Television Systems." *Proceedings of the IEEE* 73, no. 4 (April 1985):646–655.

TBI Television Broadcasting International, 1995 (Washington, D.C.: National Association of Broadcasters, 1995).

"TCI survey tracks DBS awareness." *Specs* (Louisville, CO: CableLabs, April 1995):3–4.

Tedesco, Richard. "The main menu: who's on first?" *Broadcasting & Cable* (April 10, 1995):58–60.

"Time Inc. says it will launch 3 magazines on CompuServe." *Information &Interactive Services Report* (Oct. 7, 1994):6.

Tobenkin, David. "The wireless system that could." *Broadcasting & Cable* (May 1, 1995):20.

Tobenkin, David. "Robertson close to deal on Vietnam cable." *Broadcasting & Cable* (Jan., 1995).

Tobenkin, David. "The wireless system that could." *Broadcasting & Cable* (May 1, 1995):20.

Travers, Jeffrey and Milgram, Stanley. "An experimental study of the 'Small World Problem'." *Sociometry* 32 (1969):425–443.

Tristam, Claire. "Stream on: video servers in the real world." *NewMedia* (April 1995):46–56.

Truxal, John G. *The Age of Electronic Messages*. New York: McGraw-Hill Publishing Co., 1990:309.

Tucker, T. "From the boardroom to the desktop." *Teleconnect* (Sep. 1993):50–53.

Turing, Alan M. "Computing machinery and intelligence." *Mind 59* (1953):11–35.

TV Answer press release. "Sportsticker and TV Answer team up to bring sports information to interactive television." June 9, 1993. TV Answer, 1941 Roland Clarke Place, Reston VA, 22091.

Twain, Mark. *Pudd'nHead Wilson & those extraordinary twins*. New York: Harper, 1922.

"Two new DBS ventures set to challenge for piece of increasingly crowded sky." *Video Technology News* 8:7 (March 27, 1995):1, 3.

Tyler Eastman, Susan. *Broadcast/Cable Programming*, 4th ed. Belmont: Wadsworth, 1993.

Tyrer, Daniel. "The high definition television programme in Europe." *European Trends*, no. 4 (4th quarter, 1991):77–81.

U.S. Congress, Office of Technology Assessment. *Global standards: building blocks for the future, TCT-512*. Washington, D.C.: U.S. Government Printing Office, March 1992:8.

"U.S. Staging Comeback in Technology." *Los Angeles Times* (September 19, 1994):D3.

Uyttendaele, A. G. Digital Video/Video Compression, draft of contribution to ISOG from NANBA Technical Committee, available from A.G. Uyttendaele, Cap Cities/ABC, Inc., 77th W. 66th Street, New York, NY 10023-6298.

Van Meter, Kenneth. Of Tele-TV, speaking at a seminar at Digital World, June 18, 1995, Los Angeles.

Van Tassel, Joan. "Compressed Video: Today and Tomorrow." *Videomaker* 12 (April 1994):73–77.

Van Tassel, Joan. "Santa Monica's Public Electronic Network." *WiReD* (March 1994).

Van Tassel, Joan. "Silicon Graphics online studio debuts." *Malibu Times* (Oct. 18, 1994):A-11.

Van Tassel, Joan. "Test Patterns." *The Hollywood Reporter* (May 18, 1995):S-9.

Van Tassel, Joan. "The WWWorld's Fair." *WiReD* (Aug. 1995):43.

Vane, E. T. and Gross, L. S. *Programming for TV, Radio and Cable.* Newton, MA: Butterworth-Heinemann, 1994.

Vanston, Lawrence K. "ADSL/HDSL: Dangerous temptation." *New Telecom Quarterly* (3rd quarter, 1994):53–61.

Veilleux, C. T. "Primestar in more Wal-Marts." *HFN* (May 8, 1995):9.

"Virtual libraries get a boost from Feds." *Chronicle of Higher Education* (Oct. 5, 1994):A26.

Vittore, Vince. "They'll like it! They'll really like it!" *Supercomm '95 Show Daily* (March 22, 1995):19.

Vorhaus, John. "Two way tube talk." *Los Angeles Times Magazine.* (Feb. 16, 1995): 14.

Waldron, B. and Harrison, J. "Re-engineering existing computer systems for fiber." *Communications Technology* (December 1994):74–79.

Walker, Gerald M. "CBS television: implementing the future." *World Broadcast News* (April 1995):32–37.

Walker, J. R. and Bellamy, R. V., Jr., eds. *The Remote Control in the New Age of Television.* Westport, CT: Praeger, 1993.

Walley, Wayne. "Latin TV sets plans for DBS." *Electronic Media* (March 20, 1995):40, 44.

Warner, R. M. "Speaker, partner and observer evaluations of affect during social interaction as a function of interaction tempo." *Journal of Language and Social Psychology* 11:4 (1992):253–266.

Waterman, David. Professor at Indiana University, electronic document, posted to discussion group on telecom regulation, Aug. 10, 1995.

Weick, Karl. *The Social Psychology of Organizing.* New York: Addison-Wesley, 1969.

Weinman, Richard. "Anytime, anywhere communication." *New Telecom Quarterly* (4th Quarter, 1994):18–22.

Weiss, S. Merrill. "Looking back at how far we've come." *TV Technology* (Sep., 1994):33, 36, 67.

West, Don and Jessell, Harry. "The way uphill: How cable's making it on the infohighway." *Broadcasting & Cable* (Nov. 28, 1994):34–44.

West, Don and McClellan, Steve. "Bob Wright and the NBCs nobody knows." *Broadcasting & Cable* (March 6, 1995):37–47.

West, Don. "HDTV gauntlet thrown down in Montreux." *Broadcasting & Cable* (June 19, 1995):37–38.

West, Don. "Wiley talks technical." *Broadcasting & Cable* (June 19, 1995):39.

Wharton, Dennis. "HDTV org threatened by flexibility." *Daily Variety* (August 8, 1994):32.

White, George. "On-line mice aren't stirring." *Los Angeles Times* (Nov. 25, 1994):D1, D2.

"Why Nintendo is not going away . . . (anytime soon)." *DFC Interactive* (April 1995):1, 2–3.

Wiley, Richard E. "High Tech and the Law." *American Lawyer* (July 26, 1994):6.

Wiley, Richard. "Entertainment for Tonight (and tomorrow too)." *The Recorder* (July 26, 1994):6–10.

"Will shift to digital HDTV, Japan firm says." *Los Angeles Times* (June 9, 1994):D3.

Willman, Chris. "Practicality? Just give me my virtual Ringo." *Los Angeles Times Calendar* (Oct. 14, 1994):28.

Wilson, Carol. "Don't know much about history?" *Inter@ctive Week* (Nov. 7, 1994):19.

Wilson, Carol. "Telco networks take the fast lane." *Inter@ctive Week* 2:9 (May 8, 1995):35.

Wilson, Peter R. "Standards: past tense and future perfect?" *IEEE Computer Graphics & Applications* (Jan., 1991):47.

Winston, B. "HDTV in Hollywood." *Gannett Center Journal* 3, no. 3 (1989):123–137.

Wirth, Michael O. "The emergence of the alternative media revenue streams." Paper delivered at the Broadcast Education Annual Convention, Las Vegas, NV, April 7, 1995. Michael Wirth is professor and chair, Department of Mass Communications, University of Denver, Denver, CO.

Yarnall, Louise. "Cyber-cinema." *Los Angeles Times Westside Magazine* (Feb. 13, 1995):12–14.

Young, Jeffrey. "World Class." *The Hollywood Reporter Special Issue on International Interactive* (Jan. 11, 1994):S8, S22.

Zukav, Gary. *The Dancing Wu Li Masters.* New York: William Morrow & Co., 1979:12–13.

List of Interviews

Robert Alexander
Gary Arlen
Dave Banks (U.S.West)
Mary Barnsdale (AT&T)
Jim Bloom (Sony)
Jim Boyle (YCTV)
Mike Brand (Ameritech)
Alan Brightman (Apple)
James Bromley
Jerry Brown (U.S. West)
Nancy Buskin (Viacom)
James Carlson (Jones Intercable)
Jim Chiddix (Time Warner)
Scott Cooper (Vela Research)
Anita Corona (IN)
Dr. Sadie Decker (TCI)
Joe DeMauro (NYNEX)
Marcia DeSonne (National Association of Broadcasters)
Richard Doherty (Envisioneering)
Bob Doyle (Digital Video Group)
Kevin Doyle (BellSouth)
Ellen East (Cox Cable)
Scott Evans (Interaxx)
Peter Fannon (Advanced Television Testing Center)
Tom Feige (Time Warner)
Bob Ferguson (Southwestern Bell)
Ginger Fisk (Bell Atlantic)
Dr. Joseph Flaherty (CBS)
Jack Galmiche (InTouch)

Jonathan Gill
August Grant
Virginia Gray (Southern New England Telephone Multimedia)
Dick Jones (GTE)
Dr. Richard Green (CableLabs)
Vincent Grosso (AT&T)
Mike Gwartney (Family Channel)
Tom Hagopian (ESPN)
David Harrah (then at IBM)
Jerry Heller
Ed Horowitz (Viacom)
Gerry Kaufhold (ThorKa Research)
Marty Lafferty (EON)
Steve Lange (U.S. West)
Sylvan LeClerc(Groupe Videotron)
Michael Liebhold (then at Apple)
Kenneth Locker (Worlds, Inc.)
Jim Mitchell (TCI)
Bob Myers (Viacom)
Dr. Woo Paik (General Instrument)
Ken Phillips
Larry Plumb (Bell Atlantic)
Susan Portwood (U.S. West)
Robert Rast (General Instrument)
Steve Rose (Viaduct Corporation)
Lee Rosenberg (William Morris Agency)
Mark Rosenthal (MTV)
William Samuels (ACTV)
Peter Samuelson (Starbright)
Joseph Schmitz (University of Tulsa)
Jonathan Seybold
Steven Spielberg
Paul Sturiale (EON)
Bill Sullivan (Prevue)
John Taylor (Zenith)
Gary Teegarden (U.S. West)
Tamiko Thiel (Worlds, Inc.)
Antoon Uyttendaele (ABC)
Ellen Van Buskirk (Sega)
Joseph Widoff (Advanced Television Test Center)
Patty Zebrowski (PacTel)

Index